John Minot Rice, William Woolsey Johnson

An Elementary Treatise on the Differential Calculus founded on the

Method of Rates or Fluxions

John Minot Rice, William Woolsey Johnson

An Elementary Treatise on the Differential Calculus founded on the Method of Rates or Fluxions

ISBN/EAN: 9783743335554

Manufactured in Europe, USA, Canada, Australia, Japa

Cover: Foto ©ninafisch / pixelio.de

Manufactured and distributed by brebook publishing software (www.brebook.com)

John Minot Rice, William Woolsey Johnson

An Elementary Treatise on the Differential Calculus founded on the Method of Rates or Fluxions

AN

ELEMENTARY TREATISE

ON THE

DIFFERENTIAL CALCULUS

FOUNDED ON THE

METHOD OF RATES OR FLUXIONS

BY

JOHN MINOT RICE
PROFESSOR OF MATHEMATICS IN THE UNITED STATES NAVY

AND

WILLIAM WOOLSEY JOHNSON
PROFESSOR OF MATHEMATICS IN SAINT JOHN'S COLLEGE ANNAPOLIS MARYLAND

ABRIDGED EDITION

THIRD THOUSAND.

NEW YORK:
JOHN WILEY AND SONS,
53 EAST TENTH STREET,
1893.

PREFACE.

IN preparing this abridgment of their treatise on the Differential Calculus, the authors have endeavored to adapt it to the wants of those instructors who find the larger work too extensive for the time allotted to this subject.

J. M. R.
W. W. J.

ANNAPOLIS, MARYLAND,
August, 1880.

CONTENTS.

CHAPTER I.

FUNCTIONS, RATES, AND DERIVATIVES.

I.
	PAGE
Functions	1
Implicit functions	3
Inverse functions	4
Classification of functions	4
Expressions involving an unknown function	5
Examples I	6

II.
Rates	9
Constant rates	10
Variable velocities	11
Illustration by means of Attwood's machine	11
The measure of a variable rate	12
Differentials	12
The differentials of polynomials	13
The differential of mx	14
Examples II	15

III.
The differentials of functions	16
The derivative—its value independent of dx	17
The geometrical meaning of the derivative	19
Examples III	21

CHAPTER II.

THE DIFFERENTIATION OF ALGEBRAIC FUNCTIONS.

IV.
The square	23
The square root	25
Examples IV	26

V.

	PAGE
The product	29
The reciprocal	30
The quotient	31
The power	32
Examples V	34

CHAPTER III.

THE DIFFERENTIATION OF TRANSCENDENTAL FUNCTIONS.

VI.

The logarithm	37
The Napierian base	39
The logarithmic curve ($y = \log_e x$)	40
Logarithmic differentiation	41
Differentials of algebraic functions deduced by logarithmic differentiation	42
Exponential functions	43
Examples VI	44

VII.

Trigonometric or circular functions	47
The sine and the cosine	48
The tangent and the cotangent	49
The secant and the cosecant	50
The versed sine	50
Examples VII	51

VIII.

Inverse circular functions—their primary values	54
The inverse sine and the inverse cosine	56
The inverse tangent and the inverse cotangent	57
The inverse secant and the inverse cosecant	58
The inverse versed-sine	59
Examples involving trigonometric reductions	59
Examples VIII	60

IX.

Differentials of functions of two variables	62
Examples IX	64
Miscellaneous examples of differentiation	65

CHAPTER IV.

Successive Differentiation.

X.

	PAGE
Velocity and acceleration	67
Component velocities and accelerations	69
Examples X	70

XI.

Successive derivatives	73
The geometrical meaning of the second derivative	73
Points of inflexion	74
Successive differentials	75
Equicrescent variables	75
Examples XI	76

CHAPTER V.

The Evaluation of Indeterminate Forms.

XII.

Indeterminate or illusory forms	79
Evaluation by differentiation	80
Examples involving decomposition	82
Examples XII	84

XIII.

The form $\frac{\infty}{\infty}$	87
Derivatives of functions which assume an infinite value	89
The form $0 \cdot \infty$	90
The form $\infty - \infty$	90
Examples XIII	91

XIV.

Functions whose logarithms take the form $0 \cdot \infty$	93
The form 1^{∞}	93
The form 0^0	94
Examples XIV	95

CHAPTER VI.

MAXIMA AND MINIMA OF FUNCTIONS OF A SINGLE VARIABLE.

XV.
                                                                                PAGE

Conditions indicating the existence of maxima and minima.................. 97
Maxima and minima of geometrical magnitudes 99
    Examples XV.. 101

XVI.

Method of discriminating between maxima and minima 103
Alternate maxima and minima. 104
The employment of a substituted function....................... 106
    Examples XVI... 107

XVII.

Employment of derivatives higher than the first 109
Complete criterion for a maximum or a minimum.......................... 111
Infinite values of the derivative................................... 113
    Examples XVII.. 114
    Miscellaneous examples of maxima and minima................ 115

CHAPTER VII.

THE DEVELOPMENT OF FUNCTIONS IN SERIES.

XVIII.

The nature of an infinite series... 119
Convergent and divergent series....................................... ... 121
Taylor's theorem... 122
Lagrange's expression for the remainder................................. 124
The binomial theorem............. 126
    Examples XVIII... 127

XIX.

Maclaurin's theorem.. 129
The exponential series and the value of e................................ 129
Logarithmic series 131
Computation of Napierian logarithms 132
The modulus of tabular logarithms....................................... 134
The developments of sin x and of cos x................................. 134
    Examples XIX 135

CHAPTER VIII.

Curve Tracing.

XX.

	PAGE
Equations in the form $y = f(x)$	138
Asymptotes parallel to the coordinate axes	138
Minimum ordinates and points of inflexion	140
Oblique asymptotes	141
Curvilinear asymptotes	143
Examples XX	144

XXI.

Curves given by polar equations	146
Asymptotes determined by means of polar equations	148
Asymptotic circles	149
Examples XXI	150

XXII.

The parabola of the nth degree	151
The cubical and the semicubical parabolas	152
The cissoid of Diocles	153
The cardioid	154
The lemniscata of Bernoulli	154
The logarithmic or equiangular spiral	155
The loxodromic curve	155
The cycloid	157
The epicycloid	159
The hypocycloid	160
The four-cusped hypocycloid	161

CHAPTER IX.

Applications of the Differential Calculus to Plane Curves.

XXIII.

The equation of the tangent	162
The equation of the normal	163
Subtangents and subnormals	164
The perpendicular from the origin upon a tangent	165
Examples XXIII	166

XXIV.

	PAGE
Polar coordinates	167
Polar subtangents and subnormals	169
The perpendicular from the pole upon a tangent	170
The perpendicular upon an asymptote	171
Points of inflexion	171
Examples XXIV	173

XXV.

Curvature	174
The direction of the radius of curvature	176
The radius of curvature in rectangular coordinates	177
Expressions for ρ in which x is not the independent variable	178
Examples XXV	179

XXVI.

Envelopes	180
Two variable parameters	183
Evolutes	185
Examples XXVI	188

CHAPTER X.

Functions of Two or More Variables.

XXVII.

The derivative regarded as the limit of a ratio	190
Partial derivatives	191
Examples XXVII	194

XXVIII.

The second derivative regarded as a limit	194
Higher partial derivatives	196
Examples XXVIII	198

THE
DIFFERENTIAL CALCULUS.

CHAPTER I.

FUNCTIONS, RATES, AND DERIVATIVES.

I.

Functions.

1. A QUANTITY which depends for its value upon another quantity is said to be a *function* of the latter quantity. Thus x^2, $\tan x$, $\log(a + x)$, and a^x are functions of x.

The quantity upon which the function depends must be regarded as variable, and be represented in the analytical expression for the function by an algebraic symbol. This quantity is called the *independent variable*. It is essential that variation of the independent variable should actually produce variation of the function. Thus the quantities x^0, $x^2 + (a + x)(a - x)$, and $(\tan x + \cot x)\sin 2x$ are *not* functions of x, since each admits of expression in a form which does not involve x.

2. The notation $f(x)$ is employed to denote any function of x, and, when several functions of x occur in the same in-

vestigation, such expressions as $F(x)$, $F'(x)$, $\phi(x)$, etc., are employed, the enclosed letter always denoting the independent variable. When expressions like $f(1), f(a), f(2x)$, or $f(0)$ are employed, it must be understood that the enclosed quantity is to be substituted for x in the expression which defines $f(x)$. Thus, if we have

$$f(x) = x^2 + x,$$

$$f(1) = 2, \quad f(2x) = 4x^2 + 2x, \quad \text{and} \quad f(0) = 0.$$

Again, if $\quad F(x) = \log_a x \quad (a > 1)$

$$F(1) = 0, \quad F(0) = -\infty, \quad \text{and} \quad F(a) = 1.$$

3. When x denotes the independent variable upon which a function depends, any quantity independent of x is, in contradistinction, called a *constant;* both when it is an absolute constant, like 1, $\sqrt{2}$, or π, and when it is denoted by a symbol, like a, u, or y, to which any value can be assigned. Thus, when a^x is denoted by $f(x)$, it is considered simply as a function of x, and a is regarded as a constant.

When it is desired to express that a quantity is a function of two quantities, both the symbols denoting them are placed between marks of parenthesis. Thus, since a^x is a function of x and a, we may write

$$f(x, a) = a^x.$$

Accordingly we have

$$f(y, b) = b^y, \quad f(3, 2) = 8, \quad \text{and} \quad f(2, 3) = 9.$$

4. It is often convenient to represent the value of a function of x by a single letter; thus, for example, $y = x^2$. When this notation is used, if we represent the independent variable x by the abscissa of a point, and the function y by the corresponding ordinate, a curve may be constructed which will graphically represent the function, and will serve to illustrate its peculiarities.

Rectangular coordinates are usually employed for this purpose. See diagram, Art. 10.

A function of the form

$$y = mx + b,$$

m and b being constants, is represented by a straight line. Functions of this form are, for this reason, called *linear functions*.

Implicit Functions.

5. When an equation is given involving two variables x and y, either variable is obviously a function of the other; and the former variable, when its value is not *directly* expressed in terms of the other, is said to be an *implicit* function of the latter. Thus, if we have

$$ax^2 - 3axy + y^3 - a^3 = 0,$$

either variable is an implicit function of the other.

By solving the above equation for x, we obtain

$$x = \frac{3y}{2} \pm \sqrt{\left(a^2 + \frac{9y^2}{4} - \frac{y^3}{a}\right)}.$$

In this form of the equation, x is said to be an *explicit* function of y.

This example will serve to illustrate the fact, that from a single equation involving two variables, there may be derived two or more explicit functions of the same variable. In the above case, x is said to be a *two-valued* function of y; while, since the equation is of the third degree in y, the latter is a *three-valued* function of x.

Inverse Functions.

6. If $y = f(x)$, x is some function of y; we may therefore write
$$y = f(x), \qquad \text{whence} \qquad x = \phi(y).$$

Each of the functions f and ϕ is then said to be the *inverse function* of the other. Thus, if
$$y = a^x, \qquad \text{we have} \qquad x = \log_a y;$$

hence each of these functions is the inverse of the other. So also the square and the square root are inverse functions.

7. In the case of the trigonometric functions, a peculiar notation for the inverse functions has been adopted. Thus, if we have
$$x = \sin \theta, \qquad \text{we write} \qquad \theta = \sin^{-1} x.$$

Whenever trigonometric functions are employed in the Calculus, the symbol representing the angle always denotes the *circular measure* of the angle; that is, the ratio of the arc to the radius. Hence $\sin^{-1} x$ may be read either "the inverse sine of x," or "the arc whose sine is x."

The inverse trigonometric functions are evidently many-valued. See Art. 54.

The Classification of Functions.

8. With reference to its *form*, an explicit function is either *algebraic* or *transcendental*.

An *algebraic function* is expressed by a definite combination of algebraic symbols, in which the exponents do not involve the independent variable.

§ I.]   THE CLASSIFICATION OF FUNCTIONS.   5

All functions not algebraic are classed as *transcendental*. Under this head are included *exponential* functions; that is, those in which one or more exponents are functions of the variable, as, for example, a^x, $x a^{\sqrt{x}}$, etc.: logarithmic functions: the direct and inverse trigonometric functions, and other forms which arise in the higher branches of mathematics.

9. With reference to its mode of variation, a function is said to be an *increasing function* when it increases and decreases with x; and a *decreasing function* when it decreases as x increases, and increases as x decreases. Thus, it is evident that x^3 is always an increasing function of x, while $\dfrac{1}{x}$ is always a decreasing function of x. Again, $\tan x$ is always an increasing function, but $\sin x$ is sometimes an increasing and sometimes a decreasing function of x.

10. The increase and decrease here considered are *algebraic*. For example, x^2 is an increasing function when x is positive, but when x is negative it becomes a decreasing function; for, when x is negative and algebraically increasing, x^2 is decreasing.

The curve $y = x^2$ which illustrates this function is constructed in Fig. 1. Since algebraic increase in the value of x is represented by motion from left to right, whether the moving point is on the left or on the right of the axis of y, the downward slope of the curve on the left of the origin indicates that x^2 is a decreasing function when x is negative.

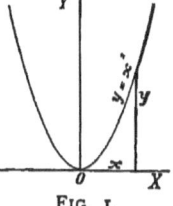

FIG. 1.

Expressions involving an Unknown Function.

11. An expression involving $f(x)$, as, for example, $x f(x)$ or $F[f(x)]$, is generally a function of x; but it may happen that such an expression has a value independent of x. Thus, suppose that, in the course of an investigation, the following equation presents itself:—

$$xf(x) = zf(z),$$

in which f denotes an unknown function, and x and z are entirely independent arbitrary quantities. When this is the case, we can make z a fixed quantity, and give to x any value whatever; that is, we can make x a variable and z a constant; but if z is a constant, $zf(z)$ is likewise a constant, we can, therefore, write

$$xf(x) = c,$$

c being an unknown constant. Hence we have

$$f(x) = \frac{c}{x}.$$

The value of the constant c is readily found, if we know the value of $f(x)$ corresponding to any one value of x.

Examples I.

1. (α) For what value of n does x^n cease to be a function of x? (β) For what values of x does it cease to be a function of n?
(α) When $n = 0$. (β) When $x = 1$, or $x = 0$.

2. If $y\left(1 - \dfrac{a-x}{a+x}\right) = x + \dfrac{ax - x^2}{a + x}$, show that y is a function of a, but not of x.

3. Show that $\sin x \tan \tfrac{1}{2}x + \cos x$ is not a function of x.

4. If $y = x + \sqrt{(1 + x^2)}$, show that $y^2 - 2xy$ is not a function of x.

5. If $f(x) = x^2$, find the value of $f(x + h)$; of $f(2x)$; of $f(x^2)$; of $f(x^2 - x)$; of $f(1)$; $f(12)$; $f[f(x)]$.

$$f(x + h) = x^2 + 2hx + h^2.$$

§ I.]   EXAMPLES OF FUNCTIONS.   7

6. If $f(\theta) = \cos\theta$, find the value of $f(0)$; of $f(\tfrac{1}{3}\pi)$; of $f(\tfrac{1}{2}\pi)$; of $f(\pi)$.

7. If $F(x) = a^x$, give the value of $F(a)$; of $F(1)$; of $F(0)$. Also show that in this case $[F(x)]^2 = F(2x)$.

8. Given $y^2 - 2ay + x^2 = 0$, make y an explicit function of x.
$$y = a \pm \sqrt{(a^2 - x^2)}.$$

9. Given $1 + \log_a y = 2 \log_a (x + a)$, make y an explicit function of x.
$$y = \frac{(x+a)^2}{a}.$$

10. Given the equations—
$$n + 1 = n(\cos^2\theta' + \cos\theta' \cos\theta + \cos^2\theta),$$
and
$$n - 1 = n(\sin^2\theta' + \sin\theta' \sin\theta + \sin^2\theta);$$

eliminate n, and make θ' an explicit function of θ. Also make n an explicit function of θ.
$$\theta' = \theta \pm \tfrac{1}{3}\pi, \text{ and } n = \mp \frac{1}{\sin\theta \cos\theta}.$$

11. Given $\sin^{-1} x + \sin^{-1} y = \alpha$, make y an explicit function of x.
$$y = \sin\alpha \sqrt{(1 - x^2)} - x \cos\alpha.$$

12. Given $\tan^{-1} x + \tan^{-1} y = \alpha$, make y an explicit function of x.
$$y = \frac{\tan\alpha - x}{1 + x \tan\alpha}.$$

13. Given $xy - 2x + y = n$, show that y is not a function of x when $n = 2$.

14. If $y = \dfrac{2x - 1}{3x - 2}$, show that the inverse function is of the same form.

15. If $y = f(x) = \dfrac{1 + x}{1 - x}$, find $z = f(y)$, and express z as a function of x.
$$z = -\frac{1}{x}.$$

16. If both f and ϕ denote increasing functions, or, if both denote decreasing functions, show that $\phi[f(x)]$ is an increasing function. Also show that the inverse of an increasing function is an increasing function.

17. Find the inverse of the function, $y = \log_e [x + \sqrt{(1 + x^2)}]$.

$$x = \tfrac{1}{2}(e^y - e^{-y}).$$

18. If $f(x)$ be an unknown function having the property

$$f(x) + f(y) = f(xy),$$

prove that $f(1) = 0$.
Put $y = 1$.

19. If $f(x)$ has the property

$$f(x + y) = f(x) + f(y),$$

prove that $f(0) = 0$. Also prove that the function has the property

$$f(px) = pf(x),$$

in which p is a positive or negative integer.
For positive integers, put $y = x$, $2x$, $3x$, etc., in the given equation; for negative integers, put $y = -x$.

20. If f denotes the same function as in Example 19, prove that

$$f(mx) = mf(x),$$

m denoting any fraction.

Solution :—

Putting $z = \dfrac{p}{q}x$, $\qquad qz = px$,

$$f(qz) = f(px);$$

hence, by Example 19, $\qquad qf(z) = pf(x),$

or $\qquad f(z) = \dfrac{p}{q}f(x),$

$\therefore \qquad f\left(\dfrac{p}{q}x\right) = \dfrac{p}{q}f(x).$

21. Given, the property of the same function proved in Example 20; viz.,

$$f(mx) = mf(x);$$

by putting z for mx, show that

$$\frac{1}{z}f(z) = \frac{1}{x}f(x),$$

and thence deduce the form of the function. *See Art.* 11.

22. Given, $[\phi(x)]^z = [\phi(z)]^x$, and $\phi(1) = \varepsilon$, determine $\phi(x)$.

$f(x) = cx.$

$\phi(x) = \varepsilon^{\frac{1}{x}}$

23. Given $\phi(x) + \phi(y) = \phi(xy)$
prove $\phi(x^m) = m\phi(x),$
and thence prove $\phi(x) = c \log x.$
Use the methods of Examples 19, 20, *and* 21.

II.

Rates.

12. In the Differential Calculus, variable quantities are regarded as undergoing continuous variation in magnitude, and the *rates* of variation, denoted by appropriate symbols, are employed in connection with the values of the variables themselves.

If a varying quantity be represented by the distance of a point moving in a straight line from a fixed origin taken on that line, the velocity of the moving point will represent the rate of increase or decrease of the varying quantity.

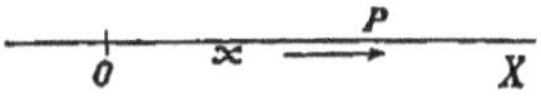

FIG. 2.

Thus O (Fig. 2) being the fixed origin and OP a variable denoted by x, P is the moving point whose velocity represents the rate of x. The velocity of P, or the rate of x, is regarded as positive when P moves in the direction in which x *increases algebraically;* thus, taking the direction OX, or toward the right, as the positive direction in laying off x, the velocity is positive when P moves toward the right, whether its position be on the right or on the left of the origin. Accordingly, a rate of *algebraic decrease* is considered as negative, and would be represented by a point moving toward the left.

Constant Rates.

13. The rate of a quantity like the velocity of a point may be either constant or variable. A velocity is uniform or constant, when the spaces passed over in any equal intervals of time are equal, or, in other words, *when the spaces passed over in any intervals of time are proportional to the intervals.*

The numerical measure of a uniform velocity *is the space passed over in a unit of time;* then if t denote the time elapsed from an assumed origin of time, and k the space passed over by a moving point in a unit of time, kt will denote the space passed over in the time t. Hence, whenever the velocity is uniform, the quotient obtained by dividing the number of units of space by the number of units of time occupied in describing this space is constant, and serves as the numerical measure of the velocity.

14. Now, if x be a quantity having a uniform rate k, it will be represented by the distance from the origin of a point having the uniform velocity k, and if a denote the value of x when t is zero, we shall have

$$x = a + kt. \quad \ldots \quad \ldots \quad (1)$$

This formula expresses a uniformly varying quantity as a function of t. When x is a uniformly decreasing quantity, k is, of course, negative.

Conversely, if x, when expressed as a function of t, is of the form (1), involving the first power only of t, then x is a quantity having a uniform rate, and the coefficient k is a measure of this rate.

Variable Velocities.

15. If the velocity of a point be *not* uniform, *its numerical measure at any instant is the number of units of space which would be described in a unit of time, were the velocity to remain constant from and after the given instant.*

Thus, when we speak of a body as having at a given instant a velocity of 32 feet per second, we mean that should the body continue to move during the whole of the next second, with the same velocity which it had at the given instant, 32 feet would be described. The *actual* space described may be greater or less, in consequence of the change in velocity which takes place during the second; it is, for instance, greater than the measure of the velocity at the beginning of the second, in the case of a falling body, because the velocity increases throughout the second.

16. Attwood's machine for determining experimentally the velocities acquired by falling bodies furnishes a familiar example of the practical application of the principle embodied in the above definition.

This apparatus consists essentially of a thread passing over a fixed pulley, and sustaining equal weights at each extremity, the pulley being so constructed as to offer but slight resistance to turning. On one of the weights a small bar of metal is placed, which, destroying the equilibrium, causes the weight to descend with an increasing velocity. To determine the value of this velocity at any point, a ring is so placed as to intercept the bar at that point, and allow the weight to pass. Thus, the sole cause of the variation of the velocity having been removed, the weight moves on uniformly with the required velocity, and the space described during the next second becomes the measure of this velocity.

Variable Rates.

17. When x is a function of t, but not of the form expressed by equation (1), Art. 14—that is, when the function is not linear—the rate of x will be variable. To obtain the measure of this rate at any given instant, we employ the same principle as in the case of a variable velocity. Thus, let x be represented by OP as in Fig. 2, Art. 12, let the symbol dt denote an assumed interval of time, and let dx denote the space which would be described in the time dt, were P to move with the velocity which it has at the given instant unchanged throughout the interval of time dt. Then the space which would be described in a unit of time is, evidently,

$$\frac{dx}{dt},$$

which is therefore the measure of the velocity of P, or the rate of x.

This ratio is in general variable, but, when x is of the form $a + kt$, it has been shown in Art. 14 that k is the measure of the rate; we therefore have

$$\frac{dx}{dt} = k, \qquad \text{when} \qquad x = a + kt.$$

Differentials.

18. The quantities dx and dt are called respectively "the differential of x" and "the differential of t."

In accordance with the definition of dx given in the preceding article, the *differential* of a variable quantity at any instant is the increment which would be received in the time dt, were the quantity to continue to increase uniformly during that interval of time with the rate it has at the given instant. *The quotient obtained by dividing the differential of any quantity by* dt *is therefore the measure of the rate of the quantity.*

The differential of a quantity is denoted by prefixing d to the symbol denoting the quantity; when the symbol denoting the quantity is not a single letter it is usually enclosed by marks of parenthesis to avoid ambiguity. Thus, $d(x^2)$, $d(xy)$, $d(\tan x)$, $d(a^x + x^n)$, etc.

The Differentials of Polynomials.

19. Let x and y denote two variable quantities, and let a and b denote particular simultaneous values of x and y, while k and k' denote corresponding values of the rates of x and y.

Now, if x and y should continue to vary with these rates, their values would (see Art. 14) be expressed by

$$x = a + kt,$$
and $$y = b + k't,$$
whence $$x + y = a + b + (k + k')t.$$

Thus the quantity $x+y$ would become a uniformly varying quantity, and, by Art. 14, its rate would be $k+k'$, which, therefore, is the measure of the rate of $x+y$ at the instant when x and y have the rates k and k'. Consequently,

$$\frac{d(x+y)}{dt} = k + k' = \frac{dx}{dt} + \frac{dy}{dt}.$$

Now, since k and k' denote any values of the rates, this equation is universally true. We have, therefore,

$$d(x+y) = dx + dy. \quad \ldots \ldots \quad (1)$$

This formula is easily extended to the sum of any number of variables. Thus,

$$d(x+y+z+\cdots) = dx + d(y+z+\cdots) = dx + dy + dz + \ldots \quad \ldots \quad (2)$$

20. The differential of a constant is evidently zero, hence

$$d(x + h) = dx. \quad \ldots \ldots \ldots \quad (3)$$

Again, if $\quad y = -x, \quad\quad y + x = 0,$

hence, by equation (1), since zero is a constant, we have

$$dy + dx = 0, \quad\quad \text{or} \quad\quad dy = -dx;$$

that is, $\quad\quad d(-x) = -dx. \quad \ldots \ldots \ldots \quad (4)$

The differential of a negative term is therefore the negative of the differential of the term taken positively.

It appears, on combining the results expressed in equations (2), (3), and (4), that *the differential of a polynomial is the algebraic sum of the differentials of its terms; and that constant terms disappear from the result.*

The Differential of a Term having a Constant Coefficient.

21. Let the term be denoted by mx, m denoting a constant.

Resuming equation (2), Art. 19; viz.,

$$d(x + y + z + \cdots) = dx + dy + dz + \cdots,$$

and denoting the number of terms by p, we put

$$x = y = z = \cdots,$$

thus obtaining $\quad\quad d(px) = p\,dx, \quad \ldots \ldots \ldots \quad (1)$

p denoting an integer.

§ II.] THE DIFFERENTIAL OF (mx).

To extend equation (1) to the case in which m denotes a fraction, let

$$z = \frac{p}{q} x, \qquad \text{then} \qquad qz = px.$$

By applying equation (1) we obtain

$$q\,dz = p\,dx, \qquad \text{or} \qquad dz = \frac{p}{q} dx;$$

that is,

$$d\left(\frac{p}{q} x\right) = \frac{p}{q} dx.$$

Hence generally, when m is positive,

$$d(mx) = m\,dx. \quad \ldots \ldots \ldots (2)$$

Since $d(-x) = -dx$, this equation is true likewise when m is negative.

It therefore follows that *the differential of a term having a constant coefficient is equal to the product of the differential of the variable factor by the constant coefficient.*

Examples II.

1. Find the differential of $\frac{2x}{3a}$, and of $\frac{x}{m-2}$.

   $\frac{2\,dx}{3a}$, and $\frac{dx}{m-2}$.

2. Find the differential of $\frac{x-a}{m^2}$, and of $\frac{a-x}{m^2}$.

   $\frac{dx}{m^2}$, and $-\frac{dx}{m^2}$.

3. Find the differential of $\frac{a+b+(a-b)x}{a^2-b^2}$.

   $\frac{dx}{a+b}$.

4. Find the differential of $\frac{a+x}{a+b}$, and of $\frac{b(x+y)}{a(a+b)}$.

   $\frac{dx}{a+b}$, and $\frac{b(dx+dy)}{a(a+b)}$.

5. Given $ay + bx + 2cx + ab = 0$, to find $\frac{dy}{dx}$.

$$\frac{dy}{dx} = -\frac{b+2c}{a}.$$

6. Given $y \log a + x \sin \alpha - y \cos \alpha - ax + \tan \alpha = 0$, to find $\frac{dy}{dx}$.

$$\frac{dy}{dx} = \frac{a - \sin \alpha}{\log a - \cos \alpha}.$$

7. Given $ay \cos^2 \alpha - 2b(1 - \sin \alpha)x = b(a - x \cos^2 \alpha)$, to find $\frac{dy}{dx}$.

$$\frac{dy}{dx} = \frac{b(1 - \sin \alpha)}{a(1 + \sin \alpha)}.$$

8. Given $a^2 + 2(1 + \cos \alpha)y = (x + y)\sin^2 \alpha$, to find $\frac{dy}{dx}$.

$$\frac{dy}{dx} = \tan^2 \frac{\alpha}{2}.$$

9. Given $\frac{x}{a} + \frac{y}{b} + \frac{z}{c} = 1$, to express dz in terms of dx and dy.

$$dz = -\frac{c}{a} dx - \frac{c}{b} dy.$$

10. A man whose height is 6 feet walks directly away from a lamp-post at the rate of 3 miles an hour. At what rate is the extremity of his shadow travelling, supposing the light to be 10 feet above the level pavement on which he is walking?

Draw a figure, and denote the variable distance of the man from the lamp-post by x, and the distance of the extremity of his shadow from the post by y.   7½ miles per hour.

11. At what rate does the man's shadow (Ex. 10) increase in length?

III.

Differentials of Functions of an Independent Variable.

22. When the variables involved in any mathematical investigation are functions of an independent variable x, the latter may be assumed to have a rate denoted by $\frac{dx}{dt}$, in which

dx is arbitrary. So also the corresponding rate of y will be denoted by $\dfrac{dy}{dt}$, and, if y is a function of x, the value of dy will depend in part upon the assumed value of dx.

To *differentiate* a function of x is to express its differential in terms of x and dx.

It is to be understood, of course, that the differentials involved in an equation are all taken with reference to the same value of dt.

If two quantities are always equal, their simultaneous rates are evidently equal; and hence their differentials are likewise equal. We can therefore *differentiate an equation;* that is, express the equality of the differentials of its members; provided the equation is true for all values of the variables involved. Thus, from the identical equation

$$(x+h)^2 = x^2 + 2hx + h^2,$$

it follows that $\quad d[(x+h)^2] = d(x^2) + 2h\, dx$.

The Derivative.

23. Before proceeding to the differentiation of the various functions of x, it is necessary to show that, if

$$y = f(x), \quad \dots \dots \dots (1)$$

the ratio $\qquad \dfrac{dy}{dx}$

has a definite value for each value of x, *independent of the assumed value of* dx.

Let a particular value of x be denoted by a, and let the corresponding value of dx be an arbitrary quantity.

Now, although dx is arbitrary, since dt is likewise arbitrary, the rate of x, that is, the ratio

$$\dfrac{dx}{dt}, \quad \dots \dots \dots (2)$$

may be assumed to have a certain *fixed value* at the instant when $x = a$. The corresponding value of the rate of y, denoted by

$$\frac{dy}{dt}, \quad \ldots \ldots \ldots \quad (3)$$

evidently depends solely upon the rate of x and upon the form of the function f in equation (1). Hence, when the value of the rate (2) is fixed, the value of (3) is also definitely fixed.

Denoting these fixed values by k and k', we have, when $x = a$,

$$\frac{dx}{dt} = k, \text{ and } \frac{dy}{dt} = k', \text{ whence } \frac{dy}{dx} = \frac{k'}{k}.$$

Hence, corresponding to a particular value a of x, there exists a determinate value $\frac{k'}{k}$, of the ratio $\frac{dy}{dx}$, notwithstanding the fact that dx has an arbitrary value; in other words, *the value of the ratio $\frac{dy}{dx}$ is independent of the arbitrary value of* dx.

24. It is obvious that, in general, this ratio will have different values corresponding to different values of x, and hence that it may be expressed as a function of x, and denoted by $f'(x)$; thus,—

$$\frac{dy}{dx} = f'(x). \quad \ldots \ldots \quad (1)$$

The form of this new function f' will evidently depend upon that of the given function f.

The function $f'(x)$ is called the *derivative* of $f(x)$, and, since equation (1) may be written in the form

$$dy = f'(x)\,dx,$$

it is also called the *differential coefficient* of y regarded as a function of x.

When, however, the given function $f(x)$ is of the linear form
$$y = mx + b,$$
the derivative is no longer a function of x, but is a constant, since the value of y gives

$$dy = m\, dx,$$

or $\qquad\dfrac{dy}{dx} = m.$

The Geometrical Meaning of the Derivative.

25. Representing the corresponding values of x and y by the rectangular coordinates of a moving point, if this point move in a uniform direction, so as to describe a straight line,— that is, if y be a linear function of x,—the value of $\dfrac{dy}{dx}$ will be constant, by the preceding article. Hence, in the general case, when this ratio is variable, the point will move in a variable direction.

If we denote the inclination of this direction to the axis of x by ϕ, the value of ϕ will vary with the value of x, and the point will describe a curve.

The tangent line to a curve is defined as follows:—

The tangent to a curve at any point is the straight line which passes through the point, and has the direction of the curve at that point.*

Hence, for any point of the curve, ϕ denotes the inclination to the axis of x of the tangent line at that point.

---

* It will be shown hereafter (Art. 49) that, in the case of the circle, this general definition of a tangent line agrees with that usually given in Plane Geometry.

26. Now, if a point, at first moving in the curve, should, after passing the point whose abscissa is a, so move that the rates $\dfrac{dx}{dt}$ and $\dfrac{dy}{dt}$ retain the values which they had at the instant of passing the given point, the direction of its motion will become constant, and the point will describe a straight line tangent to the curve at the given point.

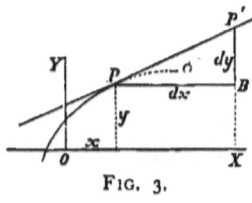

Fig. 3.

The value of dx may be represented by an arbitrary increment of x as in Fig. 3; the value of dy will then be represented by the corresponding increment which would be received by y, were the point moving in the tangent line, as indicated in the diagram. Hence

$$\frac{dy}{dx} = \tan \phi,$$

which is evidently independent of the assumed value of dx.* It follows that the value of the derivative of $f(x)$, for any value of x, is represented by the trigonometric tangent of the inclination to the axis of x of the curve $y = f(x)$, at the point corresponding to the given value of x.

27. The moving point, which is conceived to describe the curve, may pass over it in either of two directions differing by 180°. The two corresponding values of ϕ give, however, the same value of $\tan \phi$, since $\tan (\phi \pm 180°) = \tan \phi$.

Thus, in Fig. 3, the point P may be regarded as moving so as to increase x and y, in which case both dx and dy will be positive, and ϕ will be in the first quadrant; or P may

---

* In other words, the value of the derivative is determined by the form of the function f which determines the curve, and the value of x which fixes the position of P.

§ III.] *GEOMETRICAL MEANING OF THE DERIVATIVE.* 21

move in the opposite direction, making dx and dy negative, and placing ϕ in the third quadrant. In either case, $\dfrac{dy}{dx}$ or $\tan\phi$ is positive.

28. It is evident that when $f(x)$ is an increasing function, as in Fig. 3, $\dfrac{dy}{dx}$ is positive, and that when it is a decreasing function, $\dfrac{dy}{dx}$ is negative.

Thus the sign of $f'(x)$ for any value of x is positive or negative according as $f(x)$ is, for that value of x, an increasing or a decreasing function. For example, it is evident that the value of the derivative of $\sin x$ must be positive when x is between 0 and $\tfrac{1}{2}\pi$, negative when x is between $\tfrac{1}{2}\pi$ and $\tfrac{3}{2}\pi$, and so on.

When the notation $\dfrac{dy}{dx}$ is used, the value of the derivative corresponding to a particular value a of x is expressed by $\left.\dfrac{dy}{dx}\right]_a$, which is equivalent to $f'(a)$. See Art. 2.

Examples III.

1. If a point move in the straight line $2y - 7x - 5 = 0$, so that its ordinate decreases at the rate of 3 units per second, at what rate is the point moving in the direction of the axis of x?

$$\frac{dx}{dt} = -\frac{6}{7}.$$

2. If a point starting from $(0, b)$ move so that the rates of its co-ordinates are k and k', show that its path is $y = mx + b$, m being equal to $\dfrac{k'}{k}$.

Express x and y in terms of t (Art. 14), and eliminate t.

3. If a point moving in a curve passes through the point (5, 3)

moving at equal rates upward and toward the left, find the value of $\left.\dfrac{dy}{dx}\right]_0$, also the equation of the tangent line to the curve at the given point. $\left.\dfrac{dy}{dx}\right]_0 = -1$, and $y+x=8$.

4. If a point is moving in the straight line

$$x \cos \alpha + y \sin \alpha = p,$$

its rate in the positive direction of the axis of x being $l \sin \alpha$, what is its rate of motion in the direction of the axis of y?

$$-l \cos \alpha.$$

5. Given $ay \sin \alpha - ax + ax \cos \alpha - b^2 \sec \alpha = 0$; show that ϕ is constant and equal to $\frac{1}{2}\alpha$.

6. If $f(x) = \tan x$, show that $f'(x)$ must always be positive.

7. Show, by tracing the curve, that if $y = x^3$, $\dfrac{dy}{dx}$ can never be negative.

CHAPTER II.

The Differentiation of Algebraic Functions.

IV.

The Square.

29. In establishing the formulas for the differentiation of the simple algebraic functions of an independent variable, we find it convenient to begin with the square. The object of this article is, therefore, to express $d(x^2)$ in terms of x and dx.

We first deduce a relation between two values of the derivative of the function and the corresponding values of the independent variable; for this purpose, we assume two values of the variable having a constant ratio m. Thus, if

$$z = m x, \qquad\qquad z^2 = m^2 x^2.$$

Differentiating by equation (2), Art. 21,

$$dz = m\, dx, \qquad \text{and} \qquad d(z^2) = m^2 d(x^2);$$

dividing, we obtain

$$\frac{d(z^2)}{dz} = m \frac{d(x^2)}{dx}.$$

Whence, dividing by $z = m x$ to eliminate m, we have

$$\frac{1}{z} \cdot \frac{d(z^2)}{dz} = \frac{1}{x} \cdot \frac{d(x^2)}{dx}. \quad \ldots \ldots (1)$$

The derivatives $\dfrac{d(z^2)}{dz}$ and $\dfrac{d(x^2)}{dx}$ are, by Art. 23, functions of z and of x respectively, independent of the values of dz and dx; moreover, equation (1) is true for all values of x and z, these quantities being entirely independent of each other, since the arbitrary ratio m has been eliminated. Therefore, either of these quantities may be assumed to have a fixed value, while the other is variable; hence it follows that the value of each member of this equation must be a fixed quantity, independent of the value of x or of z. Denoting this fixed value by c, we therefore write

$$\frac{1}{x} \cdot \frac{d(x^2)}{dx} = c,$$

or $\qquad\qquad d(x^2) = c\, x\, dx \quad \ldots\ \ldots\ \ldots\ (2)$

30. *To determine the unknown constant c,* we apply this result to the identity

$$(x + h)^2 = x^2 + 2hx + h^2.$$

Differentiating each member (Art. 22) by equation (2), we have

$$c\,(x + h)\, d(x + h) = c\, x\, dx + 2h\, dx\,;$$

since $d(x + h) = dx$, this equation reduces to

$$c\, h\, dx = 2h\, dx,$$
or $\qquad\qquad (c - 2)\, h\, dx = 0.$

Now, since h and dx are arbitrary quantities, this equation gives

$$c = 2\,;$$

this value of c substituted in (2) gives

$$d(x^2) = 2x\, dx. \quad \ldots\ \ldots\ \ldots\ (a)$$

That is, *the differential of the square of a variable equals twice the product of the variable and its differential.*

31. Employing the derivative notation, this result may also be expressed thus:—

If $\qquad f(x) = x^2, \qquad f'(x) = 2x.$

This derivative is negative for negative values of x, therefore, for these values, x^2 is a decreasing function, as already mentioned (Art. 10) in connection with the curve illustrating this function.

Since x and dx are arbitrary, we may substitute for them any variable and its differential. Equation (a) therefore enables us to differentiate the square of any variable whose differential is known. Thus,—

$$d(5x - 3)^2 = 2(5x - 3)5dx = 10(5x - 3)dx.$$

Again, $\qquad d(ax^2 + bx)^2 = 2(ax^2 + bx)\,d(ax^2 + bx)$
$$= 2(ax^2 + bx)(2ax + b)\,dx.$$

The Square Root.

32. To derive the differential of the *square root*, we put

$$y = \sqrt{x},$$

whence $\qquad y^2 = x;$

differentiating by (a), $\quad 2y\,dy = dx,$

or $\qquad dy = \dfrac{dx}{2y}.$

But $y = \sqrt{x}, \therefore \qquad d(\sqrt{x}) = \dfrac{dx}{2\sqrt{x}}. \quad \ldots \ldots \ldots (b)$

That is, *the differential of the square root of a variable is equal to the quotient arising from dividing the differential of the variable by twice the given square root.*

Thus, $\quad d[\sqrt{(a^2 - x^2)}] = \dfrac{-x\,dx}{\sqrt{(a^2 - x^2)}}$,

or, using derivatives,

$$\dfrac{d[\sqrt{(a^2 - x^2)}]}{dx} = -\dfrac{x}{\sqrt{(a^2 - x^2)}}.$$

Examples IV.

1. Differentiate $(2x + 3)^2$, and find the numerical value of its rate, when x has the value 8, and is decreasing at the rate of 2 units per second.

The differential required is denoted by $d[(2x + 3)^2]$, and the rate by $\dfrac{d[(2x + 3)^2]}{dt}$; the given rate $\dfrac{dx}{dt} = -2$. $\qquad$ 152 units per second.

2. Find the numerical value of the rate of $(x^2 - 2x)^2$, when $x = 3$, and is increasing at the rate of $\tfrac{1}{2}$ of one unit per second.

Differentiate the given expression before substituting.

$\qquad$ 12 units per second.

3. Find the numerical value of the rate of $\sqrt{(y^2 + x^2)}$, when $y = 7$ and $x = -7$, if y is increasing at the rate of 12 units per second, and x at the rate of 4 units per second. $\qquad$ $4\sqrt{2}$ units per second.

4. If $f(x) = x - \sqrt{(x^2 - a^2)}$, find $f'(x)$, and show that $f(x)$ is a decreasing function. $\qquad f'(x) = 1 - \dfrac{x}{\sqrt{(x^2 - a^2)}}$.

5. Differentiate the identity $(\sqrt{x} + \sqrt{a})^2 = x + a + 2\sqrt{ax}$, and show that the result is an identity.

6. Differentiate $\sqrt{\left(\dfrac{x^2 - 2ax}{a^2 - 2ab}\right)}$.

The constant factor $\dfrac{1}{\sqrt{(a^2 - 2ab)}}$ should be separated from the variable factor before differentiation. $\qquad \dfrac{1}{\sqrt{(a^2 - 2ab)}} \cdot \dfrac{x - a}{\sqrt{(x^2 - 2ax)}} dx$.

§ IV.]   EXAMPLES.   27

7. If $f(x) = (1 + x^2)^{\frac{1}{2}}$, $f'(x) = \dfrac{x}{(1 + x^2)^{\frac{1}{2}}}.$

8. If $f(x) = \sqrt{(a^3 + 2b^2 x + c x^2)}$, $f'(x) = \dfrac{b^2 + c x}{\sqrt{(a^3 + 2b^2 x + c x^2)}}.$

9. If $f(x) = \sqrt{[x + \sqrt{(1 + x^2)}]}$, $f'(x) = \dfrac{\sqrt{[x + \sqrt{(1 + x^2)}]}}{2\sqrt{(1 + x^2)}}.$

10. If $f(x) = \dfrac{a^2}{x - \sqrt{(x^2 - a^2)}}$, $f'(x) = 1 + \dfrac{x}{\sqrt{(x^2 - a^2)}}.$
Rationalize the denominator before differentiating.

11. Given $\dfrac{x^2}{a^2} + \dfrac{y^2}{b^2} = 1$, express $\dfrac{dy}{dx}$ in terms of x, and give the values of $\left.\dfrac{dy}{dx}\right]_0$ and $\left.\dfrac{dy}{dx}\right]_a$. $\dfrac{dy}{dx} = \mp \dfrac{b}{a} \cdot \dfrac{x}{\sqrt{(a^2 - x^2)}}.$

12. Given $y^2 = 4a x$, express $\dfrac{dy}{dx}$ in terms of x, also in terms of y, and give the values of $\left.\dfrac{dy}{dx}\right]_a$ and $\left.\dfrac{dy}{dx}\right]_{4a}$. $\dfrac{dy}{dx} = \sqrt{\dfrac{a}{x}} = \dfrac{2a}{y}.$

13. A man is walking on a straight path at the rate of 5 ft. per second; how fast is he approaching a point 120 ft. from the path in a perpendicular, when he is 50 ft. from the foot of the perpendicular?

Solution:—

Let x denote the variable distance of the man from the foot of the perpendicular, so that $\dfrac{dx}{dt}$ may denote the known velocity of the man, and let a denote the length of the perpendicular (120 ft.); then the distance of the man from the point is $\sqrt{(a^2 + x^2)}$, of which the rate of change is denoted by

$$\dfrac{d[\sqrt{(a^2 + x^2)}]}{dt} = \dfrac{x}{\sqrt{(a^2 + x^2)}} \dfrac{dx}{dt}.$$

At the instant considered, $x = 50$ ft., while $a = 120$ ft., and $\dfrac{dx}{dt} = -5$ ft.

per second. By substituting these values, we obtain $-1\frac{4}{5}$. Hence his distance from the point is diminishing (that is, he is approaching it) at the rate of $1\frac{4}{5}$ ft. per second.

14. If the side of an equilateral triangle increase uniformly at the rate of 3 ft. per second, at what rate per second is the area increasing, when the side is 10 ft.? $15\sqrt{3}$ sq. ft.

15. A stone dropped into still water produces a series of continually enlarging concentric circles; it is required to find the rate per second at which the area of one of them is enlarging, when its diameter is 12 inches, supposing the wave to be then receding from the centre at the rate of 3 inches per second. 36π

16. If a circular disk of metal expand by heat so that the area A of each of its faces increases at the rate of 0.01 sq. ft. per second, at what rate per second is its diameter increasing?

17. A man standing on the edge of a wharf is hauling in a rope attached to a boat at the rate of 4 ft. per second. The man's hands being 9 ft. above the point of attachment of the rope, how fast is the boat approaching the wharf when she is at a distance of 12 ft. from it?   5 ft. per second.

18. A ladder 25 ft. long reclines against a wall; a man begins to pull the lower extremity, which is 7 ft. distant from the bottom of the wall, along the ground at the rate of 2 ft. per second; at what rate per second does the other extremity *begin* to descend along the face of the wall?   7 inches.

19. One end of a ball of thread is fastened to the top of a pole 35 ft. high; a man holding the ball 5 ft. above the ground moves uniformly from the bottom at the rate of five miles an hour, allowing the thread to unwind as he advances. What is the man's distance from the pole when the thread is unwinding at the rate of one mile per hour? $\frac{5}{2}\sqrt{6}$ ft.

20. A vessel sailing due south at the uniform rate of 8 miles per hour is 20 miles north of a vessel sailing due east at the rate of 10 miles an

§ IV.] EXAMPLES. 29

hour. At what rate are they separating—(α) at the end of $1\frac{1}{2}$ hours?
(β) at the end of $2\frac{1}{2}$ hours?
Express the distances in terms of the time.   (α) $5\frac{1}{17}$ miles per hour.

21. When are the two ships mentioned in the preceding example neither receding from nor approaching each other?
Put the expression for their rate of separation equal to zero.
When $t = \frac{49}{34}$ of an hour.

22. Derive, by the method employed in Art. 29 to determine the differential of the square, the result $d\left(\dfrac{1}{x}\right) = \dfrac{c\,dx}{x^2}$, c being an unknown constant.

V.

The Product.

33. Let x and y denote any two variables; in order to derive the differential of their product, we express xy by means of squares, since we have already obtained a formula for the differentiation of the square. From the identity

$$(x+y)^2 = x^2 + 2xy + y^2,$$

we derive

$$xy = \tfrac{1}{2}(x+y)^2 - \tfrac{1}{2}x^2 - \tfrac{1}{2}y^2.$$

Differentiating, $d(xy) = (x+y)(dx+dy) - x\,dx - y\,dy,$

therefore, $d(xy) = y\,dx + x\,dy.$ . . . . (*c*)

Since x and y denote any variables whatever, and dx and dy their differentials, we can substitute for x and y any variable expressions, and for dx and dy the corresponding differentials. Thus,

$$d[(1+x^2)\sqrt{(a^2-x^2)}] = \sqrt{(a^2-x^2)}\,2x\,dx - \dfrac{(1+x^2)x\,dx}{\sqrt{(a^2-x^2)}}$$

$$= \dfrac{2a^2 - 3x^2 - 1}{\sqrt{(a^2-x^2)}}\,x\,dx.$$

34. Formula (c) is readily extended to products consisting of any number of factors. Thus let $x_1\, x_2\, x_3\, \ldots\, x_p$ denote the product of p variable factors, then

$$d(x_1 x_2 x_3 \cdots x_p) = x_2 x_3 \cdots x_p\, dx_1 + x_1\, d(x_2 x_3 \cdots x_p)$$
$$= x_2 x_3 \cdots x_p\, dx_1 + x_1 x_3 \cdots x_p\, dx_2 + x_1 x_2\, d(x_3 \cdots x_p)$$
$$= x_2 x_3 \cdots x_p\, dx_1 + x_1 x_3 \cdots x_p\, dx_2 \cdots + x_1 x_2 \cdots x_{p-1}\, dx_p. \quad . \;(c')$$

The Reciprocal.

35. The differential of the reciprocal may now be obtained by means of the implicit form of this function. Denoting the function by y, we have

$$y = \frac{1}{x} \quad \therefore \quad xy = 1.$$

Differentiating the latter equation by formula (c), we obtain

$$y\, dx + x\, dy = 0,$$

whence
$$dy = -\frac{y\, dx}{x};$$

substituting the value of y,

$$d\!\left(\frac{1}{x}\right) = -\frac{dx}{x^2}. \quad \ldots \ldots \;(d)$$

Formula (d) enables us to differentiate any fraction of which the denominator alone is variable; thus,

$$d\!\left(\frac{a+b}{a+x}\right) = -(a+b)\frac{dx}{(a+x)^2}.$$

The Quotient.

36. By the term *quotient*, as used in this article, we mean a fraction whose numerator and denominator are both variable. In deriving its differential, the quotient is regarded as the product of its numerator by the reciprocal of its denominator. Thus, applying formulas (*c*) and (*d*),

$$d\left(\frac{x}{y}\right) = d\left(x\frac{1}{y}\right) = \frac{1}{y}dx + x\,d\left(\frac{1}{y}\right)$$
$$= \frac{dx}{y} - \frac{x\,dy}{y^2},$$
$$\therefore \quad d\left(\frac{x}{y}\right) = \frac{y\,dx - x\,dy}{y^2}. \quad \ldots \ldots \ldots (e)$$

It will be noticed that the negative sign belongs to the term which contains the differential of the denominator.

As an illustration of the application of this formula, we have

$$d\left(\frac{2x-a}{x^2+b}\right) = \frac{2(x^2+b) - 2x(2x-a)}{(x^2+b)^2}dx = 2\frac{b+ax-x^2}{(x^2+b)^2}dx.$$

Formula (*e*) is to be used *only when both terms of the fraction are variable;* for, when the numerator is constant, the fraction is equivalent to the product of a constant and the reciprocal of a variable, and, when the denominator is constant, it is equivalent to the product of a constant by a variable factor. Thus, if it be required to differentiate the fraction $\frac{x^2+a^2}{ax}$, the use of formula (*e*) may be avoided by first making the transformation,

$$\frac{x^2+a^2}{ax} = \frac{x}{a} + \frac{a}{x};$$

since, in this form, one term of each fraction is constant. Hence,

$$d\left(\frac{x^2+a^2}{ax}\right) = \frac{dx}{a} - \frac{a\,dx}{x^2}.$$

The Power.

37. To obtain the differential of the power when the exponent is a positive integer, suppose each of the variables $x_1, x_2, x_3, \cdots x_p$ in formula (c'), Art. 34, to be replaced by x. The first member contains p factors, and the second p terms; the equation therefore reduces to

$$d(x^p) = p\,x^{p-1}\,dx. \quad \ldots \ldots \quad (1)$$

Next, when the exponent is a fraction, let

$$y = x^{\frac{p}{q}}, \qquad \text{then} \qquad y^q = x^p;$$

differentiating by (1), p and q being positive integers, we have

$$q\,y^{q-1}\,dy = p\,x^{p-1}\,dx,$$

therefore, $$dy = \frac{p}{q} \cdot \frac{x^{p-1}}{y^{q-1}}\,dx.$$

Substituting the value of y,

$$d(x^{\frac{p}{q}}) = \frac{p}{q} \cdot \frac{x^{p-1}}{x^{p-\frac{p}{q}}}\,dx = \frac{p}{q}\,x^{\frac{p}{q}-1}\,dx. \quad \ldots \ldots \quad (2)$$

Again, when the exponent is negative, we have

$$x^{-m} = \frac{1}{x^m}.$$

Differentiating by formula (d), Art. 35, we obtain

$$d(x^{-m}) = -\frac{d(x^m)}{x^{2m}},$$

and, since m is positive, we have, by (1) or (2),

$$d(x^{-m}) = -\frac{mx^{m-1}dx}{x^{2m}} = -mx^{-m-1}dx. \quad \ldots \quad (3)$$

Equations (1), (2), and (3) show that, for all values of n,

$$d(x^n) = nx^{n-1}dx. \quad \ldots \quad \ldots \quad (f)$$

By giving to n the values 2, $\frac{1}{2}$, and -1, successively, it is readily seen that this more general formula includes formulas (a), (b) and (d).

38. It is frequently advantageous to transform a given expression by the use of fractional or negative exponents, and employ formula (f) instead of formulas (b) and (d). Thus,

$$d\left[\frac{1}{(a^2-2x^2)^2}\right] = d(a^2-2x^2)^{-2} = 8(a^2-2x^2)^{-3}x\,dx,$$

and $\quad d\left[\frac{1}{\sqrt{(a+x)^3}}\right] = d(a+x)^{-\frac{3}{2}} = -\frac{3}{2}(a+x)^{-\frac{5}{2}}dx.$

When the derivative of a function is required, it may be written at once instead of first writing the differential, since the former differs from the latter only in the omission of the factor dx, which must necessarily occur in every term. Thus, given

$$y = \frac{x}{\sqrt{(1+x^2)}} = x(1+x^2)^{-\frac{1}{2}},$$

we derive $\dfrac{dy}{dx} = (1+x^2)^{-\frac{1}{2}} - \frac{1}{2}x(1+x^2)^{-\frac{3}{2}} \cdot 2x = \dfrac{1}{(1+x^2)^{\frac{3}{2}}}.$

Examples V.

1. From the identity $xy = \frac{1}{4}(x+y)^2 - \frac{1}{4}(x-y)^2$ derive the formula for differentiating the product.

2. Differentiate $\dfrac{a+bx+cx^2}{x}$.

   Put the expression in the form $\dfrac{a}{x} + b + cx$. $\qquad \left(c - \dfrac{a}{x^2}\right)dx.$

3. Find the derivative of

   $y = \dfrac{a^2 - b^2}{a^2 - x^2}$. $\quad$ See remark, Art. 35. $\qquad \dfrac{dy}{dx} = (a^2 - b^2)\dfrac{2x}{(a^2 - x^2)^2}$.

4. $y = \sqrt{(x^3 - a^2)}$. $\qquad \dfrac{dy}{dx} = \dfrac{3x^2}{2\sqrt{(x^3 - a^2)}}$.

5. $y = \dfrac{2x^4}{a^2 - x^2}$. $\qquad \dfrac{dy}{dx} = \dfrac{4x^3(2a^2 - x^2)}{(a^2 - x^2)^2}$.

6. $y = (1 + 2x^2)(1 + 4x^3)$. $\qquad \dfrac{dy}{dx} = 4x(1 + 3x + 10x^2)$.

7. $y = (a^2 + x^2)(b^2 + 3x^2)$. $\qquad \dfrac{dy}{dx} = 3(5x^3 + b^2x + 2a^2)x$.

8. $y = (1 + x)^4(1 + x^2)^3$. $\qquad \dfrac{dy}{dx} = 4(1 + x)^3(1 + x^2)(1 + x + 2x^2)$.

9. $y = (1 + x^m)^n + (1 + x^n)^m$.

   $\dfrac{dy}{dx} = mn[(1 + x^m)^{n-1}x^{m-1} + (1 + x^n)^{m-1}x^{n-1}]$.

10. $y = \dfrac{x^2 - 2a^2}{x - a}$. $\qquad \dfrac{dy}{dx} = 1 + \dfrac{a^2}{(x-a)^2}$.

11. $y = \dfrac{a-x}{\sqrt{x}}$. $\qquad \dfrac{dy}{dx} = -\dfrac{a+x}{2x^{\frac{3}{2}}}$.

§ V.] EXAMPLES. 35

12. $y = \dfrac{\sqrt{(x^2 - a^2)}}{x}$. $\dfrac{dy}{dx} = \dfrac{a^2}{x^2 \sqrt{(x^2 - a^2)}}$.

13. $y = \dfrac{ab}{cx\sqrt{(x^2-a^2)}}$. See Art. 38. $\dfrac{dy}{dx} = -\dfrac{ab}{c} \cdot \dfrac{2x^2 - a^2}{x^2(x^2-a^2)^{\frac{3}{2}}}$.

14. $y = \dfrac{1}{\sqrt{(1+x)}} + \dfrac{1}{\sqrt{(1-x)}}$. $\dfrac{dy}{dx} = \tfrac{1}{2}[(1-x)^{-\frac{3}{2}} - (1+x)^{-\frac{3}{2}}]$.

15. $y = (1+x)\sqrt{(1-x)}$. $\dfrac{dy}{dx} = \dfrac{1-3x}{2\sqrt{(1-x)}}$.

16. $y = (a+x)^3(b-x)^4 x^2$.
$\dfrac{dy}{dx} = x(a+x)^2(b-x)^3[2ab + (5b-6a)x - 9x^2]$.

17. $y = \dfrac{x^n + 1}{x^n - 1}$. $\dfrac{dy}{dx} = -\dfrac{2n\,x^{n-1}}{(x^n - 1)^2}$.

18. $y = (3b + 2ax)^{\frac{3}{2}}(b - ax)$. $\dfrac{dy}{dx} = -5a^2 x \sqrt{(3b + 2ax)}$.

19. $y = \dfrac{a^2 - b^2}{(2ax - x^2)^{\frac{3}{2}}}$.

Put in the form $(a^2 - b^2)(2ax - x^2)^{-\frac{3}{2}}$. $\dfrac{dy}{dx} = 3(a^2 - b^2)\dfrac{x - a}{(2ax - x^2)^{\frac{5}{2}}}$.

20. $y = \dfrac{x}{\sqrt{(a^2 - x^2)}}$. $\dfrac{dy}{dx} = \dfrac{a^2}{(a^2 - x^2)^{\frac{3}{2}}}$.

21. $y = \dfrac{bx}{\sqrt{(2ax - x^2)}}$. $\dfrac{dy}{dx} = \dfrac{abx}{(2ax - x^2)^{\frac{3}{2}}}$.

22. $y = \sqrt{\dfrac{1+x}{1-x}}$. $\dfrac{dy}{dx} = \dfrac{1}{(1-x)\sqrt{(1-x^2)}}$.

23. $y = \dfrac{x}{\sqrt{(a^2 + x^2)} - x}$.

Rationalize the denominator. $\dfrac{dy}{dx} = \dfrac{1}{a^2}\left[\dfrac{a^2 + 2x^2}{\sqrt{(a^2 + x^2)}} + 2x\right]$.

24. Two locomotives are moving along two straight lines of railway which intersect at an angle of 60°; one is approaching the intersection at the rate of 25 miles an hour, and the other is receding from it at the rate of 30 miles an hour; find the rate per hour at which they are separating from each other when each is 10 miles from the intersection.

$2\frac{1}{2}$ miles.

25. A street-crossing is 10 ft. from a street-lamp situated directly above the curbstone, which is 60 ft. from the vertical walls of the opposite buildings. If a man is walking across to the opposite side of the street at the rate of 4 miles an hour, at what rate per hour does his shadow move upon the walls—(α) when he is 5 ft. from the curbstone? (β) when he is 20 ft. from the curbstone?

(α) 96 miles; (β) 6 miles.

26. Assuming the volume of a tree to be proportional to the cube of its diameter, and that the latter increases uniformly; find the ratio of the rate of its volume when the diameter is 6 inches to the rate when the diameter is 3 ft. $\frac{1}{36}$.

27. If an ingot of silver in the form of a parallelopiped expand $\frac{1}{1000}$ part of each of its linear dimensions for each degree of temperature, at what rate per degree of temperature is its volume increasing when the sides are respectively 2, 3, and 6 inches?

If x denote a side, dx may be assumed to denote the rate per degree of temperature. $\frac{27}{250}$ of a cubic inch.

28. Prove generally that, if the coefficient of expansion of each linear dimension of a solid is k, its coefficient of expansion in volume is $3k$.

Solution :—

Let x denote any side; then, if V denote the volume, we shall have $V = cx^3$; c being a constant dependent on the shape of the body.

Therefore $\quad dV = 3cx^2\, dx\,;$

or, since $\quad dx = kx,$

$\quad dV = 3kcx^3 = 3kV.$

CHAPTER III.

The Differentiation of Transcendental Functions.

VI.

The Logarithmic Function.

39. In this chapter, the formulas for the differentiation of the simple transcendental functions are to be established.

We begin by deducing the differential of the logarithmic function, employing the method exemplified in Art. 29.

The symbol $\log x$ is used in this article to denote the logarithm of x to any base, and $\log_b x$ is used when we wish to designate a particular base b.

Let $\quad z = mx, \quad \therefore \quad \log z = \log m + \log x,$

differentiating by Art. 21,

$$dz = m\, dx, \quad \text{and} \quad d(\log z) = d(\log x);$$

whence
$$\frac{d(\log z)}{dz} = \frac{d(\log x)}{m\, dx}$$

Multiplying by $z = mx$, to eliminate m, we obtain

$$z\,\frac{d(\log z)}{dz} = x\,\frac{d(\log x)}{dx}. \quad \ldots \ldots (1)$$

The derivatives, $\dfrac{d(\log z)}{dz}$ and $\dfrac{d(\log x)}{dx}$, are, by Art. 23, functions of z and of x respectively, independent of the values of dz and dx; moreover, equation (1) is true for all values of x and z, these quantities being entirely independent of each other, since the arbitrary ratio m has been eliminated. Hence, in equation (1), one of the quantities, x or z, may be assumed to have a fixed value, while the other is variable; whence it follows that the members of this equation have a fixed value independent of the values of x and z; we therefore write

$$x \frac{d(\log x)}{dx} = \text{a constant.} \quad \ldots \ldots (2)$$

This constant, although independent of x, may be dependent on the value of the base of the system of logarithms under consideration. Denoting the base of the system by b, we therefore denote the constant by B, and write equation (2) thus,—

$$d(\log_b x) = \frac{B\,dx}{x}. \quad \ldots \ldots \ldots (3)$$

40. *To determine the value of B*, we establish a relation between two values of the base and the corresponding values of this unknown quantity.

Denoting another value of the base by a, and the corresponding value of the unknown constant by A, we have

$$d(\log_a x) = \frac{A\,dx}{x}. \quad \ldots \ldots \ldots (4)$$

The relation sought may now be obtained by differentiating, by means of (3) and (4), the identical equation

$$\log_a x = \log_a b \, \log_b x,* \quad \ldots \ldots (5)$$

---

* This identity is most readily obtained thus,—by definition

$$x = b^{\log_b x}.$$

thus obtaining $$\frac{A\,dx}{x} = \log_a b \frac{B\,dx}{x},$$

or $$B \log_a b = A,$$

hence $$\log_a b^B = A,$$

that is, A is the logarithm to the base a of b^B; whence we have
$$b^B = a^A. \quad \ldots \ldots \ldots \quad (6)$$

Now, it is obvious that the value of a^A cannot depend upon b, hence equation (6) shows that the value of b^B likewise cannot depend upon b; b^B must, therefore, have a value entirely independent of b. Denoting this constant value by ε, we write
$$b^B = \varepsilon. \quad \ldots \ldots \ldots \quad (7)$$

Adopting this constant as a base, and taking the logarithms of each member of equation (7), we have
$$B \log_\varepsilon b = 1,$$

whence $$B = \frac{1}{\log_\varepsilon b}.$$

Introducing this value of B in equation (3), we obtain
$$d(\log_b x) = \frac{dx}{\log_\varepsilon b \cdot x} \quad \ldots \ldots \quad (g)$$

In this equation, the differential of a logarithm to any given base is expressed by the aid of the unknown constant ε.

41. The constant ε is employed as the base of a system of

---

taking the logarithm to the base a of each member, we have

$$\log_a x = \log_b x \, \log_a b.$$

logarithms, sometimes called *natural* or *hyperbolic*, but more commonly *Napierian* logarithms, from the name of the inventor of logarithms. Hence ε is known as the *Napierian base*.

Putting $b = \varepsilon$ in formula (g) we derive

$$d(\log_\varepsilon x) = \frac{dx}{x}. \qquad \ldots \ldots (g')$$

The logarithms employed in analytical investigations are almost exclusively Napierian. Whenever it is necessary, for the purpose of obtaining numerical results, these logarithms may be expressed in terms of the common tabular logarithms by means of the formula,

$$\log_{10} x = \log_{10} \varepsilon \; \log_\varepsilon x,$$

which is derived from equation (5), Art. 40, by writing 10 for a and ε for b. The value of the constant $\log_{10} \varepsilon$ will be computed in a subsequent chapter.

Hereafter, whenever the symbol log is employed without the subscript, $\log_\varepsilon$ is to be understood.

The Logarithmic Curve.

42. The curve, corresponding to the equation

$$y = \log_\varepsilon x \qquad \ldots \ldots (1)$$

is called the *logarithmic curve*.

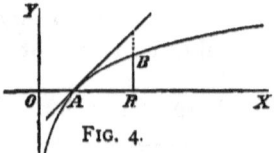

Fig. 4.

The shape of this curve is indicated in Fig. 4. It passes through the point A whose coordinates are (1, 0), since

$$\log 1 = 0.$$

Since we have, from formula (g'),

§ VI.] LOGARITHMIC DIFFERENTIATION. 41

$$\tan\phi = \frac{dy}{dx} = \frac{1}{x}, \quad \ldots \ldots (2)$$

the value of $\tan\phi$ at the point A is unity, and therefore the tangent line at this point cuts the axis of x at an angle of 45°, as in the diagram. We have from equation (2),

when $\quad\quad x > 1 \quad\quad \tan\phi < 1,$
and when $\quad x < 1 \quad\quad \tan\phi > 1$;

the curve, therefore, lies below this tangent, as shown in Fig. 4.

The point $(\varepsilon, 1)$ is a point of the curve; let B, Fig. 4, be this point, then OR will represent the Napierian base, and $BR = 1$. Since

$$OA = 1, \quad\quad \text{and} \quad\quad AR > BR,$$
$$OR > 2;$$

that is, the Napierian base ε is somewhat greater than 2.

The quantity ε is incommensurable: the method of computing its value to any required degree of accuracy is given in a subsequent chapter.

Logarithmic Differentiation.

43. The differential of the Napierian logarithm of the variable x, that is the expression $\dfrac{dx}{x}$, is called the *logarithmic differential* of x.

When x has a negative value, the expression $\log x$ has no real value; in this case, however, $\log(-x)$ is real, and we have

$$d[\log(-x)] = \frac{d(-x)}{-x} = \frac{dx}{x}.$$

This expression therefore, in the case of a negative quantity, is identical with the logarithmic differential of the positive quantity having the same numerical value.

44. The process of taking logarithms and differentiating the result is called *logarithmic differentiation*. By means of this method, all the formulas for the differentiation of algebraic functions may be derived.

In the following logarithmic equations, it is to be understood that that sign is taken in each case which will render the logarithm real.

By differentiating the formulas,—

$$\log(\pm xy) = \log(\pm x) + \log(\pm y),$$

$$\log\left(\pm \frac{x}{y}\right) = \log(\pm x) - \log(\pm y),$$

$$\log(\pm x^n) = n \log(\pm x),$$

we obtain

$$\frac{d(xy)}{xy} = \frac{dx}{x} + \frac{dy}{y},$$

$$\frac{y}{x} d\left(\frac{x}{y}\right) = \frac{dx}{x} - \frac{dy}{y},$$

$$\frac{d(x^n)}{x^n} = n \frac{dx}{x}.$$

These formulas are evidently equivalent to (*c*), (*e*), and (*f*), of which we thus have an independent proof.

45. The method of logarithmic differentiation may frequently be used with advantage in finding the derivatives of complicated algebraic expressions. For example, let us take

$$u = \frac{\sqrt{(2x)}(1-x^2)^{\frac{1}{4}}}{(x-2)^{\frac{3}{2}}}. \quad \ldots \ldots \quad (1)$$

Hence, we derive

§ VI.]   THE EXPONENTIAL FUNCTION.   43

$$\log u = \tfrac{1}{2} \log (2x) + \tfrac{3}{4} \log (1 - x^2) - \tfrac{2}{3} \log (x - 2), \quad . \quad . \quad (2)$$

differentiating,

$$\frac{du}{u\,dx} = \frac{1}{2x} - \tfrac{3}{2}\frac{x}{1 - x^2} - \tfrac{2}{3}\frac{1}{x - 2}, \quad . \quad . \quad . \quad (3)$$

adding and reducing,

$$\frac{du}{u\,dx} = \frac{-8x^3 + 24x^2 - x - 6}{6(1 - x^2)(x - 2)x};$$

therefore

$$\frac{du}{dx} = \frac{-8x^3 + 24x^2 - x - 6}{3(2x)^{\frac{1}{2}}(1 - x^2)^{\frac{1}{4}}(x - 2)^{\frac{2}{3}}}.$$

For certain values of x, one or more of the quantities whose logarithms appear in equation (2) become negative. When this is the case these logarithms should, strictly speaking, be replaced by the logarithms of the numerical values of the quantities in question; this change however would not affect the form of equation (3). See Art. 43.

Exponential Functions.

46. An *exponential function* is an expression in which an exponent is a function of the independent variable. The quantity affected by the exponent may be constant or variable. In the first case, let the function be denoted by

$$y = a^x. \quad . \quad . \quad . \quad . \quad . \quad . \quad (1)$$

If a is negative, a^x cannot denote a continuously varying quantity. We therefore exclude the case in which a has a negative value, and regard a^x as a continuously varying positive quantity.

Taking Napierian logarithms of both members of equation (1), we have

$$\log y = x \log a;$$

differentiating by (g'),

$$\frac{dy}{y} = \log a \, . \, dx \, ;$$

hence $\quad dy = \log a \, . \, y \, dx,$

or $\quad d(a^x) = \log a \, . \, a^x \, dx. \quad \ldots \ldots \quad (h)$

Exponential functions of the form ε^x are of frequent occurrence. Putting $a = \varepsilon$ in formula (h), we have

$$d(\varepsilon^x) = \varepsilon^x \, dx \, ; \quad \ldots \ldots \ldots \quad (h')$$

hence the derivative of the function ε^x is identical with the function itself. This function is the inverse of the Napierian logarithm; it has been proposed to denote it by the symbol $\exp x$.

47. When both the exponent and the quantity affected by it are variable, the method of logarithmic differentiation may be employed. Thus, if the given function be

$$z = (n\,x)^{x^2},$$

we shall have $\quad \log z = x^2 \log (n\,x) \, ;$

differentiating, $\quad \dfrac{dz}{z} = x^2 \dfrac{dx}{x} + 2x \log (n\,x) \, dx,$

hence $\quad d[(n\,x)^{x^2}] = (n\,x)^{x^2} x[1 + 2 \log (n\,x)] \, dx.$

Examples VI.

1. Given the function $y = \log_b x$; show that $\left.\dfrac{dy}{dx}\right]_\varepsilon = \dfrac{\log_b \varepsilon}{\varepsilon}$, and hence prove that the tangent to the corresponding curve, at the point whose abscissa is ε, passes through the origin.

Put $a = x = \varepsilon$ in equation 5, Art. 40.

§ VI.]   EXAMPLES.   45

2. $y = x^n \log x$. $\dfrac{dy}{dx} = x^{n-1}(1 + n \log x)$.

3. $y = \log(\log x)$. $\dfrac{dy}{dx} = \dfrac{1}{x \log x}$.

4. $y = \log[\log(a + bx^n)]$. $\dfrac{dy}{dx} = \dfrac{nb\, x^{n-1}}{(a + bx^n)\log(a + bx^n)}$.

5. $y = \sqrt{x} - \log(\sqrt{x} + 1)$. $\dfrac{dy}{dx} = \dfrac{1}{2(\sqrt{x}+1)}$.

6. $y = \log \dfrac{\sqrt{a} + \sqrt{x}}{\sqrt{a} - \sqrt{x}}$. $\dfrac{dy}{dx} = \dfrac{\sqrt{a}}{(a-x)\sqrt{x}}$.

Put in the form, $\log(\sqrt{a} + \sqrt{x}) - \log(\sqrt{a} - \sqrt{x})$.

7. $y = \log[\sqrt{(x-a)} + \sqrt{(x-b)}]$. $\dfrac{dy}{dx} = \dfrac{1}{2\sqrt{[(x-a)(x-b)]}}$.

8. $y = \log[x + \sqrt{(x^2 \pm a^2)}]$. $\dfrac{dy}{dx} = \dfrac{1}{\sqrt{(x^2 \pm a^2)}}$.

9. $y = \log \dfrac{x}{\sqrt{(1 + x^2)}}$. $\dfrac{dy}{dx} = \dfrac{1}{x(1 + x^2)}$.

10. $y = \log \dfrac{\sqrt{(1+x)} + \sqrt{(1-x)}}{\sqrt{(1+x)} - \sqrt{(1-x)}}$. $\dfrac{dy}{dx} = -\dfrac{1}{x\sqrt{(1-x^2)}}$.

11. $y = \log[x + \sqrt{(a^2 - x^2)}]$. $\dfrac{dy}{dx} = \dfrac{\sqrt{(a^2 - x^2)} - x}{\sqrt{(a^2 - x^2)}[x + \sqrt{(a^2 - x^2)}]}$.

12. $y = \log \dfrac{x}{\sqrt{(x^2 + a^2)} - x}$. $\dfrac{dy}{dx} = \dfrac{1}{x} + \dfrac{1}{\sqrt{(x^2 + a^2)}}$.

13. $y = \log[\sqrt{(1 + x^2)} + \sqrt{(1 - x^2)}]$. $\dfrac{dy}{dx} = \dfrac{1}{x} - \dfrac{1}{x\sqrt{(1 - x^4)}}$.

14. $y = \log(x - a) - \dfrac{a(2x - a)}{(x - a)^2}$. $\dfrac{dy}{dx} = \dfrac{x^2 + a^2}{(x - a)^3}$.

15. $y = a^{x^2}$. $\frac{dy}{dx} = 2\log a \cdot a^{x^2} x$.

16. $y = e^{\frac{1}{1+x}}$. $\frac{dy}{dx} = -\frac{1}{(1+x)^2} \cdot e^{\frac{1}{1+x}}$.

17. $y = e^x(1-x^3)$. $\frac{dy}{dx} = e^x(1-3x^2-x^3)$.

18. $y = (x-3)e^{2x} + 4xe^x$. $\frac{dy}{dx} = (2x-5)e^{2x} + 4(x+1)e^x$.

19. $y = \dfrac{e^x - e^{-x}}{e^x + e^{-x}}$. $\frac{dy}{dx} = \dfrac{4}{(e^x + e^{-x})^2}$.

20. $y = b^{a^x}$. $\frac{dy}{dx} = \log a \cdot \log b \cdot b^{a^x} \cdot a^x$.

21. $y = a^{x^n}$. $\frac{dy}{dx} = n a^{x^n} \cdot x^{n-1} \cdot \log a$.

22. $y = \dfrac{x}{e^x - 1}$. $\frac{dy}{dx} = \dfrac{e^x(1-x) - 1}{(e^x - 1)^2}$.

23. $y = \log(e^x + e^{-x})$. $\frac{dy}{dx} = \dfrac{e^x - e^{-x}}{e^x + e^{-x}}$.

24. $y = a^{\log x}$. $\frac{dy}{dx} = \dfrac{1}{x} \log_e a \cdot a^{\log x}$.

25. $y = \log \dfrac{e^x}{1 + e^x}$. $\frac{dy}{dx} = \dfrac{1}{1 + e^x}$.

26. $y = x^x$. $\frac{dy}{dx} = x^x(1 + \log x)$.

27. $y = \dfrac{(x-1)^{\frac{2}{3}}}{(x-2)^{\frac{3}{4}}(x-3)^{\frac{7}{5}}}$. $\frac{dy}{dx} = -\dfrac{(x-1)^{\frac{2}{3}}(7x^2 + 30x - 97)}{12(x-2)^{\frac{7}{4}}(x-3)^{\frac{19}{5}}}$.

See *Art.* 45.

28. $y = \dfrac{\sqrt{[ax(x-3a)]}}{\sqrt{(x-4a)}}$. $\dfrac{dy}{dx} = \dfrac{\sqrt{a(x^2-8ax+12a^2)}}{2[x(x-3a)]^{\frac{1}{2}}(x-4a)^{\frac{3}{2}}}$.

29. $y = \dfrac{(x+1)^{\frac{1}{2}}(x+3)^{\frac{2}{3}}}{(x+2)^{\frac{1}{6}}}$. $\dfrac{dy}{dx} = \dfrac{x^2(x+3)^{\frac{1}{3}}}{(x+2)^{\frac{7}{6}}(x+1)^{\frac{1}{2}}}$.

VII.

The Trigonometric or Circular Functions.

48. In deriving the differentials of the trigonometric functions of a variable angle, we employ the *circular measure* of the angle, and denote it by θ. Thus, let s denote the length of the arc subtending the angle in the circle whose radius is a, then

$$\theta = \frac{s}{a}.$$

In Fig. 5, let OA be a fixed line, and OP an equal line rotating about the origin O; then P will describe the circle whose equation (the coordinates being rectangular) is

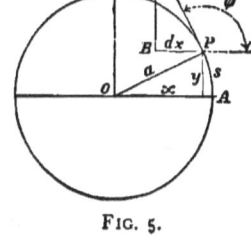

FIG. 5.

$$x^2 + y^2 = a^2.$$

The velocity of the point P is the rate of s, and (see Art. 17) is denoted by $\dfrac{ds}{dt}$, which has a positive value when P moves so as to increase θ. Let PP', taken in the direction of the motion of P, represent ds; then, according to the definition given in Art. 25, PP' is a tangent line, and PB and BP' will represent dx and dy, as in Art. 26.

49. We have first to show that the line PP', which is a tangent to the curve according to the general definition (Art. 25), is perpendicular to the radius.

Differentiating the equation of the circle, we have

$$x\,dx + y\,dy = 0;$$

whence
$$\tan\phi = \frac{dy}{dx} = -\frac{x}{y}.$$

Now (see Fig. 5), $\quad \dfrac{y}{x} = \tan\theta,$

therefore, $\quad \tan\phi = -\cot\theta = \tan(\theta \pm \tfrac{1}{2}\pi),$

or, $\quad \phi = \theta \pm \tfrac{1}{2}\pi;$

hence the tangent line is perpendicular to the radius. Assuming ϕ to be the angle between the positive directions of x and ds, we have

$$\phi = \theta + \tfrac{1}{2}\pi.$$

The Sine and the Cosine.

50. From Fig. 5, it is evident that

$$\sin\theta = \frac{y}{a}, \quad \text{and} \quad \cos\theta = \frac{x}{a},$$

therefore $\quad d(\sin\theta) = \dfrac{dy}{a}, \quad$ and $\quad d(\cos\theta) = \dfrac{dx}{a}. \quad \ldots \quad (1)$

In equations (1) we have to express dy and dx in terms of θ and $d\theta$.

§ VII.] *THE TANGENT AND THE COTANGENT.* 49

Again, from the figure, we have

$$dy = \sin\phi\, ds, \quad \text{and} \quad dx = \cos\phi\, ds\,;*$$

substituting in equations (1), we obtain

$$d(\sin\theta) = \sin\phi\,\frac{ds}{a}, \quad \text{and} \quad d(\cos\theta) = \cos\phi\,\frac{ds}{a}. \quad \ldots \quad (2)$$

Since $\quad \phi = \theta + \tfrac{1}{2}\pi, \quad$ and $\quad \dfrac{s}{a} = \theta,$

$$\sin\phi = \cos\theta, \quad \cos\phi = -\sin\theta, \quad \text{and} \quad \frac{ds}{a} = d\theta.$$

Substituting these values in equations (2), we obtain

$$d(\sin\theta) = \cos\theta\, d\theta, \quad \ldots \ldots \quad (i)$$

and $\qquad d(\cos\theta) = -\sin\theta\, d\theta. \ldots \ldots \quad (j)$

The Tangent and the Cotangent.

51. The differential of $\tan\theta$ is found by applying formula (e) to the equation

$$\tan\theta = \frac{\sin\theta}{\cos\theta};$$

thus, $\qquad d(\tan\theta) = \dfrac{\cos\theta\, d(\sin\theta) - \sin\theta\, d(\cos\theta)}{\cos^2\theta},$

or $\qquad d(\tan\theta) = \dfrac{d\theta}{\cos^2\theta} = \sec^2\theta\, d\theta. \quad \ldots \ldots \quad (k)$

---

* In Fig. 5, dx is negative; but, ϕ being in the second quadrant, $\cos\phi$ is likewise negative.

The differential of $\cot\theta$ is found by applying formula (k) to the equation

$$\cot\theta = \tan(\tfrac{1}{2}\pi - \theta);$$

whence $\quad d(\cot\theta) = -\dfrac{d\theta}{\sin^2\theta} = -\csc^2\theta\, d\theta.$ . . . (l)

The Secant and the Cosecant.

52. The differential of $\sec\theta$ is found by applying formula (d) to the equation

$$\sec\theta = \frac{1}{\cos\theta};$$

whence $\quad d(\sec\theta) = \dfrac{\sin\theta\, d\theta}{\cos^2\theta} = \sec\theta\tan\theta\, d\theta.$ . . (m)

The differential of $\csc\theta$ is found by applying formula (m) to the equation

$$\csc\theta = \sec(\tfrac{1}{2}\pi - \theta);$$

whence $\quad d(\csc\theta) = -\dfrac{\cos\theta\, d\theta}{\sin^2\theta} = -\csc\theta\cot\theta\, d\theta.$. (n)

The Versed-Sine.

53. The *versed-sine* is defined by the equation

$$\operatorname{vers}\theta = 1 - \cos\theta;$$

therefore $\quad d(\operatorname{vers}\theta) = \sin\theta\, d\theta.$ . . . . . . (o)

Examples VII.

1. The value of $d(\sin\theta)$ being given, derive that of $d(\cos\theta)$ from the formula
$$\cos\theta = \sin(\tfrac{1}{2}\pi - \theta);$$
also from the identity
$$\cos^2\theta = 1 - \sin^2\theta.$$

2. From the identity $\sec^2\theta = 1 + \tan^2\theta$, derive the differential of $\sec\theta$.

3. From the identity $\sin 2\theta = 2\sin\theta\cos\theta$, derive another by taking derivatives. $\cos 2\theta = \cos^2\theta - \sin^2\theta.$

4. From the identity $\sin(\theta \pm \tfrac{1}{4}\pi) = \tfrac{1}{2}\sqrt{2}\,(\sin\theta \pm \cos\theta)$, derive another by taking derivatives. $\cos(\theta \pm \tfrac{1}{4}\pi) = \tfrac{1}{2}\sqrt{2}\,(\cos\theta \mp \sin\theta).$

5. Prove the formulas:—

$$d(\log \sin\theta) = -d(\log \csc\theta) = \cot\theta\, d\theta;$$
$$d(\log \cos\theta) = -d(\log \sec\theta) = -\tan\theta\, d\theta;$$
$$d(\log \tan\theta) = -d(\log \cot\theta) = (\tan\theta + \cot\theta)\, d\theta.$$

6. Obtain an identity by taking derivatives of both members of the equation
$$\tan \tfrac{1}{2}\theta = \frac{1-\cos\theta}{\sin\theta}.$$
$$\tfrac{1}{2}\sec^2 \tfrac{1}{2}\theta = \frac{1-\cos\theta}{\sin^2\theta}.$$

7. $y = \theta + \sin\theta\cos\theta$. $\quad\dfrac{dy}{d\theta} = 2\cos^2\theta$

8. $y = \sin\theta - \tfrac{1}{3}\sin^3\theta$. $\quad\dfrac{dy}{d\theta} = \cos^3\theta.$

9. $y = \dfrac{\sin\theta}{\sqrt{(\cos\theta)}}$. $\quad\dfrac{dy}{d\theta} = \dfrac{1 + \cos^2\theta}{2(\cos\theta)^{\frac{3}{2}}}.$

52 TRANSCENDENTAL FUNCTIONS. [Ex. VII.

10. $y = \frac{1}{3}\tan^3\theta - \tan\theta + \theta.$ $\qquad \dfrac{dy}{d\theta} = \tan^4\theta.$

11. $y = \frac{1}{3}\tan^3\theta + \tan\theta.$ $\qquad \dfrac{dy}{d\theta} = \sec^4\theta.$

12. $y = \sin e^x.$ $\qquad \dfrac{dy}{dx} = e^x \cos e^x.$

13. $y = x \sin x^2.$ $\qquad \dfrac{dy}{dx} = \sin x^2 + 2x^2 \cos x^2.$

14. $y = a^{\sin x}.$ $\qquad \dfrac{dy}{dx} = \log a \cdot a^{\sin x} \cos x.$

15. $y = \tan^2\theta + \log(\cos^2\theta).$ $\qquad \dfrac{dy}{d\theta} = 2\tan^3\theta.$

16. $y = \log(\tan\theta + \sec\theta).$ $\qquad \dfrac{dy}{d\theta} = \sec\theta.$

17. $y = \log\tan(\tfrac{1}{4}\pi + \tfrac{1}{2}\theta).$ $\qquad \dfrac{dy}{d\theta} = \dfrac{1}{\cos\theta}.$

18. $y = x + \log\cos(\tfrac{1}{4}\pi - x).$ $\qquad \dfrac{dy}{dx} = \dfrac{2}{1 + \tan x}.$

19. $y = \log\sqrt{(\sin x)} + \log\sqrt{(\cos x)}.$ $\qquad \dfrac{dy}{dx} = \cot 2x.$

20. $y = \sin n\theta (\sin\theta)^n.$ $\qquad \dfrac{dy}{d\theta} = n(\sin\theta)^{n-1}\sin(n+1)\theta.$

21. $y = \dfrac{\sin x}{1 + \tan x}.$ $\qquad \dfrac{dy}{dx} = \dfrac{\cos^2 x - \sin^3 x}{(\sin x + \cos x)^2}.$

22. $y = e^{ax}\cos bx.$ $\qquad \dfrac{dy}{dx} = e^{ax}(a\cos bx - b\sin bx).$

23. $y = \log\sqrt{\dfrac{a\cos x - b\sin x}{a\cos x + b\sin x}}.$ $\qquad \dfrac{dy}{dx} = \dfrac{-ab}{a^2\cos^2 x - b^2\sin^2 x}.$

24. $y = \epsilon^x (\cos x - \sin x)$. $\dfrac{dy}{dx} = -2\epsilon^x \sin x$.

25. The crank of a small steam-engine is 1 foot in length, and revolves uniformly at the rate of two turns per second, the connecting rod being 5 ft. in length; find the velocity per second of the piston when the crank makes an angle of 45° with the line of motion of the piston-rod; also when the angle is 135°, and when it is 90°.

Solution:—

Let a, b, and x denote respectively the crank, the connecting-rod, and the variable side of the triangle; and let θ denote the angle between a and x.

We easily deduce

$$x = a\cos\theta + \sqrt{(b^2 - a^2 \sin^2\theta)};$$

whence $\dfrac{dx}{dt} = -\left(a\sin\theta + \dfrac{a^2 \sin\theta \cos\theta}{\sqrt{(b^2 - a^2 \sin^2\theta)}}\right)\dfrac{d\theta}{dt}$.

In this case, $\dfrac{d\theta}{dt} = 4\pi$, $a = 1$, and $b = 5$.

When $\theta = 45°$, $\dfrac{dx}{dt} = -\dfrac{16\pi\sqrt{2}}{7}$ ft.

26. An elliptical cam revolves at the rate of two turns per second about a horizontal axis passing through one of the foci, and gives a reciprocating motion to a bar moving in vertical guides in a line with the centre of rotation: denoting by θ the angle between the vertical and the major axis, find the velocity per second with which the bar is moving when $\theta = 60°$, the eccentricity of the ellipse being $\tfrac{1}{3}$, and the semi-major axis 9 inches. Also find the velocity when $\theta = 90°$.

The relation between θ and the radius vector is expressed by the equation

$$r = \dfrac{a(1 - e^2)}{1 - e\cos\theta}.$$

When $\theta = 60°$, $\dfrac{dr}{dt} = -12\sqrt{3}\,\pi$ inches.

27. Find an expression in terms of its azimuth for the rate at which the altitude of a star is increasing.

Solution:—

Let h denote the altitude and A the azimuth of the star, p its polar distance, t the hour angle, and L the latitude of the observer; the formulas of spherical trigonometry give

54   TRANSCENDENTAL FUNCTIONS.   [Ex. VII.

$$\sin h = \sin L \cos p + \cos L \sin p \cos t, \quad \ldots \quad (1)$$
and
$$\sin p \sin t = \sin A \cos h. \quad \ldots \ldots (2)$$

Differentiating (1), p and L being constant,

$$\cos h \frac{dh}{dt} = -\cos L \sin p \sin t,$$

whence, substituting the value of $\sin p \sin t$, from equation (2),

$$\frac{dh}{dt} = -\cos L \sin A.$$

It follows that $\frac{dh}{dt}$ is greatest when $\sin A$ is numerically greatest; that is, when the star is on the prime vertical. In the case of a star that never reaches the prime vertical, the rate is greatest when A is greatest.

VIII.

The Inverse Circular Functions.

54. It is shown in Trigonometry that, if

$$x = \sin \theta,$$

the expressions

$$2n\pi + \theta \qquad \text{and} \qquad (2n+1)\pi - \theta, \quad \ldots (1)$$

in which n denotes zero or any integer, include all the arcs of which the sine is x; hence each of these arcs is a value of the *inverse function*

$$\sin^{-1} x.$$

Among these values, there is always one, and *only one*, which falls between $-\tfrac{1}{2}\pi$ and $+\tfrac{1}{2}\pi$; since, while the arc passes from the former of these values to the latter, the sine passes from -1 to $+1$; that is, it passes once through all its possible values.

Let θ, in the expressions (1), denote this value, which we shall call the *primary* value of the function.

55. In a similar manner, if

$$x = \cos\theta,$$

each of the arcs included in the expression

$$2n\pi \pm \theta \quad . \quad . \quad . \quad . \quad . \quad . \quad . \quad (2)$$

is a value of the inverse function

$$\cos^{-1} x.$$

One of these values, and only one, falls between 0 and π; since, while the arc passes from the former of these values to the latter, its cosine passes from $+1$ to -1; that is, once through all its possible values. In expression (2), let θ denote this value, which we shall call the *primary* value of this function.

56. In the case of the function

$$\csc^{-1} x,$$

the definition of the *primary* value that was adopted in the case of $\sin^{-1} x$, and the same general expressions (1) for the values of the function, are applicable.

In the case of the function

$$\sec^{-1} x,$$

the definition of the *primary* value adopted in the case of

$\cos^{-1} x$ and expression (2) for the general value of the function are applicable.

Finally, in the case of each of the functions

$$\tan^{-1} x \qquad \text{and} \qquad \cot^{-1} x$$

the *primary* value (θ) is taken between $-\tfrac{1}{2}\pi$ and $+\tfrac{1}{2}\pi$, and the general expression for the value of the function is

$$n\pi + \theta. \qquad \dots \dots \dots \quad (3)$$

The Inverse Sine and the Inverse Cosine.

57. To find the differential of the inverse sine, let

$$\theta = \sin^{-1} x;$$

then $\qquad x = \sin\theta, \qquad$ and $\qquad dx = \cos\theta\, d\theta,$

or $$d\theta = \frac{dx}{\cos\theta}.$$

Now, $\qquad \cos\theta = \pm \sqrt{(1 - \sin^2\theta)} = \pm \sqrt{(1 - x^2)},$

hence $$d(\sin^{-1} x) = \frac{dx}{\pm \sqrt{(1 - x^2)}}. \qquad \dots \quad (1)$$

If θ denotes the primary value of this function; that is, the value between $-\tfrac{1}{2}\pi$ and $+\tfrac{1}{2}\pi$, $\cos\theta$ is positive. Hence the upper sign in this ambiguous result belongs to the differential of the primary value of the function; it is therefore usual to write

$$d(\sin^{-1} x) = \frac{dx}{\sqrt{(1 - x^2)}}. \qquad \dots \quad (p)$$

Since we have, from expressions (1), Art. 54,

$$d(2n\pi + \theta) = d\theta, \qquad \text{and} \qquad d[(2n+1)\pi - \theta] = -d\theta,$$

§ VIII.] *THE INVERSE SINE AND INVERSE TANGENT.* 57

it is evident that the positive sign in equation (1) belongs not only to the differential of the primary value of $\sin^{-1}x$, but likewise to the differentials of all the values included in $2n\pi + \theta$; and that the negative sign belongs to the differentials of the values of $\sin^{-1}x$ included in $(2n+1)\pi - \theta$.

58. Similarly, if

$$\theta = \cos^{-1}x, \qquad x = \cos\theta;$$

whence

$$d\theta = \frac{dx}{-\sin\theta},$$

or

$$d(\cos^{-1}x) = \frac{dx}{\mp \sqrt{(1-x^2)}} \quad \ldots \quad (1)$$

If θ denote the primary value of the function which in this case is between 0 and π, $\sin\theta$ is positive; hence the upper sign in this ambiguous result belongs to the differential of the primary value. It is therefore usual to write

$$d(\cos^{-1}x) = \frac{dx}{-\sqrt{(1-x^2)}}. \quad \ldots \quad (q)$$

Since, from expression (2), Art. 55, we have

$$d(2n\pi \pm \theta) = \pm d\theta;$$

it is evident that the upper and lower signs in equation (1) correspond to the upper and lower signs, respectively, in the general expression $2n\pi \pm \theta$.

The Inverse Tangent and the Inverse Cotangent.

59. Let

$$\theta = \tan^{-1}x, \qquad \text{then} \qquad x = \tan\theta;$$

differentiating, we derive,

$$d\theta = \frac{dx}{\sec^2\theta}.$$

But $\sec^2\theta = 1 + \tan^2\theta = 1 + x^2$, therefore,

$$d(\tan^{-1} x) = \frac{dx}{1+x^2}. \quad \ldots \ldots \quad (r)$$

No ambiguity arises in the value of the differential of this function; since, from expression (3), Art. 56, we have

$$d(n\pi + \theta) = d\theta.$$

Similarly, putting

$$\theta = \cot^{-1} x,$$

we derive $\quad d(\cot^{-1} x) = -\dfrac{dx}{1+x^2}. \quad \ldots \ldots \quad (s)$

The Inverse Secant and Inverse Cosecant.

60. Let

$$\theta = \sec^{-1} x, \qquad \text{then} \qquad x = \sec\theta;$$

differentiating, we derive

$$d\theta = \frac{dx}{\sec\theta \tan\theta}.$$

But $\sec\theta = x$, and $\tan\theta = \pm \sqrt{(\sec^2\theta - 1)} = \pm \sqrt{(x^2 - 1)}$, therefore,

$$d(\sec^{-1} x) = \frac{dx}{\pm x \sqrt{(x^2-1)}}.$$

If x is positive, and if θ denotes the primary value of the function, $\tan\theta$ is positive. Hence it is usual to write

$$d(\sec^{-1} x) = \frac{dx}{x\sqrt{(x^2-1)}}. \quad \ldots \ldots \quad (t)$$

When x is negative, if θ denotes the primary value of the function, which in this case is in the second quadrant, $\tan\theta$ is negative; consequently the radical must be taken with the negative sign. Hence, since x is also negative, the value of the differential is positive, when the arc is taken in the second quadrant.

In like manner we derive

$$d(\operatorname{cosec}^{-1}x) = -\frac{dx}{x\sqrt{(x^2-1)}}. \quad \ldots \quad (u)$$

Similar remarks apply also to this differential when x is negative.

The Inverse Versed-Sine.

61. Let

$$\theta = \operatorname{vers}^{-1}x, \quad \text{then} \quad x = \operatorname{vers}\theta = 1 - \cos\theta,$$

and $\quad 1 - x = \cos\theta, \qquad \therefore \qquad d\theta = \dfrac{dx}{\sin\theta}.$

But $\sin\theta = \sqrt{(1-\cos^2\theta)} = \sqrt{(2x-x^2)}$, therefore,

$$d(\operatorname{vers}^{-1}x) = \frac{dx}{\sqrt{(2x-x^2)}}. \quad \ldots \quad (v)$$

Illustrative Examples.

62. It is sometimes advantageous to transform a given function before differentiating, by means of one of the following formulas:—

$$\sin^{-1}\frac{\alpha}{\beta} = \operatorname{cosec}^{-1}\frac{\beta}{\alpha}, \quad \cos^{-1}\frac{\alpha}{\beta} = \sec^{-1}\frac{\beta}{\alpha}, \quad \tan^{-1}\frac{\alpha}{\beta} = \cot^{-1}\frac{\beta}{\alpha}.$$

Thus, let
$$y = \tan^{-1} \frac{\varepsilon^x \cos x}{1 + \varepsilon^x \sin x},$$

then
$$y = \cot^{-1}(\varepsilon^{-x} \sec x + \tan x).$$

By formula (s),

$$\frac{dy}{dx} = -\frac{\varepsilon^{-x} \sec x \tan x - \varepsilon^{-x} \sec x + \sec^2 x}{\sec^2 x + 2\varepsilon^{-x} \sec x \tan x + \varepsilon^{-2x} \sec^2 x};$$

multiplying both terms by $\varepsilon^{2x} \cos^2 x$,

$$\frac{dy}{dx} = \frac{\varepsilon^x(\cos x - \sin x - \varepsilon^x)}{1 + 2\varepsilon^x \sin x + \varepsilon^{2x}}.$$

63. Trigonometric substitutions may sometimes be employed with advantage. Thus, let

$$y = \tan^{-1} \frac{x}{\sqrt{(1 + x^2)} + 1}.$$

If in this example we put $x = \tan \theta$, we have

$$y = \tan^{-1} \frac{\tan \theta}{\sec \theta + 1} = \tan^{-1} \frac{\sin \theta}{1 + \cos \theta}$$

$$= \tan^{-1}(\tan \tfrac{1}{2}\theta) = \tfrac{1}{2}\theta = \tfrac{1}{2}\tan^{-1} x.$$

Examples VIII.

1. Derive from (p), (r), and (t) the formulas:—

$$d\left(\sin^{-1}\frac{x}{a}\right) = \frac{dx}{\sqrt{(a^2 - x^2)}};$$

$$d\left(\tan^{-1}\frac{x}{a}\right) = \frac{a\,dx}{a^2 + x^2};$$

$$d\left(\sec^{-1}\frac{x}{a}\right) = \frac{a\,dx}{x\sqrt{(x^2-a^2)}}.$$

2. Derive $d(\sec^{-1}x)$ from the equation $\sec^{-1}x = \cos^{-1}\frac{1}{x}$.

3. Derive $d\left(\cot^{-1}\frac{x}{a}\right)$ from the equation $\cot^{-1}\frac{x}{a} = \tan^{-1}\frac{a}{x}$.

4. $y = \sin^{-1}(2x^2)$. $\qquad\qquad \dfrac{dy}{dx} = \dfrac{4x}{\sqrt{(1-4x^4)}}$.

5. $y = \sin^{-1}(\cos x)$. $\qquad\qquad \dfrac{dy}{dx} = -1$.

6. $y = \sin(\cos^{-1}x)$. $\qquad\qquad \dfrac{dy}{dx} = -\dfrac{x}{\sqrt{(1-x^2)}}$.

7. $y = \sin^{-1}(\tan x)$. $\qquad\qquad \dfrac{dy}{dx} = \dfrac{\sec^2 x}{\sqrt{(1-\tan^2 x)}}$.

8. $y = \cos^{-1}(2\cos x)$. $\qquad\qquad \dfrac{dy}{dx} = +\dfrac{2\sin x}{\sqrt{(1-4\cos^2 x)}}$.

9. $y = x\sin^{-1}x + \sqrt{(1-x^2)}$. $\qquad \dfrac{dy}{dx} = \sin^{-1}x$.

10. $y = \tan^{-1}e^x$. $\qquad\qquad \dfrac{dy}{dx} = \dfrac{1}{e^x + e^{-x}}$.

11. $y = (x^2+1)\tan^{-1}x - x$. $\qquad \dfrac{dy}{dx} = 2x\tan^{-1}x$.

12. $y = a^2\sin^{-1}\dfrac{x}{a} + x\sqrt{(a^2-x^2)}$. $\qquad \dfrac{dy}{dx} = 2\sqrt{(a^2-x^2)}$.

13. $y = \tan^{-1}\dfrac{mx}{1-x^2}$. $\qquad\qquad \dfrac{dy}{dx} = \dfrac{m(1+x^2)}{1+(m^2-2)x^2+x^4}$.

14. $y = \sin^{-1}\dfrac{x+1}{\sqrt{2}}$. $\qquad\qquad \dfrac{dy}{dx} = \dfrac{1}{\sqrt{(1-2x-x^2)}}$.

15. $y = \tan^{-1}\dfrac{x}{\sqrt{(1-x^2)}}$. $\dfrac{dy}{dx} = \dfrac{1}{\sqrt{(1-x^2)}}$.

16. $y = \sec^{-1}\dfrac{a}{\sqrt{(a^2-x^2)}}$. $\dfrac{dy}{dx} = \dfrac{1}{\sqrt{(a^2-x^2)}}$.

17. $y = \sin^{-1}\dfrac{x}{\sqrt{(x^2+a^2)}}$. $\dfrac{dy}{dx} = \dfrac{a}{a^2+x^2}$.

18. $y = \sin^{-1}\sqrt{(\sin x)}$. $\dfrac{dy}{dx} = \tfrac{1}{2}\sqrt{(1+\operatorname{cosec} x)}$.

19. $y = \sqrt{(1-x^2)}\sin^{-1}x - x$. $\dfrac{dy}{dx} = -\dfrac{x\sin^{-1}x}{\sqrt{(1-x^2)}}$.

20. $y = \tan^{-1}\dfrac{m+x}{1-mx}$. $\dfrac{dy}{dx} = \dfrac{1}{1+x^2}$.

21. $y = \cos^{-1}\dfrac{1-x^2}{1+x^2}$. $\dfrac{dy}{dx} = \dfrac{2}{1+x^2}$.

22. $y = \tan^{-1}\sqrt{\dfrac{1-\cos x}{1+\cos x}}$. $\dfrac{dy}{dx} = \tfrac{1}{2}$.

23. $y = \dfrac{x\sin^{-1}x}{\sqrt{(1-x^2)}} + \log\sqrt{(1-x^2)}$. $\dfrac{dy}{dx} = \dfrac{\sin^{-1}x}{(1-x^2)^{\frac{3}{2}}}$.

24. $y = (x+a)\tan^{-1}\sqrt{\dfrac{x}{a}} - \sqrt{(ax)}$. $\dfrac{dy}{dx} = \tan^{-1}\sqrt{\dfrac{x}{a}}$.

IX.

Differentials of Functions of Two Variables.

64. The formulas already deduced enable us to differentiate any function of two variables, expressed by elementary functional symbols; the application of these formulas is, however, sometimes facilitated by a general principle which will now be shown to be applicable to such functions.

The formulas mentioned above involve differential factors of the first degree only. It follows, therefore, that the differentials resulting from their application consist of terms each of which contains the first power of the differential of one of the variables. In other words, if

$$u = f(x,y),$$

$$du = \phi(x,y)dx + \psi(x,y)dy. \quad \ldots \quad (1)$$

Now, if y were constant, we should have $dy = 0$, and the value of du would reduce to that of the first term in the right-hand member of (1); hence this term may be found by *differentiating* u *on the supposition that* y *is constant*, and in like manner the second term can be found by differentiating u on the supposition that x is constant. The sum of the results thus obtained is therefore the required value of du.

65. As an example, let

$$z = u^v.$$

Were v constant, we should have for the value of dz, by formula (f), Art. 37,

$$v u^{v-1} du;$$

and, were u constant, we should have, by formula (h), Art. 46,

$$\log u \cdot u^v dv;$$

whence, adding these results,

$$dz = u^{v-1}(v\, du + u \log u\, dv).$$

Although this result has been obtained on the supposition that u and v are independent variables, it is evident that any two functions of a single variable may be substituted for u and v. Thus, if

$$u = nx \quad \text{and} \quad v = x^2,$$

we have
$$z = (nx)^{x^2},$$

and, on substituting,

$$dz = (nx)^{x^2-1}(x^2 n\, dx + nx \log(nx) \cdot 2x\, dx),$$
$$= x(nx)^{x^2}[1 + 2\log(nx)]dx,$$

which is identical with the expression obtained in Art. 47, for the differential of this function.

Examples IX.

1. $u = xy\, e^{x+2y}$. $\quad du = e^{x+2y}[y(1+x)dx + x(1+2y)dy]$.

2. $u = \log \tan \dfrac{x}{y}$. $\quad du = 2\dfrac{y\, dx - x\, dy}{y^2 \sin 2\dfrac{x}{y}}$.

3. $u = \log \tan^{-1} \dfrac{x}{y}$. $\quad du = \dfrac{y\, dx - x\, dy}{(x^2 + y^2) \tan^{-1}\dfrac{x}{y}}$.

4. $u = \dfrac{\sqrt{x} + \sqrt{y}}{x + y}$.

$\quad du = \dfrac{[y - x - 2\sqrt{(xy)}]\sqrt{y}\, dx + [x - y - 2\sqrt{(xy)}]\sqrt{x}\, dy}{2\sqrt{(xy)}(x+y)^2}$.

5. $u = \dfrac{e^x y}{(x^2 + y^2)^{\frac{1}{2}}}$. $\quad du = \dfrac{y\, e^x dx}{(x^2+y^2)^{\frac{1}{2}}} + \dfrac{x(x\, dy - y\, dx)\, e^x}{(x^2+y^2)^{\frac{3}{2}}}$.

6. $u = \tan^{-1} \dfrac{x-y}{x+y}$. $\quad du = \dfrac{y\, dx - x\, dy}{x^2 + y^2}$.

7. $u = \sqrt{\dfrac{x^2 - y^2}{x^2 + y^2}}.$ $\qquad du = \dfrac{2xy(y\,dx - x\,dy)}{(x^2 + y^2)^{\frac{3}{2}} \sqrt{(x^2 - y^2)}}.$

8. $u = \log \dfrac{x + \sqrt{(x^2 - y^2)}}{x - \sqrt{(x^2 - y^2)}}.$ $\qquad du = \dfrac{2(y\,dx - x\,dy)}{y\sqrt{(x^2 - y^2)}}.$

9. Given $x = r\cos\theta$, and $y = r\sin\theta$; eliminate θ and find dr; also eliminate r and find $d\theta$.

$$dr = \dfrac{x\,dx + y\,dy}{\sqrt{(x^2 + y^2)}}, \text{ and } d\theta = \dfrac{x\,dy - y\,dx}{x^2 + y^2}.$$

Miscellaneous Examples.

1. $y = \dfrac{x}{\sqrt{(1 + x)}}.$ $\qquad \dfrac{dy}{dx} = \dfrac{x + 2}{2(1 + x)^{\frac{3}{2}}}.$

2. $y = \sqrt{\dfrac{a^2 - x^2}{b^2 - x^2}}.$ $\qquad \dfrac{dy}{dx} = \dfrac{(a^2 - b^2)x}{(a^2 - x^2)^{\frac{1}{2}}(b^2 - x^2)^{\frac{3}{2}}}.$

3. $y = \dfrac{\sqrt{(a + x)}}{\sqrt{a} + \sqrt{x}}.$ $\qquad \dfrac{dy}{dx} = \dfrac{\sqrt{a}(\sqrt{x} - \sqrt{a})}{2\sqrt{x}\sqrt{(a + x)}(\sqrt{a} + \sqrt{x})^2}.$

4. $y = (\sqrt{x} - 2\sqrt{a})\sqrt{(\sqrt{a} + \sqrt{x})}.$ $\qquad \dfrac{dy}{dx} = \dfrac{3}{4\sqrt{(\sqrt{a} + \sqrt{x})}}.$

5. $y = \dfrac{(x - 1)(e^x + 1)e^x}{e^x - 1}.$ $\qquad \dfrac{dy}{dx} = \dfrac{e^x(xe^{2x} - 2xe^x + 2e^x - x)}{(e^x - 1)^2}.$

6. $y = \log \dfrac{(1 + x^2)^{\frac{1}{4}}}{(1 + x)^{\frac{1}{2}}} + \tfrac{1}{2}\tan^{-1}x.$ $\qquad \dfrac{dy}{dx} = \dfrac{x}{(1 + x)(1 + x^2)}.$

7. $y = \log\left(\dfrac{1 + x}{1 - x}\right)^{\frac{1}{4}} - \tfrac{1}{2}\tan^{-1}x.$ $\qquad \dfrac{dy}{dx} = \dfrac{x^2}{1 - x^4}.$

8. $y = \log[x + \sqrt{(x^2 - a^2)}] + \sec^{-1}\dfrac{x}{a}.$ $\qquad \dfrac{dy}{dx} = \dfrac{1}{x}\sqrt{\left(\dfrac{x + a}{x - a}\right)}.$

9. $y = \dfrac{2\sin^{-1}x}{\sqrt{(1 - x^2)}} + \log\dfrac{1 - x}{1 + x}.$ $\qquad \dfrac{dy}{dx} = \dfrac{2x\sin^{-1}x}{(1 - x^2)^{\frac{3}{2}}}.$

10. $y = \dfrac{(1 + 3x + 3x^2)^{\frac{1}{3}}}{x}$. $\dfrac{dy}{dx} = -\dfrac{(1+x)^2}{x^2(1+3x+3x^2)^{\frac{2}{3}}}$.

11. $y = a \log \dfrac{a + \sqrt{(a^2 - x^2)}}{x} - \sqrt{(a^2 - x^2)}$.

$$\dfrac{dy}{dx} = -\dfrac{\sqrt{(a^2 - x^2)}}{x}.$$

12. $y = \dfrac{(1 - x^2)^{\frac{3}{2}} \sin^{-1} x}{x}$.

$$\dfrac{dy}{dx} = \dfrac{1 - x^2}{x} - \dfrac{1 + 2x^2}{x^2} \sqrt{(1 - x^2)} \sin^{-1} x.$$

13. $y = \log \sqrt{\dfrac{1 - \cos x}{1 + \cos x}}$. $\dfrac{dy}{dx} = \dfrac{1}{\sin x}$.

14. $y = \tan^{-1}\left[\sqrt{\dfrac{a-b}{a+b}} \cdot \tan \dfrac{x}{2}\right]$. $\dfrac{dy}{dx} = \dfrac{\sqrt{(a^2 - b^2)}}{2(a + b\cos x)}$.

15. $y = \sec^{-1} \dfrac{1}{2x^2 - 1}$. $\dfrac{dy}{dx} = -\dfrac{2}{\sqrt{(1 - x^2)}}$.

16. $y = \cos^{-1} \dfrac{x^{2n} - 1}{x^{2n} + 1}$. $\dfrac{dy}{dx} = -\dfrac{2nx^{n-1}}{x^{2n} + 1}$.

17. $y = a \cos^{-1} \dfrac{a - x}{b} - \sqrt{[b^2 - (a - x)^2]}$.

$$\dfrac{dy}{dx} = \dfrac{x}{\sqrt{[b^2 - (a - x)^2]}}.$$

18. $y = \cos^{-1} x - 2\sqrt{\dfrac{1-x}{1+x}}$. $\dfrac{dy}{dx} = \dfrac{\sqrt{(1 - x)}}{(1 + x)^{\frac{3}{2}}}$.

19. $y = \dfrac{ax - 1}{\sqrt{(1 + x^2)}} e^{a\tan^{-1}x}$. $\dfrac{dy}{dx} = \dfrac{(1 + a^2)x}{(1 + x^2)^{\frac{3}{2}}} e^{a\tan^{-1}x}$.

Use logarithmic differentials.

CHAPTER IV.

SUCCESSIVE DIFFERENTIATION.

X.

Velocity and Acceleration.

66. IF the variable quantity x represent the distance of a point, moving in a straight line, from a fixed origin taken on the line, the rate of x will represent the velocity of the point.

Denoting this velocity by v_x we have, in accordance with the definition given in Art. 17,

$$v_x = \frac{dx}{dt}. \quad \ldots \ldots \ldots (1)$$

In this expression the arbitrary interval of time dt is regarded as constant, while dx, and consequently v_x, is in general variable. Differentiating equation (1) we have, since dt is constant,

$$dv_x = \frac{d(dx)}{dt}.$$

The differential of dx, denoted above by $d(dx)$, is called the *second differential* of x; it is usually written in the abbreviated form d^2x, and read "d-second x." The rate of v_x is therefore expressed thus:—

$$\frac{dv_x}{dt} = \frac{d^2x}{(dt)^2}.$$

The rate of the velocity of a point is called its *acceleration*, and is usually denoted by α; hence we write

$$\alpha_x = \frac{dv_x}{dt} = \frac{d^2x}{dt^2}, \quad \ldots \ldots (2)$$

the marks of parenthesis being usually omitted in the denominator of this expression.

67. When the space x described by a moving point is a given function of the time t, the derivative of this function is, by equation (1), an expression for the velocity in terms of t. The derivative of the latter expression, which is called the *second derivative* of x, is therefore, by equation (2), an expression for the acceleration in terms of t.

A *positive* value of the acceleration α indicates an *algebraic increase* of the velocity v, whether the latter be positive or negative; and, on the other hand, a negative value of α indicates an *algebraic decrease* of the velocity.

68. As an illustration, let x denote the space which a body falling freely describes in the time t. A well-known mechanical formula gives

$$x = \tfrac{1}{2}gt^2. \quad \ldots \ldots \ldots (1)$$

Hence we derive $\quad v_x = \dfrac{dx}{dt} = gt, \quad \ldots \ldots \ldots (2)$

and $\quad \alpha_x = \dfrac{dv_x}{dt} = \dfrac{d^2x}{dt^2} = g. \quad \ldots \ldots (3)$

In this case, therefore, the acceleration is constant and positive, and accordingly v_x, which is likewise positive, is numerically increasing.

69. When the velocity is given in terms of x, the acceleration can readily be expressed in terms of the same variable, as in the following example.

§ X.] VELOCITY AND ACCELERATION. 69

Given $v_x = 2 \sin x$;

whence $\dfrac{dv_x}{dt} = 2 \cos x \dfrac{dx}{dt}$;

that is, $a_x = 2 \cos x \cdot v_x = 4 \cos x \sin x = 2 \sin 2x$.

The general expression for a_x, when v_x is given in terms of x, is

$$a_x = \frac{dv_x}{dt} = \frac{dv_x}{dx}\frac{dx}{dt} = v_x \frac{dv_x}{dx} = \frac{1}{2}\frac{d(v_x^2)}{dx} \quad \ldots \quad (1)$$

Component Velocities and Accelerations.

70. When the motion of a point is not rectilinear but is nevertheless confined to a plane, its position is referred to co-ordinate axes; the coordinates, x and y, are evidently functions of t, and the derivatives $\dfrac{dx}{dt}$ and $\dfrac{dy}{dt}$, which denote the rates of these variables, are called the *component* or *resolved velocities* in the directions of the axes. Denoting these component velocities by v_x and v_y, we have

$$v_x = \frac{dx}{dt}, \text{ and } v_y = \frac{dy}{dt}.$$

Again, denoting by s the actual space described, as measured from some fixed point of the path, s will likewise be a function of t, and the derivative $\dfrac{ds}{dt}$ will denote the actual velocity of the point. (Compare Art. 48.) Now, the axes being rectangular, and ϕ denoting the inclination of the direction of the motion to the axis of x, we have

$$dx = ds \cos \phi, \text{ and } dy = ds \sin \phi.$$

Hence, $\dfrac{dx}{dt} = \dfrac{ds}{dt} \cos \phi, \text{ and } \dfrac{dy}{dt} = \dfrac{ds}{dt} \sin \phi;$

or $\quad\quad v_x = v\cos\phi,\ \text{and}\ v_y = v\sin\phi.$

Squaring and adding,

$$v_x^2 + v_y^2 = v^2.$$

The last equation enables us to determine from the component velocities the actual velocity in the curve.

71. If we represent the accelerations of the resolved motions in the directions of the axes by α_x and α_y, we shall have, by Art. 66,

$$\alpha_x = \frac{d^2x}{dt^2}\ \text{and}\ \alpha_y = \frac{d^2y}{dt^2}.$$

These accelerations, α_x and α_y, will be positive when the resolved motions are *accelerated in the positive directions of the corresponding axes;* that is, when they increase a positive resolved velocity, or numerically decrease a negative resolved velocity.

Examples X.

1. The space in feet described in the time t by a point moving in a straight line is expressed by the formula

$$x = 48t - 16t^2;$$

find the acceleration, and the velocity at the end of $2\frac{1}{2}$ seconds; also find the value of t for which $v = 0$.

$\quad\quad\quad\quad \alpha = -32;\ v = 0,\ \text{when}\ t = 1\frac{1}{2}.$

2. If the space described in t seconds be expressed by the formula

$$x = 10\log\frac{4}{4+t};$$

find the velocity and acceleration at the end of 1 second, and at the end of 16 seconds. $\quad$ When $t = 1$, $v = -2$ and $\alpha = \frac{2}{5}$.

EXAMPLES.

3. If a point moves in a fixed path so that

$$s = \sqrt{t},$$

show that the acceleration is negative and proportional to the cube of the velocity. Find the value of the acceleration at the end of one second, and at the end of nine seconds. $-\frac{1}{4}$, and $-\frac{1}{108}$.

4. If a point move in a straight line so that

$$x = a \cos \tfrac{1}{2}\pi t,$$

show that $\quad \alpha = -\tfrac{1}{4}\pi^2 x.$

5. If $\quad x = a\, \varepsilon^t + b\, \varepsilon^{-t},$

prove that $\quad \alpha = x.$

6. If a point referred to rectangular coordinate axes move so that

$$x = a \cos t + b \quad \text{and} \quad y = a \sin t + c,$$

show that its velocity will be uniform. Find the equation of the path described.
Eliminate t from the given equations.

7. A projectile moves in the parabola whose equation is

$$y = x \tan \alpha - \frac{g}{2\,V^2 \cos^2\alpha} x^2,$$

(the axis of y being vertical) with a uniform horizontal velocity

$$v_x = V \cos \alpha\,;$$

find the velocity in the curve, and the vertical acceleration.

$$v = \sqrt{(V^2 - 2gy)}, \text{ and } \alpha_y = -g.$$

8. A point moves in the curve, whose equation is

$$x^{\frac{2}{3}} + y^{\frac{2}{3}} = a^{\frac{2}{3}},$$

so that v_x is constant and equal to k; find the acceleration in the direction of the axis of y.

$$\alpha_y = \frac{a^{\frac{2}{3}}k^2}{3x^{\frac{4}{3}}y^{\frac{1}{3}}}.$$

9. If a point move so that $v = \sqrt{(2gx)}$; determine the acceleration. Use equation (1), Art. 69. $\alpha = g.$

10. If a point move so that we have

$$v^2 = c - \mu \log x,$$

determine the acceleration. $\alpha = -\dfrac{\mu}{2x}.$

11. If a point move so that we have

$$v^2 = c + \frac{2\mu}{\sqrt{(x^2 + b^2)}},$$

determine the acceleration. $\alpha = -\dfrac{\mu x}{(x^2 + b^2)^{\frac{3}{2}}}.$

12. The velocity of a point is inversely proportional to the square of its distance from a fixed point of the straight line in which it moves, the velocity being 2 feet per second when the distance is six inches; determine the acceleration at a given distance s from the fixed point.

$$-\frac{1}{2s^5} \text{ feet.}$$

13. The velocity of a point moving in a straight line is m times its distance from a fixed point at the perpendicular distance a from the straight line; determine the acceleration at the distance x from the foot of the perpendicular. $\alpha = m^2 x.$

14. The relation between x and t being expressed by

$$t\sqrt{\frac{2\mu}{a}} = \sqrt{(ax - x^2)} - \tfrac{1}{2}a \operatorname{vers}^{-1}\frac{2x}{a};$$

find the acceleration in terms of x. $\alpha = -\dfrac{\mu}{x^2}.$

15. A point moves in the hyperbola

$$y^2 = p^2 x^2 + q^2$$

in such a manner that v_x has the constant value c; prove that

$$v_y{}^2 = p^2 c^2 - \frac{p^2 c^2 q^2}{y^2},$$

and thence derive α_y by equation (1), Art. 69.

$$\alpha_y = \frac{p^2 c^2 q^2}{y^3}.$$

16. A point describes the conic section

$$y^2 = 2mx + nx^2,$$

v_x having the constant value c; determine the value of α_y.
Express v_y^2 in terms of y, and proceed as in Example 15.

$$\alpha_y = -\frac{m^2 c^2}{y^3}.$$

XI.

Successive Derivatives.

72. The derivative of $f(x)$ is another function of x, which we have denoted by $f'(x)$; if we take the derivative of the latter, we obtain still another function of x, which is called the second derivative of the original function $f(x)$, and is denoted by $f''(x)$. Thus if

$$f(x) = x^3, \qquad f'(x) = 3x^2, \qquad \text{and} \qquad f''(x) = 6x.$$

Similarly the derivative of $f''(x)$ is denoted by $f'''(x)$, and is called the third derivative of $f(x)$; etc. When one of these successive derivatives has a constant value, the next and all succeeding derivatives evidently vanish. Thus, in the above example, $f'''(x) = 6$, consequently, in this case, $f^{\text{iv}}(x)$ and all higher derivatives vanish.

The Geometrical Meaning of the Second Derivative.

73. If the curve whose equation is

$$y = f(x)$$

be constructed, we have seen (Art. 26) that

$$\frac{dy}{dx} = f'(x) = \tan\phi,$$

ϕ being the inclination of the curve to the axis of x; hence

$$f''(x) = \frac{d(\tan\phi)}{dx}.$$

If now the value of this derivative be *positive*, $\tan\phi$ will be an *increasing* function of x, as in Fig. 6, in which, as we proceed toward the right, $\tan\phi$ (at first negative) increases algebraically throughout. In this case, therefore, the curve appears *concave when viewed from above*. On the other hand, if $f''(x)$ be *negative*, $\tan\phi$ will be a decreasing function of x, as in Fig. 7, in which, as we proceed toward the right, $\tan\phi$ decreases algebraically throughout, the curve appearing *convex when viewed from above*.

FIG. 6.

FIG. 7.

74. A point which separates a concave from a convex portion of a curve is called a *point of inflexion*, or a *point of contrary flexure*.

It is obvious from the preceding article that, at a point of inflexion, like P in Fig. 8, $f''(x)$ must *change sign*; hence at such a point, the value of this derivative must become either zero or infinity.

FIG. 8.

75. When a curve is described by a moving point, the character of the curvature is dependent upon the component accelerations of the motion. For, if we put

$$v_x = c, \quad \text{or} \quad dx = c\,dt,$$

c denoting a constant, we have

$$f'(x) = \frac{dy}{c\,dt};$$

and hence $$f''(x) = \frac{1}{c^2} \cdot \frac{d^2y}{dt^2} = \frac{\alpha_y}{c^2}.$$

Whence it follows that, if v_x is constant, α_y and $f''(x)$ have the same sign, and consequently that a portion of a curve which is concave when viewed from above is one in which α_y is positive when α_x is zero.

Successive Differentials.

76. The successive differentials of a *function* of x involve the successive differentials of x; thus, if

$$y = x^3,$$

we have $\quad dy = 3x^2 dx,$

$$d^2y = 6x(dx)^2 + 3x^2 d^2x,$$

and $\quad d^3y = 6(dx)^3 + 18x\, dx\, d^2x + 3x^2 d^3x.$

In general, if

$$y = f(x),$$
$$dy = f'(x)\, dx,$$
$$d^2y = f''(x)(dx)^2 + f'(x)\, d^2x,$$

and $\quad d^3y = f'''(x)(dx)^3 + 3f''(x)\, dx\, d^2x + f'(x)\, d^3x.$

Equicrescent Variables.

77. A variable is said to be *equicrescent* when its rate is constant; since dt in the expression $\dfrac{dx}{dt}$ is assumed to be constant, dx is also constant, when x is equicrescent.

In expressing the differentials of a function, it is admissible to assume the independent variable to be equicrescent, since the differential of this variable is arbitrary. This hypothesis greatly simplifies the expressions for the second and higher differentials of functions of x, inasmuch as it is evidently equivalent to making all differentials of x higher than the first vanish. Thus, in the general expressions for d^2y and d^3y given in the preceding article, all the terms except the first disappear, and it is easy to see that, in general, we shall have

$$d^n y = f^n(x)(dx)^n,$$

when x is equicrescent.

78. From the above equation we derive

$$\frac{d^n y}{dx^n} = f^n(x).$$

The expression in the first member of this equation is the usual symbol for the nth derivative of y regarded as a function of x. The nth *differential* which occurs in this symbol is always understood to denote the value which this differential assumes *when the variable indicated in the denominator is equicrescent.*

The symbol $\dfrac{d}{dx}$ is frequently used to denote the operation of taking the derivative with reference to x, and similarly the symbol $\left(\dfrac{d}{dx}\right)^n$, or $\dfrac{d^n}{dx^n}$, is used to denote the operation of taking the derivative with respect to x, n times in succession.

Examples XI.

1. Find the second derivative of $\sec x$, and distinguish the concave from the convex portions of the curve $y = \sec x$. Also show that the curve $y = \log x$ is everywhere convex.

2. Find the points of inflexion in the curve $y = \sin x$.

3. Find the point of inflexion of the curve
$$y = 2x^3 - 3x^2 - 12x + 6.$$
The point is $(\tfrac{1}{2}, -\tfrac{1}{2})$.

4. Show that the curve $y = \tan x$ is concave when y is positive, and convex when y is negative.

5. Find the points of inflexion of the curve
$$y = x^4 - 2x^3 - 12x^2 + 11x + 24.$$
The points are $(2, -2)$ and $(-1, 4)$.

6. If $f(x) = \dfrac{1+x}{1-x}$, find $f^{v}(x)$. $\qquad f^{v}(x) = \dfrac{240}{(1-x)^6}$.

7. If $f(x) = \dfrac{a}{x^n}$, find $f'''(x)$. $\qquad f'''(x) = -\dfrac{n(n+1)(n+2)a}{x^{n+3}}$.

8. If y is a function of x of the form
$$Ax^n + Bx^{n-1} + \cdots + Mx + N,$$
prove that $\qquad \dfrac{d^n y}{dx^n} = 1 \cdot 2 \cdot 3 \cdots n \, A.$

9. If $f(x) = b^{ax}$, find $f^v(x)$. $\qquad f^v(x) = a^5 (\log b)^5 b^{ax}$.

10. If $f(x) = x^3 \log(mx)$, find $f^{IV}(x)$. $\qquad f^{IV}(x) = \dfrac{6}{x}$.

11. If $f(x) = \log \sin x$, find $f'''(x)$. $\qquad f'''(x) = \dfrac{2 \cos x}{\sin^3 x}$.

12. If $f(x) = \sec x$, find $f''(x)$ and $f'''(x)$.

$f''(x) = 2 \sec^3 x - \sec x$, and $f'''(x) = \sec x \tan x (6 \sec^2 x - 1)$.

13. If $f(x) = \tan x$, find $f'''(x)$ and $f^{IV}(x)$.

$f'''(x) = 6 \sec^4 x - 4 \sec^2 x$, and $f^{IV}(x) = 8 \tan x \sec^2 x (3 \sec^2 x - 1)$.

14. If $f(x) = x^x$, find $f''(x)$. $f''(x) = x^x(1 + \log x)^2 + x^{x-1}$.

15. If $y = \varepsilon^{\frac{1}{x}}$, find $\dfrac{d^3y}{dx^3}$. $\dfrac{d^3y}{dx^3} = -\dfrac{1}{x^6}(1 + 6x + 6x^2)\varepsilon^{\frac{1}{x}}$.

16. If $y = \varepsilon^{-x^2}$, find $\dfrac{d^3y}{dx^3}$. $\dfrac{d^3y}{dx^3} = 4x(3 - 2x^2)\varepsilon^{-x^2}$.

17. If $y = \log(\varepsilon^x + \varepsilon^{-x})$, find $\dfrac{d^3y}{dx^3}$. $\dfrac{d^3y}{dx^3} = -8\dfrac{\varepsilon^x - \varepsilon^{-x}}{(\varepsilon^x + \varepsilon^{-x})^3}$.

18. If $y = \dfrac{1}{\varepsilon^x - 1}$, find $\dfrac{d^2y}{dx^2}$ and $\dfrac{d^4y}{dx^4}$.

$\dfrac{d^2y}{dx^2} = \dfrac{\varepsilon^{2x} + \varepsilon^x}{(\varepsilon^x - 1)^3}$, and $\dfrac{d^4y}{dx^4} = \dfrac{\varepsilon^x + 11\varepsilon^{2x} + 11\varepsilon^{3x} + \varepsilon^{4x}}{(\varepsilon^x - 1)^5}$.

19. If $y = \sin^{-1} x$, find $\dfrac{d^4y}{dx^4}$. $\dfrac{d^4y}{dx^4} = \dfrac{9x + 6x^3}{(1 - x^2)^{\frac{7}{2}}}$.

20. If $y = \varepsilon^{\sin x}$, find $\dfrac{d^3y}{dx^3}$.

$\dfrac{d^3y}{dx^3} = -\varepsilon^{\sin x} \cos x \sin x (\sin x + 3)$.

21. If $y = \dfrac{x}{1 + \log x}$, find $\dfrac{d^2y}{dx^2}$. $\dfrac{d^2y}{dx^2} = \dfrac{1 - \log x}{x(1 + \log x)^3}$.

22. Find the value of $d^2(\varepsilon^x)$, when x is not equicrescent.

$d^2(\varepsilon^x) = \varepsilon^x(dx)^2 + 3\varepsilon^x d^2x\, dx + \varepsilon^x d^3x$.

23. Find the value of $\dfrac{d^3}{dt^3}(\sin \theta)$, θ being a function of t.

$\dfrac{d^3}{dt^3}(\sin \theta) = -\cos \theta \left(\dfrac{d\theta}{dt}\right)^3 - 3 \sin \theta \, \dfrac{d\theta}{dt} \cdot \dfrac{d^2\theta}{dt^2} + \cos \theta \, \dfrac{d^3\theta}{dt^3}$.

CHAPTER V.

THE EVALUATION OF INDETERMINATE FORMS.

XII.

Indeterminate or Illusory Forms.

79. WHEN a function is expressed in the form of a fraction each of whose terms is variable, it may happen that, for a certain value of the independent variable, both terms reduce to zero. The function then takes the form $\frac{0}{0}$, and is said to be *indeterminate*, since its value cannot be ascertained by the ordinary process of dividing the value of the numerator by that of the denominator. The function has, nevertheless, a value as determinate for this as for any other value of the independent variable. It is the object of this chapter to show that such definite values exist, and to explain the methods by which they are determined.

The term *illusory form* is often used as synonymous with *indeterminate form*, and these terms are applied indifferently, not only to the form $\frac{0}{0}$, but also to the forms $\frac{\infty}{\infty}$, $\infty \cdot 0$, $\infty - \infty$, and to certain others whose logarithms assume the form $\infty \cdot 0$.

When a function of x takes an illusory form for $x=a$, the corresponding value of the function is sometimes called its *limiting value* as x approaches the value a.

80. The values of functions which assume illusory forms may sometimes be ascertained by making use of certain algebraic transformations. Thus, for example, the function

$$\frac{a - \sqrt{(a^2 - bx)}}{x}$$

takes the form $\frac{0}{0}$ when $x = 0$.

Multiplying both terms by the complementary surd

$$a + \sqrt{(a^2 - bx)},$$

we obtain $\qquad \dfrac{bx}{x[a + \sqrt{(a^2 - bx)}]} = \dfrac{b}{a + \sqrt{(a^2 - bx)}}.$

The last form is not illusory for the given value of x, since the factor which becomes zero has been removed from both terms of the fraction. The value of the fraction for $x = 0$ is evidently $\dfrac{b}{2a}$.

The following notation is used to indicate this and similar results; viz.,

$$\left. \frac{a - \sqrt{(a^2 - bx)}}{x} \right]_0 = \frac{b}{2a},$$

the subscript denoting that value of the independent variable for which the function is evaluated.

Evaluation by Differentiation.

81. Let $\dfrac{v}{u}$ represent a function in which both u and v are functions of x, which vanish when $x = a$; in other words, for this value of x, we have $u = 0$, and $v = 0$.

§ XII.] *EVALUATION BY DIFFERENTIATION.* 81

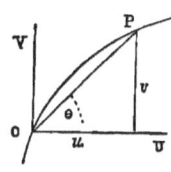

Let P be a moving point of which the abscissa and ordinate are simultaneous values of u and v (x not being represented in the figure); then, denoting the angle POU by θ, and the inclination of the motion of P to the axis of u by ϕ, we have

Fig. 9.

$$\tan \theta = \frac{v}{u}, \quad \text{and} \quad \tan \phi = \frac{dv}{du}.$$

At the instant when x passes through the value a, u and v being zero by the hypothesis, P passes through the origin; the corresponding value of θ is evidently determined by the direction in which P is moving at that instant, and is therefore equal to the value of ϕ at that point.

Hence the values of $\tan \theta$ and $\tan \phi$ corresponding to $x = a$ are equal, or

$$\left.\frac{v}{u}\right]_{x=a} = \left.\frac{dv}{du}\right]_{x=a};$$

therefore, to determine the value of $\frac{v}{u}$ for $x = a$, we substitute for it the function $\frac{dv}{du}$, whose value is the same as that of the given function, when $x = a$.

82. This result may also be expressed in the following manner: let $f(x)$ and $\phi(x)$ be two functions, such that $f(a) = 0$, and $\phi(a) = 0$; then

$$\frac{f(a)}{\phi(a)} = \frac{f'(a)}{\phi'(a)}. \quad \ldots \ldots \ldots (1)$$

As an illustration, let us take $\frac{\log x}{x-1}$. When $x = 1$, this function takes the form $\frac{0}{0}$; by the above process, we have

$$\frac{\log x}{x-1}\bigg]_1 = \frac{x^{-1}}{1}\bigg]_1 = 1,$$

the required value.

83. Since the substituted function $\dfrac{dv}{du}$ or $\dfrac{f'(x)}{\phi'(x)}$ frequently takes the indeterminate form, several repetitions of the process are sometimes requisite before the value of the function can be ascertained.

For example, the function $\dfrac{1-\cos\theta}{\theta^2}$ takes the form $\dfrac{0}{0}$ when $\theta = 0$; employing the process for evaluating, we have

$$\frac{1-\cos\theta}{\theta^2}\bigg]_0 = \frac{\sin\theta}{2\theta}\bigg]_0,$$

which is likewise indeterminate; but, by repeating the process, we obtain

$$\frac{1-\cos\theta}{\theta^2}\bigg]_0 = \frac{\sin\theta}{2\theta}\bigg]_0 = \frac{\cos\theta}{2}\bigg]_0 = \tfrac{1}{2}.$$

84. If the given function, or any of the substituted functions, contains a factor which does not take the indeterminate form, this factor may be evaluated at once, as in the following example.

The function

$$\frac{(1-x)\epsilon^x - 1}{\tan^2 x}$$

is indeterminate for $x = 0$. By employing the usual process once, we obtain

$$\frac{(1-x)\epsilon^x - 1}{\tan^2 x}\bigg]_0 = \frac{-x\epsilon^x}{2\sec^2 x \tan x}\bigg]_0,$$

which is likewise indeterminate; but, before repeating the process, we may evaluate the factor $-\dfrac{\epsilon^x}{2\sec^2 x}\bigg]_0$. The value of this factor is $-\tfrac{1}{2}$; hence we write

§ XII.] ABBREVIATED METHODS. 83

$$\left.\frac{(1-x)\varepsilon^x - 1}{\tan^2 x}\right]_0 = -\left.\frac{x\varepsilon^x}{2\sec^2 x \tan x}\right]_0 = -\tfrac{1}{2}\frac{x}{\tan x}$$

$$= -\tfrac{1}{2}\left.\frac{1}{\sec^2 x}\right]_0 = -\tfrac{1}{2}.$$

85. When the given function can be decomposed into factors each of which takes the indeterminate form, these factors may be evaluated separately. Thus, if the given function be

$$\frac{(\varepsilon^x - 1)\tan^2 x}{x^3},$$

the form $\left(\dfrac{\tan x}{x}\right)^2 \left(\dfrac{\varepsilon^x - 1}{x}\right)$

may be employed. We have

$$\left.\frac{\tan x}{x}\right]_0 = 1, \text{ and } \left.\frac{\varepsilon^x - 1}{x}\right]_0 = 1\,;$$

hence the value of the given function is unity.

When this method is used, if one of the factors is found to take the value zero while another is infinite, their product, being of the form $0 \cdot \infty$, must be treated by the usual method, since $0 \cdot \infty$ is itself an illusory form.

86. Another mode of decomposing a given function is that of separating it into parts, and substituting the values of such parts as are found on evaluation to be finite.

As an illustration, we take the expression,

$$u_0 = \left.\frac{(\varepsilon^x - \varepsilon^{-x})^2 - 2x^2(\varepsilon^x + \varepsilon^{-x})}{x^4}\right]_0.$$

Each of the fractions into which this function can be decomposed being obviously infinite, we first apply the usual process, thus obtaining

$$u_0 = \frac{2(\varepsilon^x - \varepsilon^{-x})(\varepsilon^x + \varepsilon^{-x}) - 4x(\varepsilon^x + \varepsilon^{-x}) - 2x^2(\varepsilon^x - \varepsilon^{-x})}{4x^3}\Big]_0.$$

Separating this expression into two fractions, thus,—

$$u_0 = \frac{(\varepsilon^x + \varepsilon^{-x})(\varepsilon^x - \varepsilon^{-x} - 2x)}{2x^3}\Big]_0 - \frac{\varepsilon^x - \varepsilon^{-x}}{2x}\Big]_0;$$

the latter is found on evaluation to have a finite value, and the expression reduces to

$$u_0 = \frac{\varepsilon^x - \varepsilon^{-x} - 2x}{x^3}\Big]_0 - 1.$$

Hence

$$u_0 = \frac{\varepsilon^x + \varepsilon^{-x} - 2}{3x^2}\Big]_0 - 1 = \frac{\varepsilon^x - \varepsilon^{-x}}{6x}\Big]_0 - 1 = -\tfrac{2}{3}.$$

Examples XII.

1. Prove $\dfrac{\sin x}{x}\Big]_0 = 1$, $\dfrac{\tan x}{x}\Big]_0 = 1$, and $\dfrac{\varepsilon^x - 1}{x}\Big]_0 = 1$.

These results are frequently useful in evaluating other functions. Evaluate the following functions:

2. $\dfrac{\varepsilon^x - \varepsilon^{-x}}{\log(1+x)}$, when $x = 0$.   2.

3. $\dfrac{a^n - x^n}{\log a - \log x}$, $x = a$. na^n.

4. $\dfrac{x^3 - 5x^2 + 7x - 3}{x^3 - x^2 - 5x - 3}$, $x = 3$. $\dfrac{1}{4}$.

5. $\dfrac{x^4 - 8x^3 + 22x^2 - 24x + 9}{x^4 - 4x^3 - 2x^2 + 12x + 9}$, $x = 3$. $\dfrac{1}{4}$.

6. $\dfrac{x\varepsilon^{2x} - \varepsilon^{2x} - x + 1}{\varepsilon^{2x} - 1}$, $x = 0$. -1.

§ XII.] EXAMPLES. 85

7. $\dfrac{\sin x - \cos x}{\sin 2x - \cos 2x - 1}$, when $x = \tfrac{1}{4}\pi$. $\tfrac{1}{2}\sqrt{2}$.

8. $\dfrac{\log x}{\sqrt{(1-x)}}$, $x = 1$. 0.

9. $\dfrac{a^x - b^x}{x}$, $x = 0$. $\log \dfrac{a}{b}$.

10. $\dfrac{\sqrt{(1+x^2)(1-x)}}{1-x^n}$, (See Art. 84), $x = 1$. $\dfrac{\sqrt{2}}{n}$.

11. $\dfrac{a^2 - x^2}{x^2}\left(1 - \cos\dfrac{x}{a}\right)$, $x = 0$. $\dfrac{1}{2}$.

12. $\dfrac{\varepsilon^{mx} - \varepsilon^{ma}}{x - a}$, $x = a$. $m\varepsilon^{ma}$.

13. $\dfrac{a^{\sin x} - a}{\log \sin x}$, $x = \tfrac{1}{2}\pi$. $a \log a$.

14. $\dfrac{1 - \cos x}{x \log(1+x)}$, $x = 0$. $\dfrac{1}{2}$.

15. $\dfrac{\sqrt{x}\tan x}{(\varepsilon^x - 1)^{\frac{3}{2}}}$, $x = 0$. 1.

Put in the form $\sqrt{\dfrac{x}{\varepsilon^x - 1} \cdot \dfrac{\tan x}{x} \cdot \dfrac{x}{\varepsilon^x - 1}}$. See Art. 85 and Example 1.

16. $\dfrac{\sqrt{x} - \sqrt{a} + \sqrt{(x-a)}}{\sqrt{(x^2 - a^2)}}$, $x = a$. $\dfrac{1}{\sqrt{(2a)}}$.

17. $\dfrac{x\sqrt{(3x - 2x^4)} - x^{\frac{6}{5}}}{1 - x^{\frac{2}{3}}}$, $x = 1$. $\dfrac{81}{20}$.

18. $\dfrac{(a^2 + ax + x^2)^{\frac{1}{2}} - (a^2 - ax + x^2)^{\frac{1}{2}}}{(a+x)^{\frac{1}{3}} - (a-x)^{\frac{1}{3}}}$, $x = 0$. $\sqrt{a}$.

Multiply both terms by the two complementary surds. See Art. 80.

19. $\dfrac{(a^2-x^2)^{\frac{1}{2}}+(a-x)^{\frac{3}{2}}}{(a^2-x^2)^{\frac{1}{3}}+(a-x)^{\frac{1}{3}}}$, when $x=a$. $\dfrac{\sqrt{(2a)}}{1+a\sqrt{3}}$.

Divide both terms by $(a-x)^{\frac{1}{3}}$.

20. $\dfrac{\sin x - x\cos x}{x - \sin x}$, $x=0$. 2.

21. $\dfrac{\varepsilon^x - \varepsilon^{-x} - 2x}{x - \tan x}$, $x=0$. -1.

22. $\dfrac{(x-2)\varepsilon^x + x + 2}{x(\varepsilon^x - 1)^2}$, $x=0$. $\dfrac{1}{6}$.

23. $\dfrac{x^x - x}{1 - x + \log x}$, $x=1$. -2.

24. $\dfrac{\tan x - \sin x}{x^3}$, $x=0$. $\dfrac{1}{2}$.

Put in the form $\dfrac{\sin x}{x}\bigg]_0 \cdot \dfrac{\sec x - 1}{x^2}\bigg]_0$.

25. $\dfrac{(x-1)^2 + \sin^2(x^2-1)^{\frac{1}{2}}}{(x+1)(x-1)^{\frac{3}{2}}}$, $x=1$. $\sqrt{2}$.

26. $\dfrac{1 - x + \log x}{1 - \sqrt{(2x - x^2)}}$, $x=1$. -1.

27. $\dfrac{\sin x - \log(\varepsilon^x \cos x)}{x^2}$, $x=0$. $\dfrac{1}{2}$.

28. $\dfrac{\frac{1}{4}\pi - \tan^{-1} x}{x^n - \varepsilon^{\sin(\log x)}}$, $x=1$. $\dfrac{1}{2(1-n)}$.

29. $\dfrac{\tan(a+x) - \tan(a-x)}{\tan^{-1}(a+x) - \tan^{-1}(a-x)}$, $x=0$. $(1+a^2)\sec^2 a$.

30. $\dfrac{x\sin x - \frac{1}{2}\pi}{\cos x}$, $x=\frac{1}{2}\pi$. -1.

§ XII.]   EXAMPLES.   87

✓ 31. $\dfrac{\varepsilon^x - \varepsilon^{\sin x}}{x - \sin x},$ when $x = 0$.   1.

✓ 32. $\dfrac{m^2 \sin nx - n^2 \sin mx}{\tan nx - \tan mx},$ $m = n$.

$n^{2-1}(n \cos nx - \sin nx) \cos^2 nx.$

In solving this and the following example, x *and* n *may be regarded as constants, and* m *as a variable.*

✓ 33. $\dfrac{\tan nx - \tan mx}{\sin(n^2 x - m^2 x)},$ $m = n$. $\dfrac{\sec^2 nx}{2n}.$

XIII.

The Form $\dfrac{\infty}{\infty}$.

87. Let $\dfrac{f(x)}{\phi(x)}$ denote a function which assumes the form $\dfrac{\infty}{\infty}$ when $x = a$, then we have

$$\dfrac{f(x)}{\phi(x)} = \dfrac{\dfrac{1}{\phi(x)}}{\dfrac{1}{f(x)}}. \quad \ldots \ldots \quad (1)$$

The second member of this equation takes the form $\dfrac{0}{0}$ when $x = a$; we therefore have, by equation (1) Art. 82,

$$\dfrac{f(a)}{\phi(a)} = \dfrac{\dfrac{1}{\phi(a)}}{\dfrac{1}{f(a)}} = \dfrac{-\dfrac{\phi'(a)}{[\phi(a)]^2}}{-\dfrac{f'(a)}{[f(a)]^2}} = \dfrac{\phi'(a)}{f'(a)} \left\{\dfrac{f(a)}{\phi(a)}\right\}^2; \quad \cdot \cdot \quad (2)$$

whence, if $\dfrac{f(a)}{\phi(a)}$ is neither zero nor infinity, we infer that

$$\frac{f(a)}{\phi(a)} = \frac{f'(a)}{\phi'(a)} \quad \cdots \cdots \cdots (3)$$

This formula, it will be observed, is identical with that employed when the function takes the form $\frac{0}{0}$.

88. When the value of $\frac{f(a)}{\phi(a)}$ is either zero or infinity, equation (2), Art. 87, will be satisfied independently of the existence of equation (3); we are not justified therefore, when this is the case, in deriving the latter from the former. The following demonstration shows, however, that equation (3) holds in these cases also.

First, when the value of $\frac{f(a)}{\phi(a)}$ is zero, by adding a finite quantity n to the given function, we have

$$\frac{f(a)}{\phi(a)} + n = \frac{f(a) + n\phi(a)}{\phi(a)},$$

a function which is by hypothesis finite. To this function therefore the demonstration given in Art. 87 applies; hence

$$\frac{f(a)}{\phi(a)} + n = \frac{f'(a) + n\phi'(a)}{\phi'(a)} = n + \frac{f'(a)}{\phi'(a)};$$

therefore
$$\frac{f(a)}{\phi(a)} = \frac{f'(a)}{\phi'(a)},$$
as before.

Again, if the value of $\frac{f(a)}{\phi(a)}$ is infinite, that of $\frac{\phi(a)}{f(a)}$ is zero, and, by the last result,

$$\frac{\phi(a)}{f(a)} = \frac{\phi'(a)}{f'(a)};$$

hence, in this case, likewise

$$\frac{f(a)}{\phi(a)} = \frac{f'(a)}{\phi'(a)}.$$

Derivatives of Functions which assume an Infinite Value.

89. *When* f(x) *becomes infinite, for a <u>finite</u> value* a *of the independent variable,* f'(a) *is likewise infinite.* For, let b denote a value of x so taken that $f(x)$ shall be finite for $x = b$ and for all values of x between b and a : then, as x varies from b to a, the rate of $f(x)$ must assume an infinite value, otherwise $f(x)$ would remain finite. The value of x for which the rate is infinite must be a or some value of x between b and a; that is, some value of x nearer to a than b is. Now, since b may be taken as near as we please to a, the value of x for which the rate is infinite cannot differ from a. The expression for this rate is $f'(x)\dfrac{dx}{dt}$, in which $\dfrac{dx}{dt}$ may be assumed finite, therefore $f'(x)$ must be infinite when $x = a$; in other words, f'(a) *is infinite when* f(a) *is infinite.*

90. It follows from the theorem proved in the preceding article that when a is finite the function obtained by the application of formula (3), Art. 87, takes the same form, $\dfrac{\infty}{\infty}$, as that assumed by the original function. Hence, except when the given value of x is infinite, the application of some other process, either to the original function or to one of the substituted functions, is always requisite. Thus in the example,

$$\left.\frac{\log (\sin 2x)}{\log \sin x}\right]_0 = \frac{\infty}{\infty};$$

by using the above formula we obtain

$$\frac{\log \sin 2x}{\log \sin x}\bigg]_0 = \frac{2 \cot 2x}{\cot x}\bigg]_0,$$

which takes the form $\frac{\infty}{\infty}$; but the last expression is equivalent to $2\dfrac{\sin x \cos 2x}{\sin 2x \cos x}\bigg]_0$, and is therefore easily shown to have the value unity.

The Form $0 \cdot \infty$.

91. A function which takes this form may, by introducing the reciprocal of one of the factors, be so transformed as to take either of the forms $\frac{0}{0}$ or $\frac{\infty}{\infty}$, as may be found most convenient. For example, let us take the function

$$x^{-n}\varepsilon^x,$$

which assumes the above form when $x = \infty$, n being positive. In this case it is necessary to reduce to the form $\frac{\infty}{\infty}$. Thus—

$$x^{-n}\varepsilon^x = \frac{\varepsilon^x}{x^n}\bigg]_\infty = \frac{\varepsilon^x}{nx^{n-1}}\bigg]_\infty = \frac{\varepsilon^x}{n(n-1)x^{n-2}}\bigg]_\infty, \text{ etc.}$$

By continuing this process, we finally obtain a fraction whose denominator is finite while its numerator is still infinite. Hence we have, for all finite values of n,

$$x^{-n}\varepsilon^x\big]_\infty = \infty.$$

The Form $\infty - \infty$.

92. A function which assumes this form may be so transformed as to take the form $\frac{0}{0}$. Let the given function be

$$\left[\frac{1}{x(1+x)} - \frac{\log(1+x)}{x^2}\right]_0,$$

which takes the form $\infty - \infty$, since the second term is easily shown to be infinite. But

$$\left[\frac{1}{x(1+x)} - \frac{\log(1+x)}{x^2}\right]_0 = \left[\frac{x - (1+x)\log(1+x)}{x^2(1+x)}\right]_0$$

$$= \left[\frac{x - (1+x)\log(1+x)}{x^2}\right]_0$$

$$= \left[\frac{1 - \log(1+x) - 1}{2x}\right]_0 = -\frac{1}{2}.$$

Examples XIII.

Evaluate the following functions:

1. $\dfrac{\sec x}{\sec 3x}$, when $x = \tfrac{1}{2}\pi$. $\quad -3$.

2. $\dfrac{a^x}{\operatorname{cosec}(ma^{-x})}$, $\quad x = \infty$. $\quad m$.

3. $\dfrac{\log x}{x^n}\ (n > 0)$, $\quad x = \infty$. $\quad 0$.

4. $\dfrac{\tan x}{\log(x - \tfrac{1}{2}\pi)}$, $\quad x = \tfrac{1}{2}\pi$. $\quad \infty$.

5. $\dfrac{\sec(\tfrac{1}{2}\pi x)}{\log(1 - x)}$, $\quad x = 1$. $\quad \infty$.

6. $\dfrac{\log\cos(\tfrac{1}{2}\pi x)}{\log(1 - x)}$, $\quad x = 1$. $\quad 1$.

7. $\dfrac{\tan x}{\tan 3x}$,	$x = \tfrac{1}{2}\pi.$	3.
8. $\dfrac{\log(1+x)}{x}$,	$x = \infty.$	0.
9. $\left(a^{\frac{1}{x}} - 1\right)x$,	$x = \infty.$	$\log a.$
10. $\dfrac{x^2 - a^2}{a^2} \tan \dfrac{\pi x}{2a}$,	$x = a.$	$-\dfrac{4}{\pi}.$
11. $x^m (\log x)^n$, (m and n being positive),	$x = 0.$	0.
12. $\varepsilon^x \sin \dfrac{1}{x}$,	$x = \infty.$	$\infty.$
13. $\varepsilon^{-\frac{1}{x}}(1 - \log x)$,	$x = 0.$	0.
14. $\sec \dfrac{\pi x}{2} \cdot \log \dfrac{1}{x}$,	$x = 1.$	$\dfrac{2}{\pi}.$
15. $\dfrac{\log \tan nx}{\log \tan x}$,	$x = 0.$	1.
16. $\dfrac{\log \cot \dfrac{x}{2}}{\cot x + \log x}$,	$x = 0.$	0.
17. $\sec x (x \sin x - \tfrac{1}{2}\pi)$,	$x = \tfrac{1}{2}\pi.$	$-1.$
18. $\log\left(2 - \dfrac{x}{a}\right) \tan \dfrac{\pi x}{2a}$,	$x = a.$	$\dfrac{2}{\pi}.$
19. $(1 - x) \tan(\tfrac{1}{2}\pi x)$,	$x = 1.$	$\dfrac{2}{\pi}.$
20. $\log(x - a) \tan(x - a)$,	$x = a.$	0.

XIV.

Functions whose Logarithms take the Form $\infty \cdot 0$.

93. In the case of a function of the form u^v, we have

$$\log u^v = v \log u.$$

The expression $v \log u$ takes the illusory form $0 \cdot \infty$ in two cases: first, when $v = 0$ and $\log u = \infty$; and secondly, when $v = \infty$ and $\log u = 0$.

Log u is infinite when $u = 0$, and also when $u = \infty$; therefore the first case will arise when the original function takes one of the forms ∞^0 or 0^0.

Log $u = 0$ when $u = 1$, therefore the second case will arise when the original function takes the form 1^∞.

Hence functions which take either of the three illusory forms,

$$\infty^0, \quad 0^0, \quad \text{or} \quad 1^\infty,$$

may be evaluated by first evaluating their logarithms, which take the form $0 \cdot \infty$.

It is to be noticed however that 0^∞ and ∞^∞ are not illusory forms, since their logarithms take the form $\infty(\mp \infty)$.

The Form 1^∞.

94. As an illustration of this form, we take the function $\left(1 + \dfrac{a}{x}\right)^x$, which assumes the form 1^∞ when $x = \infty$. Denoting this function by u, we have

$$\log u = x \log\left(1 + \frac{a}{x}\right) = \frac{\log\left(1 + \dfrac{a}{x}\right)}{\dfrac{1}{x}},$$

the last expression assuming the form $\dfrac{0}{0}$ when $x = \infty$.

In evaluating this logarithm, it is convenient to substitute z for $\frac{1}{x}$; then

$$\log u_\infty = \frac{\log(1+az)}{z}\bigg]_0,$$

since, when $x = \infty$, $z = 0$. Taking derivatives, we have

$$\log u_\infty = \frac{\log(1+az)}{z}\bigg]_0 = \frac{a}{1+az}\bigg]_0 = a.$$

Hence
$$u_\infty = \left(1 + \frac{a}{x}\right)^x\bigg]_\infty = \varepsilon^a.$$

95. If $a = 1$, we have

$$\left(1 + \frac{1}{x}\right)^x\bigg]_\infty = \varepsilon;$$

that is, as x increases indefinitely, the *limiting value* of the function $\left(1 + \frac{1}{x}\right)^x$ is ε. The Napierian base is often defined as the limiting value of this function, or, what is the same thing, by formula

$$\varepsilon = (1 + x)^{\frac{1}{x}}\bigg]_0.$$

The Form 0^0.

96. The function $x^x]_0$, by the aid of which many functions of similar form may be evaluated, will serve as an illustration of the form 0^0.

Let $$u = x^x;$$

then $$\log u = x \log x,$$

and $\qquad \log u \Big]_0 = \dfrac{\log x}{x^{-1}}\Big]_0 = -\dfrac{x^{-1}}{x^{-2}}\Big]_0 = 0;$

therefore $\qquad x^x \Big]_0 = \varepsilon^0 = 1.$

The value of a function which takes the form 0^0 is usually found, as in the above example, to be unity. This is not, however, *universally* true, as the function

$$x^{\frac{a+x}{\log x}}$$

(one of those earliest adduced for this purpose [*]) will show.

This function takes the form 0^0, when $x = 0$; but since its logarithm reduces to $a + x$, its value when $x = 0$ is ε^a.

Examples XIV.

1. $(\cos x)^{\cot^2 x}$, $\qquad\qquad$ when $x = 0$. $\qquad \varepsilon^{-\frac{1}{2}}$.

2. $\left(\dfrac{\tan x}{x}\right)^{\frac{1}{x^2}}$, $\qquad\qquad x = 0.$ $\qquad \sqrt[3]{\varepsilon}.$

3. $(\cos \alpha x)^{\cosec^2 \beta x}$, $\qquad\qquad x = 0.$ $\qquad \varepsilon^{-\frac{a^2}{2\beta^2}}.$

4. $\left(\dfrac{1}{x}\right)^{\tan x}$, $\qquad\qquad x = 0.$ $\qquad 1.$

5. $(\tan x)^{\tan 2x}$, $\qquad\qquad x = -\tfrac{1}{4}\pi.$ $\qquad 1.$

6. $\left(\dfrac{1}{x^n}\right)^{x^m} (m > 0),$ $\qquad\qquad x = 0.$ $\qquad 1.$

7. $(1 - x)^{\frac{1}{x}}$, $\qquad\qquad x = 0.$ $\qquad \dfrac{1}{\varepsilon}.$

8. $(\sin x)^{\sec^2 x}$, $\qquad\qquad x = \tfrac{1}{2}\pi.$ $\qquad \varepsilon^{-\frac{1}{2}}.$

[*] See *Crelle's Journal*, vol. xii, p. 293.

9. $(\cot x)^{\sin x}$, $\qquad x = 0$.

$$\text{Solution}: \ (\cot x)^{\sin x}\Big]_0 = \frac{(\cos x)^{\sin x}]_0}{(\sin x)^{\sin x}]_0} = 1. \quad (\textit{See Art. 96.})$$

10. $(\sin x)^{\tan x}$, $\qquad x = 0$. $\qquad 1$.

11. $(\sin x)^{\tan x}$, $\qquad x = \dfrac{\pi}{2}$. $\qquad 1$.

12. $x^{\frac{a}{\log \sin x}}$, $\qquad x = 0$. $\qquad \varepsilon^a$.

13. $(\sin x)^{\frac{a^2 - x^2}{\log \tan x}}$, $\qquad x = 0$. $\qquad \varepsilon^{a^2}$.

14. $x^{x^a}\ (a > 0)$, $\qquad x = 0$. $\qquad 1$.

15. $(x^2)^{\frac{(a+x)^2}{\log(x + \log \cos x)}}$, $\qquad x = 0$. $\qquad \varepsilon^{2a^2}$.

16. $x^{\frac{1}{1-x}}$, $\qquad$ when $x = 1$. $\qquad \dfrac{1}{\varepsilon}$.

17. $x^{x^x - 1}$, $\qquad x = 0$. $\qquad 1$.

18. $(\cos mx)^{\frac{n}{x^2}}$, $\qquad x = 0$. $\qquad \varepsilon^{-\frac{1}{2}nm^2}$.

19. $\left(\dfrac{\log x}{x}\right)^{\frac{1}{x}}$, $\qquad x = \infty$. $\qquad 1$.

20. $(1 \pm x)^{\frac{1}{x}}$, $\qquad x = \infty$. $\qquad 1$.

21. $x^m (\sin x)^{\tan x}\left(\dfrac{\pi - 2x}{2 \sin 2x}\right)^2$, $\qquad x = \dfrac{\pi}{2}$. $\qquad \dfrac{\pi^m}{2^{m+2}}$.

22. $\dfrac{(m^2 - 1)(a \sin x - \sin ax)^n}{x^n \sin x (\cos x - \cos ax)^n}$, $\qquad x = 0$. $\qquad \left(\dfrac{a}{3}\right)^n \log m$.

CHAPTER VI.

Maxima and Minima of Functions of a Single Variable.

XV.

Conditions Indicating the Existence of Maxima and Minima.

97. If, while the independent variable increases continuously, a function dependent on it increases up to a certain value, and then decreases, this value of the function is said to be a *maximum* value. In other words, a function $f(x)$ has a maximum value corresponding to $x = a$, if, when x increases through the value a, the function changes from an increasing to a decreasing function.

Since $f'(x)$ is positive, when $f(x)$ is an increasing function, and negative when it is a decreasing function; it is obvious that if $f(a)$ is a maximum value of $f(x)$, $f'(x)$ must *change sign*, from $+$ to $-$, as x increases through the value a.

On the other hand, a function is said to have a *minimum* value for $x = a$, if it is a decreasing function before x reaches this value and an increasing one afterward. In this case, $f'(x)$ changes sign from $-$ to $+$.

98. The derivative $f'(x)$ can only change sign on passing through zero or infinity. Hence a value of x, for which $f(x)$ is a maximum or a minimum, must satisfy one of the two following equations:

$$f'(x) = 0 \quad \text{and} \quad f'(x) = \infty.$$

The required values of x will therefore be found among the roots of these equations.

The case which usually presents itself, and which will therefore be considered first, is that in which the required value of x is a root of the equation $f'(x) = 0$.

99. As an illustration, let it be required *to divide a number into two such parts that the square of one part multiplied by the cube of the other shall give the greatest possible product.*

Denote the given number by a, and the part to be squared by x; then we have

$$f(x) = x^2 (a - x)^3.$$

It is evident that a maximum value of this function exists; for when $x = 0$ its value is zero, and when $x = a$ its value is again zero, while for intermediate values of x it is positive; hence the function must change from an increasing to a decreasing function at least once, while x passes from the value zero to the value a.

Taking the derivative of this function, the equation

$$f'(x) = 0$$

is in this case $\quad 2x(a - x)^3 - 3x^2 (a - x)^2 = 0$,

or $\quad\quad\quad x(a - x)^2 (2a - 5x) = 0.$

0 and a are roots of this equation; but, as we are in search of a value of the function corresponding to an intermediate value of x, we put

$$2a - 5x = 0,$$

and obtain $\quad\quad\quad x = \tfrac{2}{5}a.$

The corresponding value of the function is $\tfrac{108}{3125}a^5$, the maximum value sought.

Maxima and Minima of Geometrical Magnitudes.

100. When the maximum or minimum value of a geometrical magnitude limited by certain conditions is required, it is necessary to obtain an expression for the magnitude in terms of a single unknown quantity, such that the determination of the value of this quantity will constitute the solution of the problem. For example: *let it be required to determine the cone of greatest convex surface among those which can be inscribed in a sphere whose radius is* a.

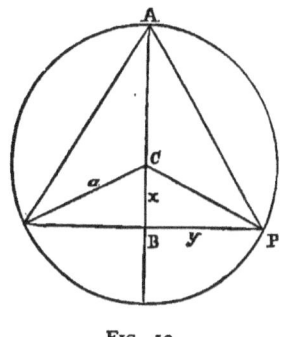

FIG. 10.

Any point A of the surface of the sphere being taken as the apex of the cone, let the diagram represent a great circle of the sphere passing through the fixed point A.

If we refer the position of the point P to rectangular coordinates, and take C as the origin, the required cone will evidently be determined when x is determined. We have now to express the convex surface S in terms of x.

The expression for the convex surface of a cone gives

$$S = \pi y \sqrt{[y^2 + (a + x)^2]}, \quad \ldots \quad (1)$$

in which the unknown quantities x and y are connected by the equation of the circle

$$x^2 + y^2 = a^2. \quad \ldots \ldots \ldots (2)$$

Substituting the value of y, we have

$$S = \pi \sqrt{(a^2 - x^2)} \sqrt{(2a^2 + 2ax)},$$

reducing, $\quad S = \pi \sqrt{(2a)} (a + x) \sqrt{(a - x)}. \quad \ldots \quad (3)$

Since the factor $\pi \sqrt{(2a)}$ is constant, we are evidently required to find the value of x for which the function

$$f(x) = (a + x)\sqrt{(a - x)}$$

is a maximum. The equation $f'(x) = 0$ is, in this case,

$$\sqrt{(a - x)} - \frac{a + x}{2\sqrt{(a - x)}} = 0;$$

whence $\quad x = \tfrac{1}{3}a.$

The altitude of the required cone is therefore $\tfrac{4}{3}a$. Substituting this value of x in equation (3), we have

$$S = \tfrac{8}{9}\sqrt{3}\cdot\pi a^2,$$

the maximum value required.

101. As a further illustration, let it be required *to determine the greatest cylinder that can be inscribed in a given segment of a paraboloid of revolution.*

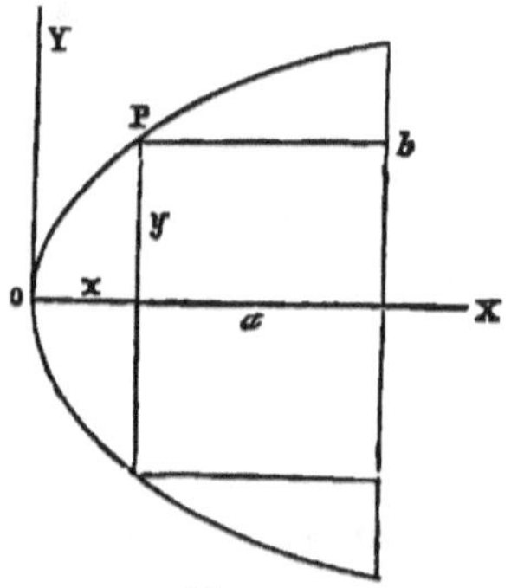

Let a denote the altitude, and b the radius of the base of the segment. The equation of the generating parabola is of the form

$$y^2 = 4cx.$$

Since (a, b) is a point of the curve, we have the condition

$$b^2 = 4ca;$$

FIG 11.

eliminating $4c$, the equation of the curve is

$$y^2 = \frac{b^2}{a}x. \qquad \ldots \ldots \ldots (1)$$

The volume V of the cylinder of which the maximum is required is expressed by

$$V = \pi y^2(a - x),$$

or, by equation (1), $\quad V = \pi \dfrac{b^2}{a} x(a - x).$

Hence we put $\quad f(x) = ax - x^2,$

and the condition $f'(x) = 0$ gives

$$x = \tfrac{1}{2}a.$$

Consequently $a - x$, the altitude of the cylinder, is one half the altitude of the segment.

Examples XV.

1. Find the sides of the largest rectangle that can be inscribed in a semicircle of radius a. The sides are $a\sqrt{2}$ and $\tfrac{1}{2}a\sqrt{2}$.

2. Determine the maximum right cone inscribed in a given sphere. The altitude is four thirds the radius of the sphere.

3. Determine the maximum rectangle inscribed in a given segment of a parabola.
The altitude of the rectangle is two thirds that of the segment.

4. Find the maximum cone of given slant height a.
The radius of the base is $\tfrac{1}{3}a\sqrt{6}$.

5. A boatman 3 miles out at sea wishes to reach in the shortest time possible a point on the beach 5 miles from the nearest point of the shore; he can pull at the rate of 4 miles an hour, but can walk at the rate of 5 miles an hour; find the point at which he must land.
Express the whole time in terms of the distance of the required point from the nearest point of the shore.
He must land one mile from the point to be reached.

6. If a square piece of sheet-lead whose side is a have a square cut out at each corner, find the side of the latter square in order that the remainder may form a vessel of maximum capacity

The side of the square is $\tfrac{1}{6}a$.

7. A given weight is to be raised by means of a lever weighing n pounds per linear inch, which has its fulcrum at one end, and at a fixed distance a from the point of suspension of the weight w; find the length of the lever in order that the power required to raise the weight may be a minimum.
$$\sqrt{\frac{2aw}{n}}$$

8. A rectangular court is to be built so as to contain a given area c^2, and a wall already constructed is available for one of the sides; find its dimensions so that the least expense may be incurred.

The side parallel to the wall is double each of the others.

9. Determine the maximum cylinder inscribed in a given cone.

The altitude of the cylinder is one third that of the cone.

10. Prove that the rectangle with given perimeter and maximum area is a square, also that the rectangle with given area and minimum perimeter is a square.

11. Find the side of the smallest square that can be inscribed in a square whose side is a.

Take as the independent variable the distance between the angles of the two squares. $\tfrac{1}{2}a\sqrt{2}$.

12. Inscribe the maximum cone in a given paraboloid, the apex of the cone being at the middle point of the base of the paraboloid.

The altitude of the cone is half that of the paraboloid.

13. Find the maximum cylinder that can be inscribed in a sphere whose radius is a. The altitude is $\tfrac{2}{3}a\sqrt{3}$.

14. Through a point whose rectangular coordinates are a and b draw a line such that the triangle formed by this line and the coordinate axes shall be a minimum.

The intercepts on the axes are $2a$ and $2b$.

15. A high vertical wall is to be braced by a beam which must pass over a parallel wall a feet high and b feet distant from the other, find the length of the shortest beam that can be used for this purpose.

Take as the independent variable the inclination of the beam to the horizon

$$\left(a^{\frac{2}{3}} + b^{\frac{2}{3}}\right)^{\frac{3}{2}}$$

16. The illumination of a plane surface by a luminous point being directly as the cosine of the angle of incidence of the rays, and inversely as the square of its distance from the point; find the height at which a bracket-burner must be placed, in order that a point on the floor of a room at the horizontal distance a from the burner may receive the greatest possible amount of illumination.

The height is $\dfrac{a}{\sqrt{2}}$.

XVI.

Methods of Discriminating between Maxima and Minima.

102. When the existence of a maximum or a minimum corresponding to a particular root a of the equation $f'(x) = 0$ is not obvious from the nature of the problem, it is necessary to determine whether $f'(x)$ *changes sign* as x passes through the value a.

If a change of sign does take place we have, in accordance with Art. 97, a *maximum* if, when x passes through the value a, the change of sign is from $+$ to $-$; that is, if $f'(x)$ is a *decreasing* function, and a *minimum* if the change of sign is from $-$ to $+$, in which case $f'(x)$ is an *increasing* function.

103. In many cases we are able to distinguish maxima from minima by examining the expression for $f'(x)$, as in the following examples.

Given
$$f(x) = \frac{x}{\log x},$$
whence
$$f'(x) = \frac{\log x - 1}{(\log x)^2};$$
$f'(x) = 0$ gives $\log x = 1$, or $x = \varepsilon$.

Since $\log x$ is an increasing function, it is obvious that, as x increases through the value ε, $f'(x)$ increases; it therefore changes sign from $-$ to $+$, and consequently $f(\varepsilon)$ is a minimum value of $f(x)$.

104. If $f'(x)$ does not change sign we have neither a maximum nor a minimum; thus, let
$$f(x) = x - \sin x,$$
whence
$$f'(x) = 1 - \cos x.$$
In this case $f'(x)$ becomes zero when $x = 2n\pi$, n being zero or any integer, but does not change sign, since $1 - \cos x$ can never be negative; consequently $f(x)$ has neither maxima nor minima values, but is an *increasing* function for all values of x.

Alternate Maxima and Minima.

105. Let the curve
$$y = f(x)$$
be constructed, and suppose it to take the form represented in Fig. 12. There is a maximum value of $f(x)$ at B, another at D, and minima values occur at A, at C, and at E.

It is obvious that in a continuous portion of the curve maxima and minima ordinates must occur alternately, and must separate the curve into segments in which the ordinate is alternately an increasing and a decreasing function; hence, if $f(x)$ has maxi-

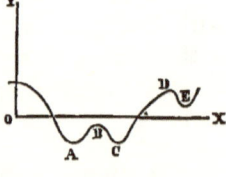

FIG. 12.

ma and minima values, they must occur alternately *unless infinite values of the function intervene.* It is also evident, with the same restriction, that a maximum is greater in value than either of the adjacent minima, but not necessarily greater than *any* other minimum; thus, in Fig. 12, the maximum at B is greater than the minima at A and C, but not greater than that at E.

106. As an illustration let us take the following function in which it is easy to discriminate between the maxima and minima values.
$$f(x) = x(x+a)^2(x-a)^3.$$
Whence,
$$f'(x) = (x+a)^2(x-a)^3 + 2x(x+a)(x-a)^3 + 3x(x+a)^2(x-a)^2,$$
$$= (x+a)(x-a)^2(6x^2 + ax - a^2).$$

a and $-a$ are evidently roots of $f'(x) = 0$; the roots derived by putting the last factor equal to zero and solving are $-\tfrac{1}{2}a$ and $\tfrac{1}{3}a$. Hence $f'(x)$ can be written in the form
$$f'(x) = 6(x+a)(x+\tfrac{1}{2}a)(x-\tfrac{1}{3}a)(x-a)^2,$$
in which the factors are so arranged that the corresponding roots are in order of magnitude.

When $x < -a$, $f'(x)$ is negative, and, if we regard x as increasing continuously, $f'(x)$ changes sign when $x = -a$, when $x = -\tfrac{1}{2}a$, and again when $x = \tfrac{1}{3}a$, but *not* when $x = a$.

Since $f'(x)$ is at first negative it changes sign from $-$ to $+$ when it first passes through zero, that is when $x = -a$; the corresponding value of $f(x)$ is therefore a minimum. Accordingly the value of $f(x)$ corresponding to the next root $x = -\tfrac{1}{2}a$ is a maximum, and that corresponding to $x = \tfrac{1}{3}a$ is another minimum; but there is neither a maximum nor a minimum corresponding to $x = a$.

107. When the function is continuous as in the above example, that is, does not become infinite for any finite value of x, it is always easy to determine by examining the function itself whether the last, or greatest value of x in question, gives a maximum or a minimum. Thus, in the above example, $f(x)$ evidently increases without limit as x increases without limit; therefore, the last value must be a minimum.

The Employment of a Substituted Function.

108. Since an increasing function of a variable increases and decreases with the variable, such a function will pass from a state of increase to a state of decrease, or the reverse, simultaneously with the variable; that is, it will reach a maximum or a minimum value at the same time with the variable.

This fact often enables us to simplify the determination of maxima and minima by substituting an increasing function of the given function for the given function itself. For example, if we have

$$f(x) = \sqrt{(b^2 + ax)} + \sqrt{(b^2 - ax)},$$

we may with advantage employ the square of the given function. The square is

$$2b^2 + 2\sqrt{(b^4 - a^2x^2)},$$

which is obviously a maximum when $x = 0$, and, since the square of a positive quantity is an increasing function, we infer that $f(x)$ is likewise a maximum for the same value of x.

109. A *decreasing* function of the given function may also be employed; but, in this case, since the substituted function decreases with the increase of the given function and increases with its decrease, a maximum of the substituted function indicates a minimum, and a minimum indicates a maximum of the given function.

Thus, if we have

$$f(x) = \frac{x}{x^2 - 3x + 1},$$

the reciprocal may be employed. The reciprocal of this function is

$$\frac{x^2 - 3x + 1}{x} = x - 3 + \frac{1}{x};$$

whence, taking the derivative, we obtain

$$1 - \frac{1}{x^2} = \frac{x^2 - 1}{x^2},$$

which vanishes when $x = \pm 1$.

Since x^2 is an increasing function when x is positive, this derivative is evidently an *increasing* function when $x = 1$. The reciprocal is therefore a *minimum* for this value of x, and consequently $f(1)$ is a *maximum* value of $f(x)$. In a similar manner it may be shown that $f(-1)$ is a minimum.

Examples XVI.

Determine the maxima and minima of the following functions:

1. $f(x) = x^x$. $\qquad$ A min. for $x = \dfrac{1}{\varepsilon}$.

2. $f(x) = \dfrac{\log x}{x^n}$. $\qquad$ A max. for $x = \varepsilon^{\frac{1}{n}}$.

3. $f(x) = \dfrac{(a-x)^2}{a-2x}$. $\qquad$ A min. for $x = \frac{1}{4}a$.

4. $f(x) = \dfrac{1+3x}{\sqrt{(1+5x)}}$. $\qquad$ A min. for $x = -\frac{1}{15}$.

5. $f(x) = \sin 2x - x$.
  A max. for $x = n\pi + \frac{1}{6}\pi$;
  a min. for $x = n\pi - \frac{1}{6}\pi$.

6. $f(x) = 2x^3 + 3x^2 - 36x + 12$.
  A max. for $x = -3$;
  a min. for $x = 2$.

7. $f(x) = x^3 - 3x^2 - 9x + 5$.
  A max. for $x = -1$;
  a min. for $x = 3$.

8. $f(x) = 3x^5 - 125x^3 + 2160x$.
  A max. for $x = -4$ and $x = 3$;
  a min. for $x = -3$ and $x = 4$.

9. $f(x) = b + c(x - a)^{\frac{5}{3}}$.
  Neither a max. nor a min.

10. $f(x) = (x - 1)^4 (x + 2)^3$.
  A max. for $x = -\frac{5}{7}$;
  a min. for $x = 1$.

11. $f(x) = (x - 9)^5 (x - 8)^4$.
  A max. for $x = 8$;
  a min. for $x = 8\frac{4}{9}$.

12. $f(x) = \dfrac{1 - x + x^2}{1 + x - x^2}$.
  A min. for $x = \frac{1}{2}$.

13. $f(x) = \dfrac{ax}{ax^2 - bx + a}$. *See Art.* 109.
  Max. for $x = 1$;
  min. for $x = -1$.

14. $f(x) = (a^{x-1})^{x+2}$.
  Min. for $x = -\frac{1}{2}$ (a being positive).

15. $f(x) = (1 + x^{\frac{2}{3}})(7 - x)^2$.
  Solve by putting $x = z^3$. For method of discriminating between maxima and minima, see Art. 107.
  Min. for $x = 0$, and $x = 7$;
  max. for $x = 1$.

16. $f(x) = 5x^6 + 12x^5 - 15x^4 - 40x^3 + 15x^2 + 60x + 27$.
  Min. for $x = -2$.

17. $f(x) = x^6 - 6x^4 + 4x^3 + 9x^2 - 12x + 3$.
  Min. for $x = -2$, and $x = 1$;
  max. for $x = -1$.

§ XVI.] EXAMPLES.

18. The top of a pedestal which sustains a statue a feet in height is b feet above the level of a man's eyes; find his horizontal distance from the pedestal when the statue subtends the greatest angle.

When the distance $= \sqrt{[b(a + b)]}$.

19. It is required to construct from two circular iron plates of radius a a buoy, composed of two equal cones having a common base, which shall have the greatest possible volume.

The radius of the base $= \frac{1}{3}a\sqrt{6}$.

20. The lower corner of a leaf of a book is folded over so as just to reach the inner edge of the page; find when the crease thus formed is a minimum.

Solution:—

Let y denote the length of the crease, x the distance of the corner from the intersection of the crease with the lower edge, and a the width of the page.

By means of the relations of similar right triangles, the following expression is deduced:

$$y = \frac{x\sqrt{x}}{\sqrt{(x - \frac{1}{2}a)}}.$$

Whence we obtain

$$x = \tfrac{3}{4}a,$$

which gives a minimum value of y.

21. Find when the area of the part folded over is a minimum.

When $x = \tfrac{2}{3}a$.

XVII.

The Employment of Derivatives Higher than the First.

110. To ascertain whether $f'(x)$ is an increasing or a decreasing function, (and thence whether $f(x)$ is a minimum or a maximum), it is frequently necessary to find the expression for its derivative, $f''(x)$. Now, if $f''(a)$ is found to have a *positive* value, it follows that $f'(x)$ is an *increasing* function when $x = a$.

and, as was shown in Art. 102, that $f(a)$ is a minimum. On the other hand, if we find that $f''(a)$ has a *negative* value, it follows that $f'(x)$ is a decreasing function, and that $f(a)$ is a maximum. To illustrate, let

$$f(x) = 3x^4 - 16x^3 - 6x^2 + 12,$$

then $\qquad f'(x) = 12x^3 - 48x^2 - 12x.$

The roots of $f'(x) = 0$ are $x = 0$, and $x = 2 \pm \sqrt{5}$.

In this case $f''(x) = 36x^2 - 96x - 12$,

hence $\qquad f''(0) = -12;$

$f(x)$ is therefore a *maximum* when $x = 0$.

It is unnecessary to find the values of $f''(x)$ for the other roots; for, since the function does not admit of infinite values, the maxima and minima occur alternately. The root $2 - \sqrt{5}$ being negative and $2 + \sqrt{5}$ positive, the root zero is intermediate in value, and therefore both the remaining roots give minima.

III. If $f'(x)$ contains a positive factor which cannot change sign, this factor may be omitted; since we can determine whether $f'(x)$ increases or decreases through zero by examining the sign of the derivative of the remaining factor. Thus, if

$$f(x) = \frac{x}{1 + x^2}, \qquad f'(x) = \frac{1 - x^2}{(1 + x^2)^2}.$$

Since $\dfrac{1}{(1 + x^2)^2}$ is always positive, we have only to determine whether the factor $1 - x^2$ changes sign. Denoting this factor by v, and putting $v = 0$, we have

$$x = \pm 1.$$

Now $\qquad \dfrac{dv}{dx} = -2x$

which is negative for $x = 1$ and positive for $x = -1$. These roots, therefore, give respectively a maximum and a minimum value of $f(x)$.

112. There may be roots of the equation $f'(x) = 0$ which correspond to neither maxima nor minima, since it is a condition essential to the existence of such values that $f'(x)$ shall change sign. When such cases arise, the form assumed by the curve $y = f(x)$ in the immediate vicinity of the point at which $x = a$ will be one of those represented at A and B in Fig. 13.

At these points the value of $\tan \phi$ or $f'(x)$ is zero, but at A it is positive on both sides of the point, and $f(x)$ or y is an increasing function, while at B $f'(x)$ is negative on both sides of the point, and $f(x)$ is a decreasing function.

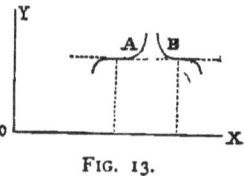

Fig. 13.

113. It is important to notice that at A the value zero assumed by $f'(x)$ constitutes a minimum value of this function, thus a root of $f'(x) = 0$ for which $f'(x)$ is a *minimum* corresponds to a case in which $f(x)$ is an *increasing* function. In like manner a root of $f'(x) = 0$ for which $f'(x)$ is a *maximum* is a case in which $f(x)$ is a *decreasing* function.

114. It follows from the preceding article and from Art. 102 that, if $f'(a) = 0$, then, of the two functions $f(x)$ and $f'(x)$, one will be a *maximum* and the other a *decreasing* function, or else one will be a *minimum* and the other an *increasing* function. Hence, if we consider the case in which the given function and several of its successive derivatives vanish for the same value of x, it is evident that when these functions are arranged in order *they will be either alternately maxima and decreasing functions, or alternately minima and increasing functions.*

115. Now suppose that $f''(x)$ is the first of these successive derivatives that does not vanish when $x = a$, then, writing the series of functions

$$f(x), \quad f'(x), \quad f''(x), \quad \cdots \cdots \quad f^{n-1}(x), \quad f^{n}(x),$$

let us assume first that $f^n(a)$ is *positive*. Then in the above series of functions $f^{n-1}(a)$, $f^{n-3}(a)$, etc., will be increasing functions while $f^{n-2}(a)$, $f^{n-4}(a)$, etc., will be minima.

Now whenever n is *odd*, the original function will belong to the first of these classes and will be an increasing function, while if n is *even* the original function will belong to the second class and will be a minimum.

On the other hand, if $f^n(a)$ has a *negative* value, the series of functions will be alternately decreasing functions and maxima; and when n is *odd* $f(a)$ will be a decreasing function, but when n is *even* $f(a)$ will be a maximum.

Thus we shall have neither maxima nor minima unless the first derivative, which does not vanish when $x = a$, is of an *even order;* but when this is the case we shall have a maximum or a minimum according as the value of this derivative is negative or positive.

116. The following function presents a case in which the above principle is advantageously employed.

$$f(x) = \varepsilon^x + \varepsilon^{-x} + 2\cos x,$$

$$f'(x) = \varepsilon^x - \varepsilon^{-x} - 2\sin x.$$

Zero is evidently a root of the equation $f'(x) = 0$.* In this case

---

* Zero is the only root of $f'(x) = 0$ in this example; for

$$f''(x) = \frac{\varepsilon^{2x} - 2\varepsilon^x \cos x + 1}{\varepsilon^x} > \frac{(\varepsilon^x - 1)^2}{\varepsilon^x}.$$

$f''(x)$ therefore cannot be negative, hence $f'(x)$ cannot again assume the value zero.

$$f''(x) = \varepsilon^x + \varepsilon^{-x} - 2\cos x \quad \therefore \quad f''(0) = 0,$$

$$f'''(x) = \varepsilon^x - \varepsilon^{-x} + 2\sin x \quad \therefore \quad f'''(0) = 0,$$

$$f^{iv}(x) = \varepsilon^x + \varepsilon^{-x} + 2\cos x \quad \therefore \quad f^{iv}(0) = 4.$$

The fourth derivative being the first that does not vanish, and having a positive value, we conclude that $x = 0$ gives a minimum value of $f(x)$.

Infinite Values of the Derivative.

117. It was shown in Art. 98 that if we have, for $x = a$,

$$f'(x) = \infty,$$

a maximum value will present itself if $f'(x)$ changes sign from $+$ to $-$, and a minimum if it changes sign from $-$ to $+$. It may, however, happen in these cases that the value of $f(a)$ is also infinite. When $f(a)$ is finite, the form of the curve

$$y = f(x)$$

in the vicinity of a maximum or a minimum ordinate of this variety is represented at A and B in Fig. 14. As an example, let

$$f(x) = (x^{\frac{1}{3}} - b^{\frac{1}{3}})^{\frac{2}{3}},$$

whence $\quad f'(x) = \tfrac{2}{3} x^{-\frac{2}{3}} (x^{\frac{1}{3}} - b^{\frac{1}{3}})^{-\frac{1}{3}}.$

$f'(x)$ is infinite when $x = 0$ and when $x = b$. When $x = 0$ $f'(x)$ does not change sign, since $x^{-\frac{2}{3}}$ cannot be negative, but when $x = b$ it changes sign from $-$ to $+$; hence $f(x)$ has a minimum value when $x = b$.

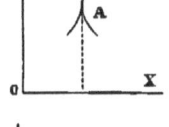
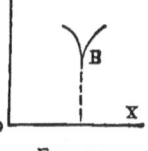

Fig. 14.

Examples XVII.

1. Show that $a\epsilon^{kx} + b\epsilon^{-kx}$ has a minimum value equal to $2\sqrt{(ab)}$.

Find the maxima and minima of the following functions:

2. $f(x) = x \sin x$.

A maximum for a value of x in the second quadrant satisfying the equation $\tan x = -x$.

3. $f(x) = \dfrac{a^2}{x} + \dfrac{b^2}{a-x}$.

The roots $x = \dfrac{a^2}{a+b}$ and $x = \dfrac{a^2}{a-b}$ give a min. and a max. if b is positive, but a max. and min. if b is negative.

4. $f(x) = 2\cos x + \sin^2 x$.

Solution:— $f'(x) = 2 \sin x (\cos x - 1)$;

rejecting the factor $2(1 - \cos x)$, which is always positive, we put

$$v = -\sin x. \quad \text{Hence } \frac{dv}{dx} = -\cos x.$$

A max. for $x = 2n\pi$;
a min. for $x = (2n+1)\pi$.

5. $f(x) = \sin x(1 + \cos x)$.

A max. for $x = \tfrac{1}{3}\pi$;
a min. for $x = -\tfrac{1}{3}\pi$;
neither for $x = \pi$.

6. $f(x) = \sec x + \log \cos^2 x$.

Multiplying the derivative by $\cos^2 x$, we obtain

$$v = \sin x (1 - 2\cos x).$$

A max. for $x = 0$, and $x = \pi$;
a min. for $x = \pm \tfrac{1}{3}\pi$.

7. $f(x) = \dfrac{\tan^2 x}{\tan 3x}$.

A min. for $x = 0$, $\tfrac{2}{8}\pi$, $\tfrac{6}{8}\pi$, and π;
a max. for $x = \tfrac{1}{8}\pi$, $\tfrac{1}{2}\pi$, $\tfrac{7}{8}\pi$, etc.

8. $f(x) = \epsilon^x + \epsilon^{-x} - x^2$.

A min. for $x = 0$.

9. Find maxima and minima of the following functions:

$f(x) = (x^{\frac{2}{3}} - b^{\frac{2}{3}})^{\frac{1}{2}}$. A min. for $x = 0$.

10. $f(x) = (x^2 - b^2)^{\frac{2}{3}}$. A max. for $x = 0$;
a min. for $x = \pm b$.

11. $f(x) = (x^2 + 3x + 2)^{\frac{2}{3}} + x^{\frac{2}{3}}$.
$f'(x) = \infty$ gives min. corresponding to $x = -2$, $x = -1$ and $x = 0$.
$f'(x) = 0$ gives two intermediate maxima.

12. $f(x) = (x^2 + 2x)^{\frac{1}{3}} - (x + 3)^{\frac{1}{3}}$. Max. for $x = \frac{1}{4}(-3 \pm \sqrt{17})$;
min. for $x = 0$ and $x = -2$.

13. $f(x) = (x - a)^{\frac{1}{3}} (x - b)^{\frac{2}{3}} + c$. A max. for $x = \dfrac{2b + a}{3}$;
min. for $x = a$ and $x = b$.

14. $f(x) = \dfrac{(x - a)(x - b)}{x^2}$. A min. for $x = \dfrac{2ab}{a + b}$.

15. $f(x) = (x - a)^{\frac{2}{3}} (x - b)^{\frac{1}{3}}$.
Solutions for $x = a$ and $x = \frac{1}{3}(2b + a)$; if $b > a$, the former gives a max. and the latter a min.

Miscellaneous Examples.

1. $f(x) = \dfrac{x - 1}{x^3 - 3x^2 + 2x + 54}$. *Use the reciprocal.*
Max. for $x = 4$.

2. $f(x) = \dfrac{x^2 - x + 1}{x^2 + x - 1}$. A max. for $x = 0$;
a min. for $x = 2$.

3. $f(x) = x^{-a} \varepsilon^{bx}$.   A min. for $x = \dfrac{a}{b}$.

4. The equation of the path of a projectile being

$$y = x \tan \alpha - \frac{x^2}{4h \cos^2 \alpha},$$

find the value of x when y is a maximum; also the maximum value of y.   Max. when $x = h \sin 2\alpha$, and $y = h \sin^2 \alpha$.

5. In a given sphere inscribe the greatest rectangular parallelopiped.

Solution :—

Regarding any one edge as of fixed length, it is easy to show that the other two edges are equal. Hence the three edges are equal.

6. In a given cone inscribe the greatest rectangular parallelopiped.

Solution :—

Regarding the parallelopiped as inscribed in a cylinder which is itself inscribed in the cone, the base is evidently a square, and the altitude is that of the maximum cylinder. See Ex. XV, 9.

7. A Norman window consists of a rectangle surmounted by a semicircle. Given the perimeter, required the height and breadth of the window when the quantity of light admitted is a maximum.

The radius of the semicircle is equal to the height of the rectangle.

8. A tinsmith was ordered to make an open cylindrical vessel of given volume, which should be as light as possible; find the ratio between the height and the radius of the base.

The height equals the radius of the base.

9. What should be the ratio between the diameter of the base and the height of cylindrical fruit-cans in order that the amount of tin used in constructing them may be the least possible?

The height should equal the diameter of the base.

§ XVII.] MISCELLANEOUS EXAMPLES. 117

10. Determine the circle having its centre on the circumference of a given circle so that the arc included in the given circle shall be a maximum.

A max. for the value of θ which is in the first quadrant.

11. Given the vertical angle of a triangle and its area; find when its base is a minimum.   The triangle is isosceles.

12. Prove that, of all circular sectors of the same perimeter, the sector of greatest area is that in which the circular arc is double the radius.

13. Find the minimum isosceles triangle circumscribed about a parabolic segment.

The altitude of the triangle is four-thirds the altitude of the segment.

14. Find the least isosceles triangle that can be described about a given ellipse, having its base parallel to the major axis.

The height is three times the minor semi-axis.

15. Inscribe the greatest parabolic segment in a given isosceles triangle.

The altitude of the segment is three-fourths that of the triangle.

16. A steamer whose speed is 8 knots per hour and course due north sights another steamer directly ahead, whose speed is 10 knots, and whose course is due west. What must be the course of the first steamer to cross the track of the second at the least possible distance from her?

$N.\ 53° 8'\ W.$

17. Determine the angle which a rudder makes with the keel of a ship when its turning effect is the greatest possible.

Solution:—

Let ϕ denote the angle between the rudder and the prolongation of the keel of the ship; then if b is the area of the rudder that of the stream of water intercepted will be $b \sin \phi$: the resulting force being decomposed, the component perpendicular to the rudder contains the factor $\sin^2 \phi$. Again decomposing this force, and taking the component that is perpendicular to the keel of the ship, which is the only part of the original force that is effective in turning the ship, the expression to be made a maximum is

$$\sin^2 \phi \cos \phi.$$

Whence we obtain

$$\tan \phi = \sqrt{2}.$$

18. The work of driving a steamer through the water being proportional to the cube of her speed, find her most economical rate per hour against a current running a knots per hour.

Solution :—

Let v denote the speed of the steamer in knots per hour. The work per hour will then be denoted by kv^3, k being a constant, and the actual distance the steamer advances per hour by $v - a$. The work per knot made good is therefore expressed by

$$\frac{kv^3}{v-a}.$$

Whence we obtain the result

$$v = \tfrac{3}{2}a.$$

CHAPTER VII.

The Development of Functions in Series.

XVIII.

The Nature of an Infinite Series.

118. A FUNCTION which can be expressed by means of a limited number of integral terms, involving powers of the independent variable with positive integral exponents only, is called a *rational integral function*.

When $f(x)$ is *not* a rational integral function, it is usually possible to derive an unlimited series of terms rational and integral with respect to x, which may be regarded as an algebraic equivalent for the function. The process of deriving this series is called the *development* of the function into an *infinite series*.

When the given function is in the form of a rational fraction, the ordinary process of division (the dividend and divisor being arranged according to ascending powers of x) suffices to effect the development. Thus—

$$\frac{1+x}{1-x} = 1 + 2x + 2x^2 + 2x^3 + \cdots,$$

a series of terms arranged according to ascending powers of x, each coefficient after the absolute term being 2.

It is to be observed, in the first place, that, owing to the indefinite number of terms in the second member, the equation as written above cannot be verified numerically for an assumed value of x. In this case, however, the process not only gives us the series, but the remainder after any number of terms. Thus carrying the quotient to the term containing x^n, and writing the remainder, we have

$$\frac{1+x}{1-x} = 1 + 2x + 2x^2 \cdots + 2x^n + \frac{2x^{n+1}}{1-x}.$$

This equation may now be verified numerically for any assumed value of x; or algebraically by multiplying each member by $1-x$, thus obtaining an identity.

The ordinary process of extracting the square root of a polynomial furnishes an example of a series which may be extended so as to include as many terms as we please; but this process gives us no expression for the remainder.

119. Assuming that $f(x)$ admits of development into a series involving ascending powers of x, and denoting the remainder after $n+1$ terms by R, we may write

$$f(x) = A + Bx + Cx^2 + \cdots + Nx^n + R, \quad . \quad . \quad (1)$$

in which $A, B, C, \ldots N$ denote coefficients independent of x, and as yet unknown; the value of R is however not independent of x. If the coefficients $B, C, \ldots N$ admit of finite values, it may be assumed that R is a function of x which vanishes when $x = 0$; and in accordance with this assumption equation (1) becomes, when $x = 0$,

$$f(0) = A, \quad . \quad . \quad . \quad . \quad . \quad . \quad . \quad (2)$$

which determines the first term of the series. If in any case the value of $f(0)$ is found to be infinite, we infer that the proposed development is impossible.

120. When the coefficients $B, C, \ldots N$ admit of finite values, and the value of the function to be developed remains

§ XVIII.] INFINITE SERIES. 121

finite, R will have a finite value. If moreover the value of R decreases as n increases, and can be made as small as we please, by sufficiently increasing n, the series is said to be *convergent*, and may be employed in finding an approximate value of the function $f(x)$; the closeness of the approximation increasing with the number of terms used. A series in which R does not decrease as n increases is said to be *divergent*.

When the successive terms of a series decrease it does not necessarily follow that the series is convergent; for the value of the equivalent function, and consequently that of R, may be infinite. To illustrate, if we put $x = 1$ in the series

$$x + \tfrac{1}{2}x^2 + \tfrac{1}{3}x^3 + \tfrac{1}{4}x^4 + \cdots,$$

we obtain the numerical series

$$1 + \tfrac{1}{2} + \tfrac{1}{3} + \tfrac{1}{4} + \tfrac{1}{5} + \cdots;$$

it can be shown that, by taking a sufficient number of terms, the sum of this series may be made to exceed any finite limit, *the value of the equivalent or generating function of the above series being in fact infinite when $x = 1$.*

121. Since R vanishes with x, every series for which finite coefficients can be determined is convergent for certain small values of x. In some cases there are limiting values of x, both positive and negative, within which the series is convergent, while for values of x without these limits the series is divergent. These values of x are called *the limits of convergence*.

---

* If we consider the first two terms separately, and regard the other terms as arranged in groups of two, four, eight, sixteen, etc., the groups will end with the terms $\tfrac{1}{4}$, $\tfrac{1}{8}$, $\tfrac{1}{16}$, $\tfrac{1}{32}$, etc. The sum of the fractions in the first group exceeds $\tfrac{2}{4}$ or $\tfrac{1}{2}$, the sum of those in the second exceeds $\tfrac{4}{8}$ or $\tfrac{1}{2}$, and so on; hence the sum of $2N$ such groups exceeds the number N, and N may be taken as large as we choose.

The generating function in this case is $\log \dfrac{1}{1-x}$, and unity is the limit of convergence.

We shall now demonstrate a theorem by which a function in the form $f(x_0 + h)$ may be developed into a series involving powers of h, and in Section XIX we shall show how this theorem is transformed so as to give the expansion of $f(x)$ in powers of x.

Taylor's Theorem.

122. A function of h of the form $f(x_0 + h)$ in general admits of development in a series involving ascending powers of h. We therefore assume

$$f(x_0 + h) = A_0 + B_0 h + C_0 h^2 + \cdots + N_0 h^n + R_0, \quad \ldots (1)$$

in which $A_0, B_0, C_0, \ldots N_0$ are independent of h, while R_0 is a function of h which vanishes when h is zero. Hence, making $h = 0$, we have

$$f(x_0) = A_0.$$

We have now to find the values of $B_0, C_0, \ldots N_0$, which are evidently functions of x_0. For this purpose we put

$$x_1 = x_0 + h, \quad \text{whence} \quad h = x_1 - x_0;$$

substituting, equation (1) takes the form

$$f(x_1) = f(x_0) + B_0(x_1 - x_0) + C_0(x_1 - x_0)^2 \cdots + N_0(x_1 - x_0)^n + R_0,$$

in which we may regard x_1 as constant and x_0 as variable. Replacing the latter by x, and its functions, $B_0, C_0, \ldots N_0$, and R_0, by $B, C, \ldots N$, and R, we have

$$f(x_1) = f(x) + B(x_1 - x) + C(x_1 - x)^2 \cdots + N(x_1 - x)^n + R. \quad . (2)$$

Taking derivatives with respect to x, we have

§ XVIII.] TAYLOR'S THEOREM. 123

$$0 = f'(x) - B + (x_1 - x)\frac{dB}{dx} - 2C(x_1 - x) + (x_1 - x)^2\frac{dC}{dx} \cdots$$

$$- nN(x_1 - x)^{n-1} + (x_1 - x)^n\frac{dN}{dx} + \frac{dR}{dx}. \quad \cdots \quad (3)$$

To render the development possible, $B, C, \ldots N$, and R must have such values as will make equation (3) *identical*, that is, true for all values of x.

123 It is evident that B may be so taken as to cause the first two terms of equation (3) to vanish, and that, this being done, C can be so determined as to cause the coefficient of $(x_1 - x)$ to vanish, D so as to make the coefficient of $(x_1 - x)^2$ vanish, and so on. The requisite conditions are

$$f'(x) - B = 0, \quad \frac{dB}{dx} - 2C = 0, \quad \frac{dC}{dx} - 3D = 0, \text{ etc.,}$$

and finally
$$(x_1 - x)^n \frac{dN}{dx} + \frac{dR}{dx} = 0.$$

From these conditions we derive

$$B = f'(x), \qquad C = \tfrac{1}{2}\frac{dB}{dx} = \tfrac{1}{2}f''(x),$$

$$D = \tfrac{1}{3}\frac{dC}{dx} = \frac{1}{1 \cdot 2 \cdot 3}f'''(x), \qquad E = \frac{1}{1 \cdot 2 \cdot 3 \cdot 4}f^{\mathrm{iv}}(x),$$

and in general $\qquad N = \dfrac{1}{1 \cdot 2 \cdots n}f^n(x).$

Putting x_0 for x, and substituting in equation (1) the values of $A_0, B_0, C_0, \ldots N_0$, we obtain

$$f(x_0 + h) = f(x_0) + f'(x_0)h + f''(x_0)\frac{h^2}{1 \cdot 2} \cdots + f^n(x_0)\frac{h^n}{1 \cdot 2 \cdots n} + R_0. \quad (4)$$

This result is called Taylor's Theorem, from the name of its discoverer, Dr. Brook Taylor, who first published it in 1715.

It is evident from equation (4) that *the proposed expansion is impossible* when the given function or any of its derived functions is infinite for the value x_0.

Lagrange's Expression for the Remainder.

124. R denotes a function of x which takes the value R_0 when $x = x_0$, and becomes zero when $x = x_1$. It has been shown in the preceding article that R must also satisfy the equation

$$(x_1 - x)^n \frac{dN}{dx} + \frac{dR}{dx} = 0,$$

or, substituting the value of N determined above,

$$\frac{dR}{dx} = -\frac{(x_1 - x)^n}{1 \cdot 2 \cdots n} f^{n+1}(x). \quad \ldots \quad (5)$$

This equation shows that $\frac{dR}{dx}$ cannot become infinite for any value of x between x_0 and x_1, provided $f^{n+1}(x)$ remains finite and real while x varies between these limits. Since it follows from the theorem proved in Art. 104 that all preceding derivatives must be likewise finite, the above hypothesis is equivalent to the assumption that $f(x)$ *and its successive derivatives to the (n + 1)th inclusive remain finite and real while* x *varies from* x_0 *to* $x_0 + h$.

125. Let P denote any assumed function of x which, like R, takes the value R_0 when $x = x_0$ and the value zero when $x = x_1$, and whose derivative $\frac{dP}{dx}$ does not become infinite or imaginary for any value of x between these limits.

§ XVIII.]   *EXPRESSIONS FOR THE REMAINDER.*   125

Then, R_0 *being assumed to be finite*, $P - R$ denotes a function of x which vanishes both when $x = x_0$ and when $x = x_1$ and whose derivative cannot become infinite for any intermediate value of x. It follows therefore that the value of this function cannot become infinite for any intermediate value of x.

Since, as x varies from x_0 to x_1, $P - R$ starts from the value zero and returns to zero again, without passing through infinity, its numerical value must pass through a maximum; hence its derivative cannot retain the same sign throughout, and as it cannot become infinite it must necessarily become zero for some intermediate value of x. Since $x_1 = x_0 + h$ this intermediate value of x can be expressed by $x_0 + \theta h$, θ being a *positive proper fraction*. It is therefore evident that at least one value of x of the form

$$x = x_0 + \theta h$$

will satisfy the equation

$$\frac{dP}{dx} - \frac{dR}{dx} = 0. \quad \ldots \quad (6)$$

126. The value of P will fulfil the required conditions if we assume

$$P = \frac{(x_1 - x)^{n+1}}{h^{n+1}} \cdot R_0,$$

for this function takes the value R_0 when $x = x_0$ and vanishes when $x = x_1$; moreover its derivative with reference to x, viz.,

$$\frac{dP}{dx} = - \frac{(n+1)(x_1 - x)^n}{h^{n+1}} \cdot R_0, \quad \ldots \quad (7)$$

does not become infinite for any intermediate value of x. Substituting in equation (6) the values of the derivatives given in equations (5) and (7), and solving for R_0, we obtain

$$R_0 = \frac{h^{n+1}}{1 \cdot 2 \cdots n \cdot (n+1)} f^{n+1}(x_0 + \theta h). \quad \ldots \quad (8)$$

This expression for the remainder was first given by Lagrange.

The series may now be written thus:

$$f(x_0 + h) = f(x_0) + f'(x_0)h + f''(x_0)\frac{h^2}{1 \cdot 2} \cdots$$

$$+ f^n(x_0)\frac{h^n}{1 \cdot 2 \cdots n} + f^{n+1}(x_0 + \theta h)\frac{h^{n+1}}{1 \cdot 2 \cdots (n+1)} \cdots \quad (9)$$

It should be noticed that the above expression for the remainder after $n + 1$ terms differs from the next, or $(n + 2)$th term of the series, simply by the addition of θh to x_0.

The Binomial Theorem.

127. We shall now apply Taylor's Theorem to the function $(a + b)^m$ in order to obtain a series involving ascending powers of b.

In this case b takes the place of h, and a that of x_0; hence

$$f(x) = x^m \qquad \therefore \ f(x_0) = x_0^m \qquad = a^m$$

$$f'(x) = mx^{m-1} \qquad \therefore \ f'(x_0) = mx_0^{m-1} \qquad = ma^{m-1}$$

$$f''(x) = m(m-1)x^{m-2} \quad \therefore \ f''(x_0) = m(m-1)x_0^{m-2} = m(m-1)a^{m-2}$$

and

$$f^n(x_0) = m(m-1)(m-2) \cdots (m-n+1)a^{m-n}.$$

Whence

$$(a + b)^m = a^m + ma^{m-1}b + \frac{m(m-1)}{1 \cdot 2}a^{m-2}b^2 \cdots$$

$$+ \frac{m(m-1)(m-2) \cdots (m-n+1)}{1 \cdot 2 \cdot 3 \cdots n}a^{m-n}b^n + \cdots$$

This result is called the *Binomial Theorem*.

Examples XVIII.

1. To expand $\log(x_0 + h)$ by Taylor's Theorem.

Solution :—

$$f(x_0 + h) = \log(x_0 + h)$$

$$f(x) = \log x \quad \therefore \quad f(x_0) = \log x_0$$

$$f'(x) = \frac{1}{x} \quad \therefore \quad f'(x_0) = \frac{1}{x_0}$$

$$f''(x) = -\frac{1}{x^2} \quad \therefore \quad f''(x_0) = -\frac{1}{x_0^2}$$

$$f'''(x) = \frac{2}{x^3} \quad \therefore \quad f'''(x_0) = \frac{1 \cdot 2}{x_0^3}$$

$$f^{iv}(x) = -\frac{2 \cdot 3}{x^4} \quad \therefore \quad f^{iv}(x_0) = -\frac{1 \cdot 2 \cdot 3}{x_0^4}$$

. . . .

$$f^n(x) = -(-1)^n \frac{1 \cdot 2 \cdots (n-1)}{x^n} \quad \therefore \quad f^n(x_0) = -(-1)^n \frac{1 \cdot 2 \cdots (n-1)}{x_0^n}.$$

By substituting in equation (4), Art. 123, we obtain

$$\log(x_0 + h) = \log x_0 + \frac{h}{x_0} - \frac{h^2}{2x_0^2} + \frac{h^3}{3x_0^3} - \frac{h^4}{4x_0^4} \cdots -(-1)^n \frac{h^n}{n x_0^n} + R_0.$$

Employing Lagrange's expression for the remainder (Art. 126) we derive

$$R_0 = (-1)^n \frac{h^{n+1}}{(n+1)(x_0 + \theta h)^{n+1}}.$$

2. Expand $a^{x_0 + h}$.

Solution :—

$$f(x_0 + h) = a^{x_0 + h}$$

$$f(x) = a^x \quad \therefore \quad f(x_0) = a^{x_0}$$
$$f'(x) = \log a \cdot a^x \quad \therefore \quad f'(x_0) = \log a \cdot a^{x_0}$$
$$f''(x) = (\log a)^2 \cdot a^x \quad \quad f''(x_0) = (\log a)^2 \cdot a^{x_0}$$
$$\cdots \cdots \cdots \cdots \cdots \cdots \cdots$$
$$f^n(x) = (\log a)^n \cdot a^x \quad \quad f^n(x_0) = (\log a)^n \cdot a^{x_0}.$$

Substituting in equation (4), Art. 123, we have

$$a^{x_0+h} = a^{x_0}\left[1 + \log a \cdot h + (\log a)^2 \frac{h^2}{1 \cdot 2} \cdots + \frac{(\log a)^n \cdot h^n}{1 \cdot 2 \cdots n}\right] + R_0.$$

3. Find the expansion of $f(x_0 + h)$, when $f(x) = x \log x - x$, writing the $(n+1)^{th}$ term of the series.

$$f(x_0 + h) = x_0 \log x_0 - x_0 + \log x_0 \cdot h + \frac{1}{x_0} \cdot \frac{h^2}{1 \cdot 2} - \frac{1}{x_0^2} \cdot \frac{h^3}{2 \cdot 3} + \cdots$$
$$+ (-1)^n \frac{1}{x_0^{n-1}} \cdot \frac{h^n}{(n-1)n} \cdots$$

4. Expand $\sin^{-1}(x_0 + h)$ to the fourth term inclusive.

$$\sin^{-1}(x_0 + h) = \sin^{-1} x_0 + \frac{h}{(1-x_0^2)^{\frac{1}{2}}} + \frac{x_0}{(1-x_0^2)^{\frac{3}{2}}} \cdot \frac{h^2}{1 \cdot 2}$$
$$+ \frac{1 + 2x_0^2}{(1-x_0^2)^{\frac{5}{2}}} \cdot \frac{h^3}{1 \cdot 2 \cdot 3} + \cdots$$

5. Prove that

$$\sin(\tfrac{1}{3}\pi + h) = \tfrac{1}{2}\left[1 + h\sqrt{3} - \frac{h^2}{1 \cdot 2} - \frac{h^3\sqrt{3}}{1 \cdot 2 \cdot 3} + \frac{h^4}{1 \cdot 2 \cdot 3 \cdot 4}\right.$$
$$\left. + \frac{h^5\sqrt{3}}{1 \cdot 2 \cdot 3 \cdot 4 \cdot 5} - \cdots\right]$$

6. Prove that

$$\tan(\tfrac{1}{4}\pi + h) = 1 + 2h + 2h^2 + \tfrac{8}{3}h^3 + \tfrac{10}{3}h^4 + \cdots$$

XIX.
Maclaurin's Theorem.

128. We shall now give a particular form of Taylor's Series, which is usually more convenient, when numerical results are to be obtained, than the general form given in the preceding section.

This form of the series is obtained by putting $x_0 = 0$ and replacing h by x in equation (4), Art. 123. Thus,

$$f(x)=f(0)+f'(0)x+f''(0)\frac{x^2}{1\cdot 2}\cdots +f^n(0)\frac{x^n}{1\cdot 2\cdots n}+R_0 \quad . . (1)$$

and, the same substitutions being made in equation (8), Art. 126, we obtain

$$R_0 = f^{n+1}(\theta x)\frac{x^{n+1}}{1\cdot 2 \cdots (n+1)}.$$

Equation (1) is called Maclaurin's Theorem: it may be used in developing any function to which Taylor's Theorem is applicable, by giving a different signification to the symbol f. Thus, if $\log(1+h)$ is to be developed by Taylor's Theorem, $f(x) = \log x$, the value of x_0 being unity; but, if $\log(1+x)$ is to be developed by Maclaurin's Theorem, we must put $f(x) = \log(1+x)$. (Compare Ex. XVIII., 1, with Art. 130.)

The Exponential Series and the Value of ε.

129. As an example of the application of the above theorem, we shall deduce the development of the function ε^x, which is called the *exponential series*, and shall thence obtain a series for computing the value of ε.

The successive derivatives of ε^x being equal to the original function, the coefficients, $f(0)$, $f'(0)$, etc., each reduce to unity;

we therefore derive, by substituting in equation (1) and introducing the value of R_0,

$$\varepsilon^x = 1 + x + \frac{x^2}{1\cdot 2} + \frac{x^3}{1\cdot 2\cdot 3} \cdots + \frac{x^n}{1\cdot 2 \cdots n} + \varepsilon^{\theta\cdot r}\cdot \frac{x^{n+1}}{1\cdot 2 \cdots (n+1)}.$$

Putting x equal to unity, we obtain the following series, which enables us to compute the value of the incommensurable quantity ε to any required degree of accuracy:

$$\varepsilon = 1 + 1 + \frac{1}{1\cdot 2} + \frac{1}{1\cdot 2\cdot 3} + \frac{1}{1\cdot 2\cdot 3\cdot 4} \cdots$$
$$+ \frac{1}{1\cdot 2\cdot 3 \cdots n} + \frac{\varepsilon^\theta}{1\cdot 2\cdot 3 \cdots (n+1)}.$$

The computation may be arranged thus, each term being derived from the preceding term by division:

$$\begin{array}{r} 2.5 \\ .16666666667 \\ 4166666667 \\ 833333333 \\ 138888889 \\ 19841270 \\ 2480159 \\ 275573 \\ 27557 \\ 2505 \\ 209 \\ 16 \\ 1 \\ \hline 2.71828182846 \end{array}$$

Since ε^θ is less than ε, the remainder (n being 14) is less than $\frac{3}{15}$ of the last term employed in the computation, and therefore cannot affect the result. Inasmuch as each term may contain a positive or negative error of one-half a unit in the last decimal place, we cannot, in general, rely upon the accuracy of the last two places of decimals, in computations involving so large a number of terms. Accordingly, this computation only justifies us in writing

$$\varepsilon = 2.718281828.$$

Logarithmic Series.

130. The logarithmic series is deduced by applying Maclaurin's Theorem to the function $\log(1 + x)$.

In this case

$$f(x) = \log(1 + x) \quad \therefore \quad f(0) = 0$$

$$f'(x) = \frac{1}{1 + x} \quad \therefore \quad f'(0) = 1$$

$$f''(x) = -\frac{1}{(1 + x)^2} \quad \therefore \quad f''(0) = -1$$

$$f'''(x) = \frac{1 \cdot 2}{(1 + x)^3} \quad \therefore \quad f'''(0) = 1 \cdot 2$$

$$f^{\text{iv}}(x) = -\frac{1 \cdot 2 \cdot 3}{(1 + x)^4} \quad \therefore \quad f^{\text{iv}}(0) = -1 \cdot 2 \cdot 3,$$

hence $\quad \log(1 + x) = x - \dfrac{x^2}{2} + \dfrac{x^3}{3} - \dfrac{x^4}{4} + \cdots \quad . \quad . \quad (1)$

Since this series is divergent for values of x greater than unity (see Art. 120), we proceed to deduce a formula for the difference of two logarithms, which may be employed in computing successive logarithms; that is, denoting the numbers corresponding to two logarithms by n and $n + h$, we derive a series for

$$\log(n + h) - \log n = \log \frac{n + h}{n}.$$

A series which could be employed for this purpose might be obtained from (1), by putting $\frac{n+h}{n}$ in the form $1 + \frac{h}{n}$. We obtain, however, a much more rapidly converging series by the process given below.

Substituting $-x$ for x in (1), we have

$$\log(1-x) = -x - \frac{x^2}{2} - \frac{x^3}{3} - \frac{x^4}{4} - \cdots \quad \cdot \quad (2)$$

Subtracting (2) from (1),

$$\log \frac{1+x}{1-x} = 2\left[x + \frac{x^3}{3} + \frac{x^5}{5} + \frac{x^7}{7} + \cdots\right], \quad \cdot \quad (3)$$

a series involving only the positive terms of series (1).

Putting $\frac{1+x}{1-x} = \frac{n+h}{n}$, we derive $x = \frac{h}{2n+h}$; substituting in (3), we have

$$\log \frac{n+h}{n} = 2\left[\frac{h}{2n+h} + \tfrac{1}{3}\frac{h^3}{(2n+h)^3} + \tfrac{1}{5}\frac{h^5}{(2n+h)^5} + \cdots\right]. \quad \cdot \quad (4)$$

The Computation of Napierian Logarithms.

131. The series given above enables us to compute Napierian logarithms. We proceed to illustrate by computing $\log_e 10$. The approximate numerical value of this logarithm could be obtained by putting $n = 1$ and $h = 9$ in (4); but, since the series thus obtained would converge very slowly, it is more convenient first to compute $\log 2$ by means of the series obtained by putting $n = 1$ and $h = 1$ in (4); thus:

$$\log_e 2 = 2\left[\frac{1}{3} + \frac{1}{3}\cdot\frac{1}{3^3} + \frac{1}{5}\cdot\frac{1}{3^5} + \frac{1}{7}\cdot\frac{1}{3^7} + \cdots\right].$$

§ XIX.] LOGARITHMIC SERIES.

We then put $n = 8$ and $h = 2$ in (4); whence

$$\log_e 10 = 3 \log_e 2 + \frac{2}{3}\left[\frac{1}{3} + \frac{1}{3}\cdot\frac{1}{3^3} + \frac{1}{5}\cdot\frac{1}{3^5} + \frac{1}{7}\cdot\frac{1}{3^7} + \cdots\right].$$

In making the computation, it is convenient first to obtain the values of the powers of $\frac{1}{3}$ which occur in the series for log 2, by successive division by 9, and afterwards to derive the values of the required terms of the series by dividing these auxiliary numbers by 1, 3, 5, 7, etc. The same auxiliary numbers are also used in the computation of $\log_e 10$. See the arrangement of the numerical work below.

$\frac{1}{3}$ 0.3333333333 : 1   0.3333333333
$(\frac{1}{3})^3$ 370370370 : 3   123456790
$(\frac{1}{3})^5$ 41152263 : 5   8230453
$(\frac{1}{3})^7$ 4572474 : 7   653211
$(\frac{1}{3})^9$ 508053 : 9   56450
$(\frac{1}{3})^{11}$ 56450 : 11   5132
$(\frac{1}{3})^{13}$ 6272 : 13   482
$(\frac{1}{3})^{15}$ 697 : 15   46
$(\frac{1}{3})^{17}$ 77 : 17   5
                                        ——————————
                                        0.3465735902
                                                   2
                       $\log_e 2 = 0.6931471804$

$\frac{1}{3}$ 0.3333333333 : 1   0.3333333333
$(\frac{1}{3})^5$ 41152263 : 3   13717421
$(\frac{1}{3})^9$ 508053 : 5   101611
$(\frac{1}{3})^{13}$ 6272 : 7   896
$(\frac{1}{3})^{17}$ 77 : 9   9
                                        ——————————
                                        0.3347153270
                                       −0.1115717757
                                        ——————————
                                        0.2231435513
                       $3 \log_e 2 = 2.0794415412$
                       $\log_e 10 = 2.30258509$

The tabular logarithms of the system of which 10 is the base, are derived from the corresponding Napierian logarithms by means of the relation

$$\log_e x = \log_e 10 \; \log_{10} x,$$

whence $\quad \log_{10} x = \dfrac{1}{\log_e 10} \cdot \log_e x = M \cdot \log_e x.$

The constant $\dfrac{1}{\log_e 10}$, denoted above by M, is called the *modulus* of common logarithms. Taking the reciprocal of $\log_e 10$, computed above, we have

$$M = 0.43429448.$$

The Developments of the Sine and the Cosine.

132. Let $\qquad f(x) = \sin x,$

then

$f'(x) = \cos x, f''(x) = -\sin x, f'''(x) = -\cos x, f^{iv}(x) = \sin x;$

f^{iv} being identical with f, it follows that these functions recur in cycles of four; their values when $x = 0$ are

$$0, \; 1, \; 0, \; -1, \text{ etc.}$$

Hence substituting in equation (1), Art. 128, we have

$$\sin x = x - \frac{x^3}{1\cdot 2\cdot 3} + \frac{x^5}{1\cdot 2\cdots 5} - \frac{x^7}{1\cdot 2\cdots 7} + \cdots \quad (1)$$

In a similar manner, we obtain

$$\cos x = 1 - \frac{x^2}{1\cdot 2} + \frac{x^4}{1\cdot 2\cdot 3\cdot 4} - \frac{x^6}{1\cdot 2\cdots 6} + \cdots \quad (2)$$

Examples XIX.

1. Expand $(1 + x)^m$.

$$(1 + x)^m = 1 + mx + \frac{m(m-1)}{1 \cdot 2} x^2 + \frac{m(m-1)(m-2)}{1 \cdot 2 \cdot 3} x^3 + \cdots$$

It is evident that no coefficient will vanish if m is negative or fractional. This is the form in which the binomial theorem is employed in computation, x being less than unity.

2. Find three terms of the expansion of $\sin^2 x$.

$$\sin^2 x = x^2 - \frac{x^4}{3} + \frac{2x^6}{3^2 \cdot 5} - \cdots$$

3. Expand $\tan x$ to the term involving x^5 inclusive.

$$\tan x = x + \frac{x^3}{3} + \frac{2x^5}{15} + \cdots$$

4. Expand $\sec x$ to the term involving x^6 inclusive.

$$\sec x = 1 + \frac{x^2}{1 \cdot 2} + \frac{5x^4}{1 \cdot 2 \cdot 3 \cdot 4} + \frac{61x^6}{1 \cdot 2 \cdots 6} + \cdots$$

5. Expand $\log \sec x$ to the term involving x^6 inclusive.

$$\log \sec x = \frac{x^2}{2} + \frac{x^4}{12} + \frac{x^6}{45} + \cdots$$

6. Find four terms of the expansion of $\varepsilon^x \sec x$.

$$\varepsilon^x \sec x = 1 + x + x^2 + \frac{2x^3}{3} + \cdots$$

7. Derive the expansion of $\log(1 - x^2)$ from the logarithmic series, and verify by adding the expansions of $\log(1 + x)$ and $\log(1 - x)$.

8. Derive the expansion of $(1 + x)\varepsilon^x$ from that of ε^x.

$$(1 + x)\varepsilon^x = 1 + 2x + \frac{3 \cdot x^2}{1 \cdot 2} \cdots + \frac{n+1}{1 \cdot 2 \cdots n} x^n.$$

9. Find, by means of the exponential series, the expansion of $x\epsilon^{2x}$, including the nth term.

10. Expand $\dfrac{\epsilon^x}{1+x}$ by division, making use of the exponential series.

$$\frac{\epsilon^x}{1+x} = 1 + \frac{x^2}{2} - \frac{x^3}{3} + \frac{3x^4}{8} - \frac{11x^5}{30} + \cdots$$

11. Find the expansion of $\epsilon^x \log(1+x)$ to the term involving x^5, by multiplying together a sufficient number of the terms of the series for ϵ^x and for $\log(1+x)$.

$$\epsilon^x \log(1+x) = x + \frac{x^2}{2} + \frac{x^3}{3} + \frac{3x^5}{40} + \cdots$$

12. Expand $\log(1+\epsilon^x)$.

$$\log(1+\epsilon^x) = \log 2 + \frac{x}{2} + \frac{x^2}{8} - \frac{x^4}{192} + \cdots$$

13. Expand $(1+\epsilon^x)^n$ to the term involving x^3 inclusive.

$$(1+\epsilon^x)^n = 2^n \left\{ 1 + \frac{n}{2}\cdot x + \frac{n(n+1)}{2^2} \cdot \frac{x^2}{1\cdot 2} \right.$$
$$\left. + \frac{n(n^2+n+2)}{2^3} \cdot \frac{x^3}{1\cdot 2\cdot 3} + \cdots \right\}.$$

14. Find the expansion of $\sqrt{(1 \pm \sin 2x)}$, employing the formula $\sqrt{(1 \pm \sin 2x)} = \cos x \pm \sin x$.

$$\sqrt{(1 \pm \sin 2x)} = 1 \pm x - \frac{x^2}{1\cdot 2} \mp \frac{x^3}{1\cdot 2\cdot 3} + \cdots$$

15. Find the expansion of $\cos^2 x$ by means of the formula $\cos^2 x = \tfrac{1}{2}(1 + \cos 2x)$.

$$\cos^2 x = 1 - x^2 + \frac{2^3 x^4}{1\cdot 2\cdot 3\cdot 4} - \frac{2^5 x^6}{1\cdot 2\cdots 6} + \cdots$$

16. Find the expansion of $\cos^3 x$, by means of the formula $\cos^3 x = \frac{1}{4}(\cos 3x + 3\cos x)$.

$$\cos^3 x = 1 - \frac{3x^2}{1\cdot 2} + \frac{3^4 + 3}{4} \cdot \frac{x^4}{1\cdot 2\cdot 3\cdot 4} \cdots + (-1)^n \frac{3^{2n} + 3}{4} \cdot \frac{x^{2n}}{1\cdot 2 \cdots 2n}.$$

17. Compute $\log_e 3$, and find $\log_{10} 3$ by multiplying by the value of M (Art. 166).

$$\log_e 3 = 1.0986123.$$
$$\log_{10} 3 = 0.4771213.$$

18. Find $\log_e 269$.

Put $n = 270 = 10 \times 3^3$, and $h = -1$.

$$\log_e 269 = 5.5947114.$$

19. Find $\log_e 7$, and $\log_e 13$.

$$\log_e 7 = 1.9459101.$$
$$\log_e 13 = 2.5649494.$$

CHAPTER VIII.

Curve Tracing.

XX.

Equations in the Form y = f(x).

133. When a curve given by its equation is to be traced, it is necessary to determine its general form especially at such points as present any peculiarity, and also the nature of those branches of the curve, if there be any, which are unlimited in extent.

The general mode of procedure, when the equation can be put in either of the forms, $y = f(x)$ or $x = \phi(y)$, is indicated in the following examples.

Asymptotes Parallel to the Coordinate Axes.

134. *Example* 1. $-a^2y - x^2y = a^3$ . . . . (1)
Solving for y, we obtain

$$y = \frac{a^3}{a^2 - x^2}. \qquad \ldots \ldots \ldots (2)$$

When $x = 0, y = a$. Numerically equal positive and negative values of x give the same values for y; the curve is therefore symmetrical with reference to the axis of y. As x increases

§ XX.] ASYMPTOTES PARALLEL TO THE AXES. 139

from zero, y increases until the denominator, $a^2 - x^2$, becomes zero, when y becomes infinite; this occurs when $x = \pm a$.

Draw the straight lines $x = \pm a$. These are lines to which the curve approaches indefinitely, for we may assign values to x as near as we please to $+ a$ or to $- a$, thus determining points of the curve as near as we please to the straight lines $x = a$ and $x = - a$. Such lines are called *asymptotes* to the curve.

When x passes the value a, y becomes negative and decreases numerically, approaching the value zero as x increases indefinitely. Hence there is a branch of the curve below the axis of x to which the lines $x = a$ and $y = 0$ are asymptotes.

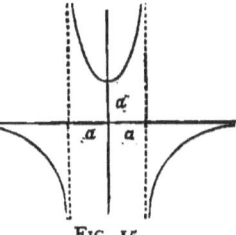

Fig. 15.

The general form of the curve is indicated in Fig. 15.

The point $(0, a)$ evidently corresponds to a minimum ordinate.

135. Example 2. $a^2 x = y(x - a)^2$. . . . . (1)
Solving for y,

$$y = \frac{a^2 x}{(x - a)^2}. \quad \cdots \cdots \cdots \quad (2)$$

When x is zero, y is zero; y increases as x increases until $x = a$, when y becomes infinite. Hence

$$x = a$$

is the equation of an asymptote. When x passes the value a, y does not change sign, but remains positive, and as x increases y diminishes, approaching zero as x increases indefinitely.

Examining now the values of y which correspond to negative values of x, we perceive that, y becoming negative, the branch which passes through the origin continues below the axis of x, and that y approaches zero as the negative value of x increases indefinitely. Hence the general form of the curve is that indicated in Fig. 17.

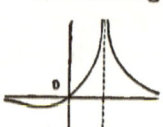

Fig. 17.

136. To determine the direction of the curve at any point, we have

$$\tan \phi = \frac{dy}{dx} = -a^2 \frac{a+x}{(x-a)^2}. \quad \ldots \quad (3)$$

The direction in which the curve passes through the origin is given by the value of $\tan \phi$ which corresponds to $x = 0$. From (3), we have

$$\left.\frac{dy}{dx}\right]_0 = 1;$$

the inclination of the curve at the origin is therefore 45°.

Minimum Ordinates and Points of Inflexion.

137. To find the minimum ordinate which evidently exists on the left of the axis of y, we put the expression for $\frac{dy}{dx}$ equal to zero, and deduce

$$x = -a.$$

The minimum ordinate is therefore found at the point whose abscissa is $-a$; its value, obtained from equation (2), is $-\frac{1}{4}a$.

A *point of inflexion* is a point at which $\frac{d^2y}{dx^2}$ changes sign (see

Art. 74); in other words, it is a point at which tan ϕ has a maximum or a minimum value. In this case there is evidently a point of inflexion on the left of the minimum ordinate. From equation (3) we derive

$$\frac{d^2y}{dx^2} = 2a^2 \frac{x+2a}{(x-a)^4},$$

putting this expression equal to zero to determine the abscissa, and deducing the corresponding ordinate from (2), we obtain, for the coordinates of the point of inflexion,

$$x = -2a, \text{ and } y = -\tfrac{2}{9}a.$$

Oblique Asymptotes.

138. *Example* 3. $x^3 - 2x^2y - 2x^2 - 8y = 0.$ . . . . . (1)
Solving this equation for y, we have

$$y = \frac{x^2}{2} \cdot \frac{x-2}{x^2+4}. \quad \ldots \quad \ldots \quad (2)$$

It is obvious from the form of equation (2) that the curve meets the axis of x at the two points (0, 0) and (2, 0). Since y is positive only when $x > 2$, the curve lies below the axis of x on the left of the origin, and also between the origin and the point (2, 0), but on the right of this point the curve lies above the axis of x.

139 Developing the second member of equation (2) into an expression involving a fraction whose numerator is lower in degree than its denominator, we have

$$y = \tfrac{1}{2}x - 1 + 2\frac{2-x}{x^2+4}. \quad \ldots \quad \ldots \quad (3)$$

The fraction in this expression decreases without limit as x increases indefinitely; hence the ordinate of the curve may, by increasing x, be made to differ as little as we please from that of the straight line

$$y = \tfrac{1}{2}x - 1.$$

This line is, therefore, an asymptote.

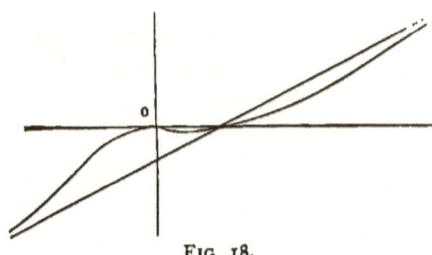

FIG. 18.

The fraction $\dfrac{2-x}{x^2+4}$ is positive for all values of x less than 2, negative for all values of x greater than 2, and does not become infinite. The curve, therefore, lies above the asymptote on the left of the point (2, 0), and below it on the right of this point, as represented in Fig. 18.

140. There is evidently a minimum ordinate between the origin and the point (2, 0).

We obtain from equation (2)

$$\frac{dy}{dx} = \frac{x}{2} \cdot \frac{x^3 + 12x - 16}{(x^2+4)^2}, \quad \ldots \quad (4)$$

and

$$\frac{d^2y}{dx^2} = 4 \cdot \frac{-x^3 + 6x^2 + 12x - 8}{(x^2+4)^3}. \quad \ldots \quad (5)$$

Putting $\dfrac{dy}{dx} = 0$, we obtain $x = 0$ and $x = 1.19$ nearly, the only real roots; the abscissa corresponding to the minimum ordinate is therefore 1.19, the value of the ordinate being about -0.11. The root zero corresponds to a maximum ordinate at the origin.

§ XX.] OBLIQUE ASYMPTOTES. 143

Putting $\frac{d^2y}{dx^2} = 0$, we obtain the three roots $x = -2$, and $x = 2(2 \pm \sqrt{3})$; the corresponding ordinates are obtained from equation (3). There are, therefore, three points of inflexion, one situated at the point $(-2, -1)$, and the others near the points $(0.54, -0.05)$, and $(7.46, 2.55)$.

The inclination of the curve is determined by means of equation (4) to be $\tan^{-1}\frac{1}{4}$ at the point $(2, 0)$, and $\tan^{-1}\frac{13}{4}$ at the left-hand point of inflexion.

Curvilinear Asymptotes.

141. *Example* 4. $x^2 - xy + 1 = 0.$ . . . . (1)
Solving for y

$$y = \frac{x^2 + 1}{x} = x + \frac{1}{x}. \quad \ldots \ldots (2)$$

In this case, on developing y in powers of x, the integral portion of its value is found to contain the second power of x; the fraction approaches zero when x increases indefinitely; hence the ordinate of this curve may be made to differ as little as we please from that of the curve

$$y = x^2. \quad \ldots \ldots \ldots (3)$$

The parabola represented by this equation is accordingly said to be a *curvilinear asymptote*. It is indicated by the dotted line in Fig. 19.

142. The sign of the fraction $\frac{1}{x}$ is always the same as that of x, and its value is infinite when x is zero; hence the curve lies below the parabola on the left of the axis of y, and above it on the right, this axis being an asymptote, as indicated in Fig. 19.

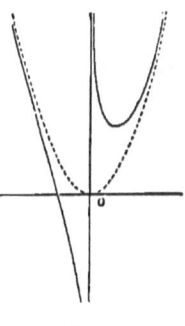

FIG. 19.

Taking derivatives, we obtain

$$\frac{dy}{dx} = 2x - \frac{1}{x^2}, \quad \ldots \ldots \quad (4)$$

and
$$\frac{d^2y}{dx^2} = 2\left(1 + \frac{1}{x^3}\right). \quad \ldots \ldots \quad (5)$$

There is a point of inflexion at $(-1, 0)$; the inclination of the curve to the axis of x at this point is $\tan^{-1}(-3)$.

There is a minimum ordinate at the point $(\frac{1}{2}\sqrt[3]{4}, \frac{3}{2}\sqrt[3]{2})$.

This cubic curve is an example of the species called by Newton *the trident*. The characteristic property of a trident is the possession of a parabolic asymptote and a rectilinear asymptote parallel to the axis of the parabola.

Examples XXVI.

1. Trace the curve $y = x(x^2 - 1)$.

Since y is an odd function of x, the curve is symmetrical with reference to the origin as a centre. Find the point of inflexion, and the minimum ordinate.

2. Trace the curve $y^2(x - 1) = x^2$.

The curve has for an asymptote the line $x = 1$; there is a minimum ordinate at $(2, 2)$, and a point of inflexion at $(4, \frac{4}{3}\sqrt{3})$.

3. Trace the curve $y^3 = x^2(x - a)$, determining its direction at the points at which it meets the axis of x.

The asymptote is found by the method of development, thus

$$y = x\left(1 - \frac{a}{x}\right)^{\frac{1}{3}} = x\left(1 - \frac{a}{3x} - \frac{a^2}{9x^2} - \text{etc.}\right);$$

the equation of the asymptote is therefore

$$y = x - \tfrac{1}{3}a,$$

4. Trace the curve $x^2 + xy + 2x - y = 0$.

5. Trace the curve $y^2 = x^2 + x^3$, and find its direction at the origin.

The curve has a maximum ordinate at $(-\frac{2}{3}, \pm \frac{2}{3}\sqrt{3})$. The value of y^2 may be taken as the function whose maximum is required. (See Art. 108.)

6. Trace the curve $y = x^3 - xy$. Find the point of inflexion, the minimum ordinate, and the asymptotes.

The curve has a rectilinear asymptote $x = -1$, and a curvilinear asymptote $y = x^2 - x + 1$. This curve is a trident. (See Art. 142.)

7. Trace the curve $y^2 = x^3 - x^4$.

Both branches of the curve are tangent to the axis of x at the origin.

8. Trace the curve $y^2 - 4y = x^3 + x^2$.

Solving for y, we obtain

$$y = 2 \pm \sqrt{(x^3 + x^2 + 4)} = 2 \pm \sqrt{[(x+2)(x^2 - x + 2)]}.$$

The factor $x^2 - x + 2$ being always positive, the curve is real on the right of the line $x = -2$.

Find the points at which the curve cuts the axis, and show that the upper branch has a maximum ordinate for $x = -\frac{2}{3}$ and a minimum ordinate for $x = 0$.

9. Trace the curve $(x - 2a)xy = a(x - a)(x - 3a)$.

10. Trace the curve $(x - 2a)xy^2 = a^2(x - a)(x - 3a)$.

11. Trace the curve $y^2 = x^4(1 - x^2)^3$: find all the points at which the tangent is parallel to the axis of x.

12. Trace the curve $6x(1 - x)y = 1 + 3x$.

This curve has a point of inflexion, determined by a cubic having only one real root, which is between -1 and -2. Find the three asymptotes, and the maximum and the minimum ordinate.

13. Trace the curve $x^2y^2 = (x+2)^2(1+x^2)$.

Solving the equation for y, we have

$$y = \pm \frac{x+2}{x}\sqrt{(1+x^2)} = \pm(2+x)\left(1+\frac{1}{x^2}\right)^{\frac{1}{2}}.$$

The asymptotes are $y = x+2, y = -x-2$, and $x=0$. The curve has a minimum ordinate corresponding to $x = \sqrt[3]{2}$; the inclination at the point at which it cuts the axis of x is $\tan^{-1}(\pm\tfrac{1}{2}\sqrt{5})$. There is a point of inflexion corresponding to the abscissa $x = -6.1$ nearly.

XXI.

Curves Given by Polar Equations.

143. The following examples will illustrate some of the methods employed when the curve is given by means of its polar equation.

Example 5. $r = a\cos\theta\cos 2\theta$. . . . . (1)

When $\theta = 0, r = a$, the generating point P therefore starts from A on the initial line. As θ increases, r decreases and becomes zero when $\theta = 45°$, P describing the half-loop in the first quadrant, and arriving at the pole in a direction having an inclination of $45°$ to the initial line. When θ passes $45°$, r becomes negative, and returns to zero again when $\theta = 90°$, P describing the loop in the third quadrant. As θ passes $90°$, r again becomes positive, but returns to zero when $\theta = 135°$, P describing the loop in the second quadrant. As θ varies from $135°$ to $180°$, r again becomes negative, P describing the half-loop in the fourth quadrant, and returning to A.

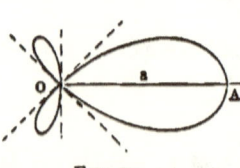

FIG. 19.

§ XXI.] *CURVES GIVEN BY POLAR EQUATIONS.* 147

In this example if we suppose θ to vary from 180° to 360°, P will again describe the same curve, and, since θ enters the equation of the curve, by means of trigonometrical functions only, it is unnecessary to consider values of θ greater than 360°.

144. Putting equation (1) in the form

$$r = a(2\cos^3\theta - \cos\theta),$$

we derive

$$\frac{dr}{d\theta} = a(-6\cos^2\theta \sin\theta + \sin\theta).$$

To determine the maxima values of r, we place this derivative equal to zero, thus obtaining the roots

$$\sin\theta = 0 \quad \text{and} \quad \cos\theta = \pm \tfrac{1}{6}\sqrt{6};$$

the former gives the point A on the initial line, and the latter gives the values of θ which determine the position of the maxima in the small loops. The corresponding values of r are $\mp \dfrac{a}{9}\sqrt{6}$.

To determine the position of the maximum ordinate, we have from (1)

$$y = r\sin\theta = \tfrac{1}{4}a\sin 4\theta.$$

The maxima values occur when $\sin 4\theta = 1$, and the minima when $\sin 4\theta = -1$; that is, we have maxima when $\theta = \tfrac{1}{8}\pi$ and when $\theta = \tfrac{5}{8}\pi$, and minima when $\theta = \tfrac{3}{8}\pi$ and $\tfrac{7}{8}\pi$.

145. In the preceding example the substitution of $\theta + \pi$ for θ changes the sign but not the numerical value of r. When this is the case, θ and $\theta + \pi$ evidently give the same point of the curve, and the complete curve is described while θ varies from 0 to π. If however this substitution changes neither the numerical value nor the sign of r, θ and $\theta + \pi$ will give points symmetrically situated with reference to the pole; that is, the curve will be symmetrical in opposite quadrants.

Again if the substitution of $-\theta$ for θ does not change the value of r, θ and $-\theta$ give points symmetrically situated with reference to the initial line, hence in this case the curve is symmetrical to this line; but, if the substitution of $-\theta$ for θ changes the sign of r without changing its numerical value, the curve is symmetrical with reference to a perpendicular to the initial line.

The Determination of Asymptotes by Means of Polar Equations.

146. When r becomes infinite for a particular value of θ the curve has an infinite branch, and, if there be a corresponding asymptote, it may be determined by means of the expression derived below.

Let θ_1 denote a value of θ for which r is infinite, and let OB be drawn through the pole, making this angle with the initial line; then, from the triangle OBP, Fig. 20, we have

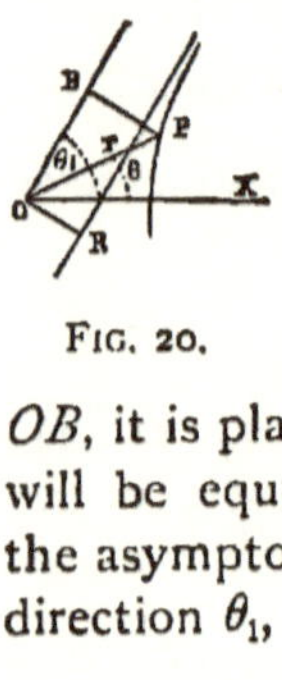

$$PB = r \sin(\theta_1 - \theta).$$

FIG. 20. Now, if the curve has an asymptote parallel to OB, it is plain that as θ approaches θ_1 the limiting value of PB will be equal to OR, the perpendicular from the pole upon the asymptote. Hence, if the curve has an asymptote in the direction θ_1, the expression

$$OR = [r \sin(\theta_1 - \theta)]_{\theta_1},$$

which takes the form $\infty \cdot 0$, will have a finite value, and this value will determine the distance of the asymptote from the pole. Fig. 20 shows that when the above expression is positive OR is to be laid off in the direction $\theta_1 - 90°$.

If upon evaluation the expression for OR is found to be infinite we infer that the infinite branch of the curve is parabolic.

147. *Example 6.* $\quad r = \dfrac{a\theta^2}{\theta^2 - 1}.$ . . . . (1)

Since r becomes infinite when $\theta = 1$, we proceed to apply the method established in the preceding article for determining the existence of an asymptote. In this case we have

$$[r \sin (\theta_1 - \theta)]_{\theta_1} = \left[\dfrac{a\theta^2}{\theta + 1} \cdot \dfrac{\sin (1 - \theta)}{\theta - 1} \right]_1 = -\tfrac{1}{2} a.$$

The angle $\theta = 1$ corresponds to $57° \ 18'$, nearly, and since the expression for the perpendicular on the asymptote is negative its direction is $\theta_1 + 90° = 147° \ 18'$; consequently, the asymptote is laid off as in Fig. 21.

Numerically equal positive and negative values of θ give the same values for r; hence the curve is symmetrical with reference to the initial line.

While θ varies from 0 to 1, r is negative and varies from 0 to ∞, giving the infinite branch in the third quadrant.

As θ passes the value unity, and increases indefinitely, r becomes positive and decreases, approaching indefinitely to the limiting value a, which we obtain from (1) by making θ infinite. Hence the curve describes an infinite number of whorls approaching indefinitely to the circle $r = a$, which is therefore called *an asymptotic circle.*

The points of inflexion in this curve are determined in Art. 175.

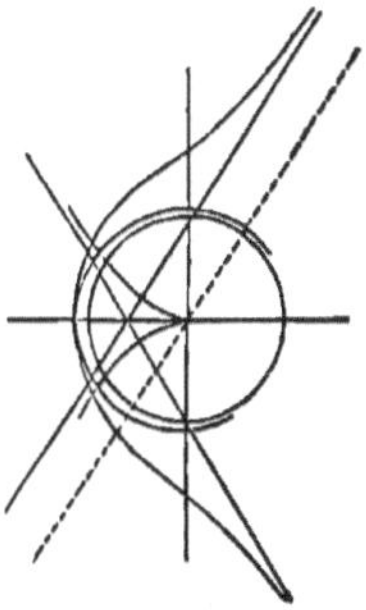

FIG. 21.

Examples XXI.

1. Trace the curve $r = a \cos^3 \tfrac{1}{3}\theta$.

Show that, to describe the curve, θ must vary from 0 to 3π; also that the curve is symmetrical to the initial line. Find the values of θ which correspond to the maxima and minima ordinates and abscissas, the initial line being taken as the axis of x.

2. Trace the curve $r = a(2 \sin \theta - 3 \sin^3\theta)$.

Show that the entire curve is described while θ varies from 0 to π, and that the curve is symmetrical with reference to a perpendicular to the initial line.

3. Trace the curve $r = 2 + \sin 3\theta$.

A maximum value of r (equal to 3) occurs at $\theta = 30°$; a minimum (equal to 1) at $\theta = 90°$. The curve is symmetrical with reference to lines inclined at the angles $30°$, $90°$, and $150°$ to the initial line.

4. Trace the curve $r = 1 + \sin 5\theta$.

The curve consists of five equal loops.

5. Trace the curve $r^2 = a^2 \sin 3\theta$.

The curve consists of three equal loops.

6. Trace the curve $r \cos \theta = a \cos 2\theta$.

The curve has an asymptote perpendicular to the initial line at the distance a on the left of the pole.

7. Trace the curve $r = 2 + \sin \tfrac{3}{2}\theta$.

A maximum value of r occurs at $\theta = 60°$, and a minimum at $\theta = 180°$. The curve has three double points, one being on the initial line.

8. Trace the curve $r \cos 2\theta = a$.

The curve is symmetrical with reference to the initial line and with reference to a perpendicular to the initial line. There are four asymptotes.

§ XXI.]   EXAMPLES.   151

9. Trace the curve $r \sin 4\theta = a \sin 3\theta$.
The curve is symmetrical to the initial line, and has three asymptotes; the minimum value of r is $\frac{3}{4}a$.

10. Trace the curve $r^2 = a^2 \cos 2\theta$.
The curve is symmetrical with respect to the pole since $r = \pm a \sqrt{(\cos 2\theta)}$: r is imaginary for values of θ between $\frac{1}{4}\pi$ and $\frac{3}{4}\pi$.

11. Trace the curve $r^{\frac{2}{3}} = a^{\frac{2}{3}} \cos \frac{2}{3}\theta$.
The curve consists of three equal loops, r being real for all values of θ.

12. Trace the curve $r^2 \cos \theta = a^2 \sin 3\theta$.
The curve consists of two loops and an infinite branch which has an asymptote perpendicular to the initial line and passing through the pole.

13. Trace the curve $r = a \dfrac{2\theta}{2\theta - 1}$.
Find the rectilinear and the circular asymptote, and also the point of inflexion.

XXII.

The Parabola of the nth Degree.

148. The term *parabola* is frequently applied to any curve in which one of the coordinates is proportional to the nth power of the other, n being greater than unity. The parabola proper is thus distinguished as the parabola of the second degree.

The general equation of the parabola of the nth degree is usually written in the homogeneous form, (a being positive)

$$a^{n-1} y = x^n.$$

The curve passes through the origin and through the point (a, a), for all values of n. Since $n > 1$, the curve is tangent to the axis of x at the origin.

149. The following three diagrams represent forms which the curve takes for different values of n. When n denotes a fraction, it is supposed to be reduced to its lowest terms.

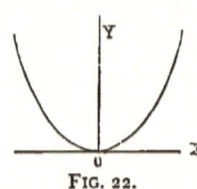

FIG. 22.

Fig. 22 represents the general shape of the curve when n is an even integer, or a fraction having an even numerator and an odd denominator.

Fig. 23 represents the form of the curve when n is an odd integer or a fraction with an odd numerator and an odd denominator, the origin being a point of inflexion.

Fig. 24 represents the form of the curve when n is a fraction having an odd numerator and an even denominator. In this case y is regarded as a two-valued function, and is imaginary when x is negative.

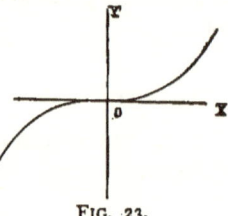

FIG. 23.

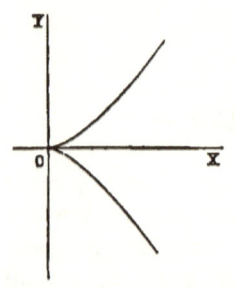

FIG. 24.

Fig. 22 is constructed for the parabola in which $n = 4$.

Fig. 23 is the *cubical parabola* in which $n = 3$.

Fig. 24 is the *semi-cubical parabola* in which $n = \frac{3}{2}$; the equation being

$$a^{\frac{1}{2}} y = \pm x^{\frac{3}{2}},$$

or $\qquad ay^2 = x^3.$

The curves corresponding to the general equation

$$y = A + Bx + Cx^2 + Dx^3 + \ldots Lx^n$$

are sometimes called *parabolic curves* of the nth degree.

The Cissoid of Diocles.

150. Let A be a point on the circumference of a circle, and BC a tangent at the opposite extremity of the diameter AB; let AC be any straight line through A, and take $CP = AD$; then the locus of P is the *cissoid*.

To find the polar equation, AB being the initial line, let DB be drawn, and denote the radius of the circle by a; then $AC = 2a \sec \theta$; and since ADB is a right angle, $AD = 2a \cos \theta$. The polar equation of the locus of P, A being the pole, is, therefore,

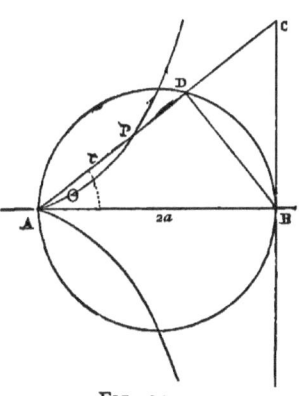

Fig. 25.

$$r = 2a(\sec \theta - \cos \theta) = 2a\frac{1 - \cos^2 \theta}{\cos \theta},$$

or
$$r = 2a\frac{\sin^2 \theta}{\cos \theta}. \quad \ldots \ldots \quad (1)$$

151. To obtain the rectangular equation, we employ the equations of transformation

$$\sin \theta = \frac{y}{r}, \quad \cos \theta = \frac{x}{r}, \quad r^2 = x^2 + y^2;$$

whence, eliminating θ we obtain

$$r = 2a\frac{y^2}{rx},$$

and thence the rectangular equation of the curve

$$x(x^2 + y^2) = 2ay^2, \quad \ldots \ldots \quad (2)$$

or
$$y^2 = \frac{x^3}{2a - x}. \quad \ldots \ldots \quad (3)$$

The Cardioid.

152. The curve defined by the polar equation

$$r = 2a(1 - \cos\theta) \quad \ldots \quad (1)$$

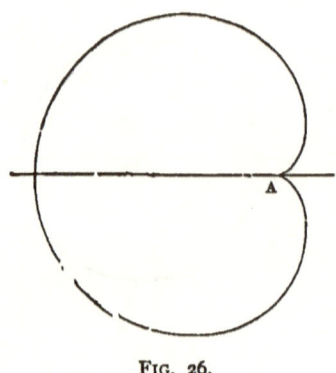

Fig. 26.

is called the *cardioid*. In Fig. 26, A denotes the pole.

The polar equation can also be written in the form

$$r = 4a \sin^2 \tfrac{1}{2}\theta \quad \ldots \quad (2)$$

Transforming equation (1) to rectangular coordinates, we have for the rectangular equation of the cardioid,

$$(x^2 + y^2)^2 + 4ax(x^2 + y^2) - 4a^2y^2 = 0 \quad \ldots \quad (3)$$

A point at which two branches of a curve have a common tangent is called a *cusp*. This curve has a cusp at the origin.

The Lemniscata of Bernoulli.

153. The curve defined by the polar equation

$$r^2 = a^2 \cos 2\theta \quad \ldots \quad (1)$$

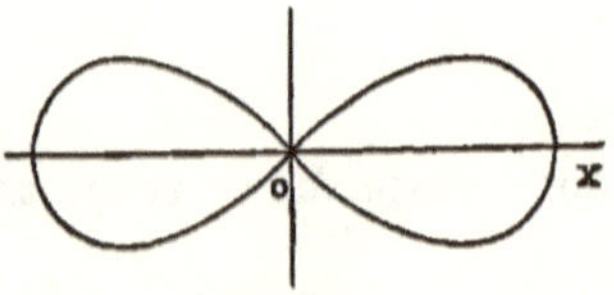

is called the *lemniscata*. In Fig. 27 O denotes the pole: a is the semi-axis of the curve.

Fig. 27.

From (1), we have

$$r^2 = a^2(\cos^2\theta - \sin^2\theta),$$

THE LEMNISCATA.

or
$$r^2 = a^2 \frac{x^2 - y^2}{r^2};$$

whence we have

$$(x^2 + y^2)^2 + a^2(y^2 - x^2) = 0, \quad \ldots \quad (2)$$

the rectangular equation of the lemniscata, referred to its centre and axis of symmetry.

If we turn the initial line back through 45°, (1) becomes

$$r^2 = a^2 \sin 2\theta, \quad \ldots \quad (3)$$

and the corresponding rectangular equation is

$$(x^2 + y^2)^2 = 2a^2xy. \quad \ldots \quad (4)$$

When the equation has this form, the coordinate axes are the tangents at the origin.

The Logarithmic or Equiangular Spiral.

154. This spiral is defined by the polar equation

$$r = a\epsilon^{n\theta}, \quad \ldots \quad (1)$$

or $\quad \log r = \log a + n\theta,$

the logarithm of the radius vector being a linear function of the vectorial angle.

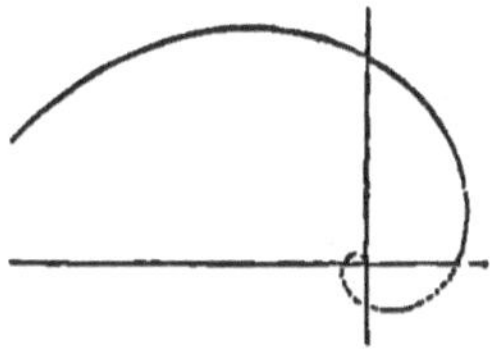

FIG. 28.

It is proved in Art. 168 that this curve cuts its radius vector at a *constant angle* whose cotangent is n; hence it is sometimes called the *equiangular spiral*.

The Loxodromic Curve.

155. The track of a ship whose course is uniform is a curve that cuts the meridians of the sphere at a constant angle, and is called a *loxodromic curve*.

If we project this curve stereographically upon the plane of the equator the meridians will project into straight lines, and, since in this projection angles are unchanged in magnitude, the projection of the curve will make a constant angle with the projections of the meridians and will therefore be an equiangular spiral.

Let θ denote the longitude of the generating point measured from the point at which the curve cuts the equator, and C the course; that is, the constant acute angle at which the curve cuts the meridians, the generating point being supposed to approach the pole as θ increases. Taking as the pole the projection of the pole of the sphere, the polar equation of the projected curve will be of the form

$$r = a\varepsilon^{n\theta}, \quad \ldots \ldots \ldots \quad (1)$$

in which a is the radius of the sphere, since $\theta = 0$ gives $r = a$; we also have

$$n = -\cot C, \quad \ldots \ldots \ldots \quad (2)$$

since the angle whose cotangent is n is the supplement of C (see the preceding article).

Denoting by ϕ the co-latitude of the projected point we have, by the mode of projection,

$$\frac{r}{a} = \tan \tfrac{1}{2}\phi; \quad \ldots \ldots \ldots \quad (3)$$

and, denoting the corresponding latitude by l,

$$\tfrac{1}{2}\phi = \tfrac{1}{4}\pi - \tfrac{1}{2}l.$$

Equation (1) is therefore equivalent to

$$\tan\left(\tfrac{1}{4}\pi - \tfrac{1}{2}l\right) = \varepsilon^{-\theta \cot C};$$

whence, solving for θ, we have

$$\theta = -\tan C \log_e \tan(\tfrac{1}{4}\pi - \tfrac{1}{2}l) = \tan C \log_e \tan(\tfrac{1}{4}\pi + \tfrac{1}{2}l),$$

or, employing common logarithms and expressing θ in degrees,

$$\theta° = 131.9284 \tan C \log_{10} \tan(45° + \tfrac{1}{2}l) \quad . \quad . \quad . \quad (4)$$

The Cycloid.

156. The path described by a point in the circumference of a circle which rolls upon a straight line is called a *cycloid*. The curve consists of an unlimited number of branches cor-

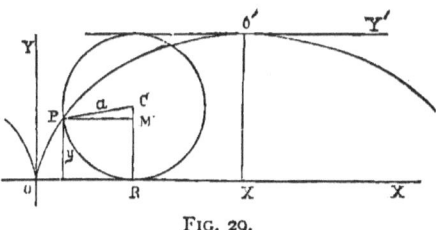

FIG. 29.

responding to successive revolutions of the generating circle; a single branch is, however, usually termed a cycloid.

Let O, the point where the curve meets the straight line, be taken as the origin, let P be the generating point of the curve, and denote the angle PCR by ψ. Since PR is equal to the line OR over which it has rolled,

$$OR = PR = a\psi,$$

and, from Fig. 29, we readily derive

$$\left.\begin{array}{l} x = a(\psi - \sin \psi) \\ y = a(1 - \cos \psi) \end{array}\right\} \quad . \quad . \quad . \quad . \quad . \quad (1)$$

157. These two equations express the values of x and y in terms of the auxiliary variable ψ, and constitute the equations of the cycloid. If desirable, ψ is easily eliminated from equations (1) and an equation between x and y obtained. Thus, from the second equation, we have

$$\cos\psi = \frac{a-y}{a}, \quad \text{whence} \quad \sin\psi = \frac{\sqrt{(2ay-y^2)}}{a};$$

and hence from the first of equations (1)

$$x = a \cos^{-1}\frac{a-y}{a} - \sqrt{(2ay-y^2)}, \quad \ldots \quad (2)$$

or
$$x = a \operatorname{vers}^{-1}\frac{y}{a} - \sqrt{(2ay-y^2)}.$$

Equations (1) will in general be found more convenient than equation (2). Thus we easily derive from (1)

$$\frac{dy}{dx} = \frac{\sin\psi\, d\psi}{(1-\cos\psi)\, d\psi} = \frac{\sin\psi}{1-\cos\psi};$$

whence

$$\frac{d^2y}{dx^2} = \frac{d}{dx}\left(\frac{dy}{dx}\right) = \frac{\cos\psi - 1}{(1-\cos\psi)^2}\frac{d\psi}{dx} = -\frac{1}{a(1-\cos\psi)^3}.$$

158. The cycloid is frequently referred to the middle point O' or vertex of the curve as an origin, the directions of the axes being turned through 90°.

Denoting the coordinates referred to the axes $O'X'$ and $O'Y'$, in Fig. 29, by x' and y', we have

$$y' = x - a\pi = a(\psi - \pi - \sin\psi),$$
$$x' = 2a - y = a(1 + \cos\psi),$$

or, denoting $\psi - \pi$ by ψ',

$$\left.\begin{array}{l} y' = a(\psi' + \sin\psi') \\ x' = a(1 - \cos\psi') \end{array}\right\} \quad \ldots \quad (3)$$

In these equations $\psi' = 0$ gives the coordinates of the vertex and $\psi' = \pm\pi$ gives those of the cusps.

The Epicycloid.

159. When a circle, tangent to a fixed circle externally, rolls upon it, the path described by a point in the circumference of the rolling circle is called an *epicycloid*.

Taking the origin at the centre of the fixed circle, and the axis of x passing through A, (one of the positions of P when in contact with the fixed circle,) a, b, ψ, and χ, being defined by the diagram, we have, evidently,

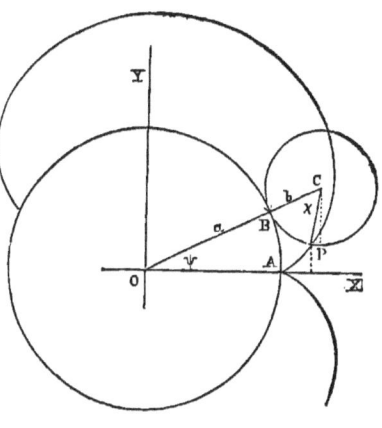

FIG. 30.

$$a\psi = b\chi \therefore \chi = \frac{a}{b}\psi.$$

The inclination of CP to the axis of x is equal to $\psi + \chi$, or to $\frac{a+b}{b}\psi$; the coordinates of P are found by subtracting the projections of CP on the axes from the corresponding projections of OC; hence

$$\left. \begin{array}{l} x = (a+b)\cos\psi - b\cos\dfrac{a+b}{b}\psi \\ y = (a+b)\sin\psi - b\sin\dfrac{a+b}{b}\psi \end{array} \right\} \quad . \quad . \quad (1)$$

These are the equations of an epicycloid *referred to an axis passing through one of the cusps.*

Were the generating point taken at the opposite extremity of a diameter passing through P in the figure, the projection of CP would be *added* to that of OC; the axis of x would in this case pass through one of the *vertices* of the curve, and the second terms in the above values of x and y would have the *positive sign*.

The Hypocycloid.

160. When the rolling circle has internal contact with the fixed circle, the curve generated by a point on the circumference is called the *hypocycloid*, whether the radius of the rolling circle be greater or less than that of the fixed circle.

Adopting the notation used in deducing the equation of the epicycloid we have (see Fig. 31),

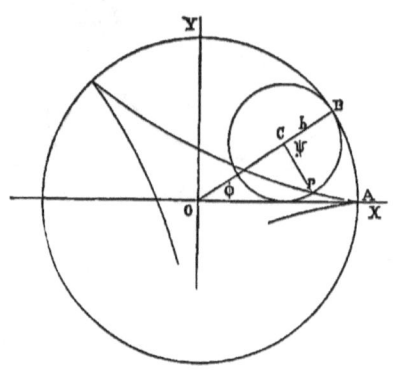

FIG. 31.

$$OC = a - b, \quad \text{and} \quad \chi = \frac{a}{b}\psi.$$

The inclination of CP to the negative direction of the axis of x is

$$\chi - \psi = \frac{a-b}{b}\psi;$$

hence the equations of the *hypocycloid* are

$$\left. \begin{array}{l} x = (a-b)\cos\psi + b\cos\dfrac{a-b}{b}\psi \\[2mm] y = (a-b)\sin\psi - b\sin\dfrac{a-b}{b}\psi \end{array} \right\} \quad \ldots \ldots (1)$$

The Four-Cusped Hypocycloid.

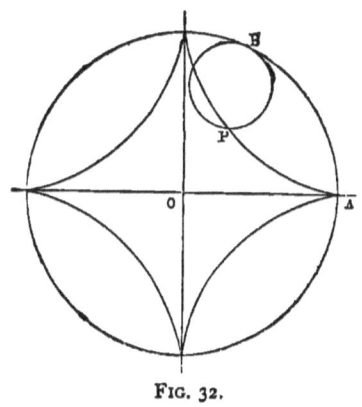

Fig. 32.

161. In the case of the hypocycloid when $b = \frac{1}{4}a$, the circumference of the rolling circle is one-fourth the circumference of the fixed circle, and the curve will have a cusp at each of the four points where the coordinate axes cut the fixed circle, as represented in Fig. 32.

On substituting $\frac{1}{4}a$ for b equations (2) Art. 160 become

$$\left. \begin{array}{l} x = \tfrac{3}{4}a \cos\psi + \tfrac{1}{4}a \cos 3\psi \\ y = \tfrac{3}{4}a \sin\psi - \tfrac{1}{4}a \sin 3\psi \end{array} \right\} \quad \ldots \ldots (1)$$

Substituting the values of $\cos 3\psi$ and $\sin 3\psi$ from the formulas,

$$\cos 3\psi = 4\cos^3\psi - 3\cos\psi,$$

and

$$\sin 3\psi = 3\sin\psi - 4\sin^3\psi,$$

we have

$$\left. \begin{array}{l} x = a\cos^3\psi \\ y = a\sin^3\psi \end{array} \right\} ; \quad \ldots \ldots (2)$$

whence $\quad x^{\frac{2}{3}} = a^{\frac{2}{3}}\cos^2\psi,\quad$ and $\quad y^{\frac{2}{3}} = a^{\frac{2}{3}}\sin^2\psi.$

Adding, we have $\quad x^{\frac{2}{3}} + y^{\frac{2}{3}} = a^{\frac{2}{3}}, \quad \ldots \ldots (3)$

the rectangular equation of the curve.

CHAPTER IX.

APPLICATIONS OF THE DIFFERENTIAL CALCULUS TO PLANE CURVES.

XXIII.

The Equation of the Tangent.

162. THE equation of the curve being given in the form $y = f(x)$, the inclination of the tangent at any point is determined by the equation

$$\tan \phi = \frac{dy}{dx} = f'(x).$$

Hence, if (x_1, y_1) be a point of the curve, the equation of a tangent at (x_1, y_1) will be found by giving to the direction-ratio m, in the general equation

$$y - y_1 = m(x - x_1),$$

the value $\left.\dfrac{dy}{dx}\right]_{x_1}$; thus

$$y - y_1 = \left.\frac{dy}{dx}\right]_{x_1} (x - x_1), \quad \ldots \ldots \quad (1)$$

or
$$y - y_1 = f'(x_1)(x - x_1). \quad \ldots \ldots \quad (2)$$

For example, in the case of the semi-cubical parabola

$$y^2 = ax^3,$$

we have
$$\frac{dy}{dx} = \tfrac{2}{3}\sqrt[3]{\frac{a}{x}}.$$

The point (a, a) is a point of this curve; the equation of the tangent at this point is, therefore,
$$3y - 2x = a.$$

The Equation of the Normal.

163. A perpendicular to the tangent at its point of contact is called a *normal* to the curve.

The coordinate axes being rectangular, the direction-ratio of the normal is the negative reciprocal of that of the tangent; for the inclination of the normal is $\tfrac{1}{2}\pi + \phi$, and
$$\tan(\tfrac{1}{2}\pi + \phi) = -\cot \phi.$$

The equation of the normal may, therefore, be written thus—

$$y - y_1 = -\left.\frac{dx}{dy}\right]_{x_1} (x - x_1), \quad \ldots \ldots (1)$$

or
$$y - y_1 = -\frac{1}{f'(x_1)}(x - x_1). \quad \ldots \ldots (2)$$

As an illustration, let us take the equation of the ellipse
$$\frac{x^2}{a^2} + \frac{y^2}{b^2} = 1;$$

whence
$$\frac{dy}{dx} = -\frac{b^2 x}{a^2 y}.$$

The equation of the normal at any point (x_1, y_1) of the ellipse is, therefore,
$$y - y_1 = \frac{a^2 y_1}{b^2 x_1}(x - x_1).$$

Subtangents and Subnormals.

164. Denoting by s the length of the arc measured from some fixed point, $\dfrac{ds}{dt}$ denotes the velocity of P, the generating point of the curve; let PT, equal to ds, be measured on the tangent at P, then PQ and QT will represent dx and dy, and the angle TPQ will be ϕ; hence

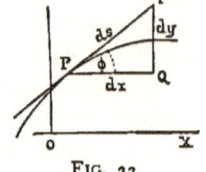

Fig. 33.

$$\cos \phi = \frac{dx}{ds}, \quad \sin \phi = \frac{dy}{ds}, \quad (1)$$

and $\quad ds = \sqrt{(dx^2 + dy^2)}. \quad . \quad . \quad (2)$

165. The distance PT (Fig. 34) on the tangent line intercepted between the point of contact and the axis of x is sometimes called the *tangent*, and in like manner the intercept PN is called the *normal*.

Fig. 34.

From the triangles PTR and NPR, we have

$$PT = y \operatorname{cosec} \phi = y \frac{ds}{dy} = y \sqrt{\left[1 + \left(\frac{dx}{dy}\right)^2\right]},$$

$$PN = y \sec \phi = y \frac{ds}{dx} = y \sqrt{\left[1 + \left(\frac{dy}{dx}\right)^2\right]}.$$

The projections of these lines on the axis of x, that is TR and RN, are called the *subtangent* and the *subnormal*.

From the same triangles, we have

§ XXIII.] SUBTANGENTS AND SUBNORMALS. 165

the subtangent, $\quad TR = y \cot \phi = y \dfrac{dx}{dy}$,

and the subnormal, $\quad RN = y \tan \phi = y \dfrac{dy}{dx}$.

The Perpendicular from the Origin upon the Tangent.

166. If a perpendicular p to the tangent PR be drawn from the origin, we have, from the triangles in Fig. 35,

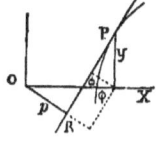

FIG. 35.

$$p = x \sin \phi - y \cos \phi, \quad \ldots \quad (1)$$

$\phi - 90°$ *being taken as the positive direction of* p. Substituting the values of $\sin \phi$ and $\cos \phi$, equation (1) becomes

$$p = \frac{xdy - ydx}{ds} = \frac{xdy - ydx}{\sqrt{(dx^2 + dy^2)}} \quad \ldots \quad (2)$$

For example, let us determine p in the case of the four-cusped hypocycloid,

$$x = a \cos^3 \psi, \qquad y = a \sin^3 \psi.$$

Differentiating,

$$dx = -3a \cos^2 \psi \sin \psi \, d\psi, \quad \text{and} \quad dy = 3a \sin^2 \psi \cos \psi \, d\psi;$$

whence $\qquad ds = 3a \sin \psi \cos \psi \, d\psi.$

Substituting in equation (2) we obtain

$$p = a \cos^3 \psi \sin \psi + a \sin^3 \psi \cos \psi = a \sin \psi \cos \psi = \sqrt[3]{(axy)}.$$

To ascertain the direction of p it is necessary to determine

ϕ. The ambiguity in the value of ϕ as determined from the equation $\tan \phi = \dfrac{dy}{dx}$ may be removed by means of one of the formulas of Art. 164. Thus, in the present case, we have

$$\tan \phi = -\tan \psi, \quad \text{whence} \quad \phi = -\psi, \quad \text{or} \quad \phi = \pi - \psi;$$

but, since
$$\cos \phi = \frac{dx}{ds} = -\cos \psi,$$

we must take
$$\phi = \pi - \psi.$$

The direction of p when positive is therefore $\frac{1}{2}\pi - \psi$.

Examples XXIII.

1. In the case of the parabola of the nth degree

$$a^{n-1} y = x^n,$$

find the equations of the tangent and the normal at the point (a, a).

2. Find the subtangent and the subnormal of the parabola

$$y^2 = 4ax.$$

3. Prove that the subtangent of the exponential curve

$$y = a^x$$

is constant, and find the ordinate of the point of contact when the tangent passes through the origin.

4. Find the subnormal of the ellipse whose equation is

$$\frac{x^2}{a^2} + \frac{y^2}{b^2} = 1.$$

5. Find the subtangent of the curve

$$a^{n-1} y = x^n.$$

6. In the case of the parabola

$$y^2 = 4ax,$$

find p in terms of x.

For the upper branch, $p = -\dfrac{x\sqrt{a}}{\sqrt{(a+x)}}$.

7. Find, in terms of ψ, the equation of the tangent to the four-cusped hypocycloid (Art. 161), and thence show that the part intercepted between the axes is of constant length.

8. In the case of the epicycloid, find the value of ds in terms of the auxiliary angle ψ. See Art. 159.

$$ds = 2(a+b)\sin\frac{a\psi}{2b}d\psi.$$

9. Determine the value of p in the case of the epicycloid employing the value of ds determined in the preceding example.

$$p = (a+2b)\sin\frac{a\psi}{2b}.$$

XXIV.

Polar Coordinates.

167. When the equation of a curve is given in polar coordinates the vectorial angle θ is usually taken as the independent variable; hence, denoting by s an arc of the curve, it is usual to assume that ds and $d\theta$ have the same sign; that is, that $\dfrac{ds}{d\theta}$ is positive.

In Fig. 36 let PT, a portion of the tangent line, represent ds; then, producing r, let the rectangle PT be completed, and let ψ denote the angle TPS; that is, the angle between the positive directions of r and s. The resolved velocities of P along and perpendicular to the radius vector are $\dfrac{dr}{dt}$ and $\dfrac{rd\theta}{dt}$, the latter being the velocity which P would have if r were constant; that is, if P moved in a circle described with r as a radius. Hence we have

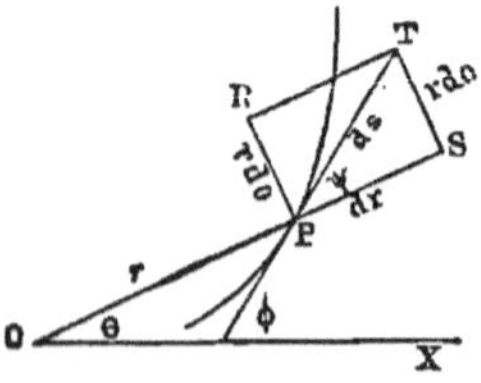

FIG. 36.

$$PS = dr \quad \text{and} \quad PR = rd\theta.$$

From the triangle PST, we derive

$$\tan\psi = \frac{rd\theta}{dr}, \quad \sin\psi = \frac{rd\theta}{ds}, \quad \cos\psi = \frac{dr}{ds}, \quad \ldots \quad (1)$$

and

$$\frac{ds}{d\theta} = \sqrt{\left[r^2 + \left(\frac{dr}{d\theta}\right)^2\right]}. \quad \ldots \ldots \quad (2)$$

168. The second of equations (1) shows that, in accordance with the assumption that ds has the sign of $d\theta$, the value of ψ will always be either in the first or in the second quadrant.

The first of equations (1) is equivalent to

$$\cot\psi = \frac{dr}{rd\theta}, \quad \ldots \ldots \quad (3)$$

which shows that $\cot\psi$ is *the logarithmic derivative of r regarded as a function of θ*. Thus in the case of the logarithmic spiral

$$r = a\varepsilon^{n\theta}$$

we have
$$\log r = \log a + n\theta,$$

hence
$$\cot\psi = n$$

§ XXIV.] POLAR COORDINATES. 169

whence it follows that, in the case of this curve, ψ is constant. See Art. 154.

169. It is frequently convenient to employ in place of the radius vector its reciprocal, which is usually denoted by u; then

$$r = \frac{1}{u}, \quad \text{and} \quad dr = -\frac{du}{u^2}. \quad \ldots \ldots (4)$$

Making these substitutions in equations (2) and (3) we have

$$\frac{ds}{d\theta} = \frac{1}{u^2}\sqrt{\left[u^2 + \left(\frac{du}{d\theta}\right)^2\right]}. \quad \ldots \ldots (5)$$

and
$$\cot \psi = -\frac{du}{u\,d\theta}. \quad \ldots \ldots (6)$$

Polar Subtangents and Subnormals.

170. Let a straight line perpendicular to the radius vector be drawn through the pole, and let the tangent and the normal meet this line in T and N respectively; then the projections of PT and PN upon this line, that is OT and ON, are called respectively the *polar subtangent* and the *polar subnormal*. In Fig. 37, $OPT = \psi$; whence

$$OT = r \tan \psi = r^2 \frac{d\theta}{dr} = -\frac{d\theta}{du},$$

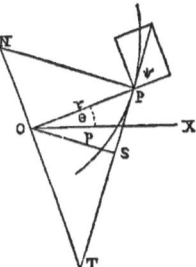

FIG. 37.

and
$$ON = r \cot \psi = \frac{dr}{d\theta} = -\frac{du}{u^2 d\theta}.$$

Fig. 37 shows that the value of OT is positive when its

direction is $\theta - 90°$; that of ON is, on the other hand, positive when its direction is $\theta + 90°$.

The Perpendicular from the Pole upon the Tangent.

171. Let p denote the perpendicular distance from the pole to the tangent; then, from Fig. 37 we obtain

$$p = r \sin \psi = r^2 \frac{d\theta}{ds} = \frac{r^2}{\sqrt{\left[r^2 + \left(\frac{dr}{d\theta}\right)^2\right]}}. \quad . \quad . \quad (1)$$

These expressions give positive values for p, because $\frac{ds}{d\theta}$ is assumed to be positive, and Fig. 37 shows that p has the direction $\phi - 90°$, ϕ being the angle which the positive direction of s makes with the initial line.

The relation between p and u is obtained thus:—from (1) we have

$$\frac{1}{p^2} = \frac{ds^2}{r^4 d\theta^2},$$

and, transforming by the formulas of Art. 169,

$$\frac{1}{p^2} = u^2 + \left(\frac{du}{d\theta}\right)^2. \quad . \quad . \quad . \quad . \quad . \quad (2)$$

172. The expression deduced below for the function $u + \frac{d^2u}{d\theta^2}$ is frequently useful.

Differentiating (2), we have

$$2u\,du + 2\frac{du}{d\theta} \cdot \frac{d^2u}{d\theta} = -\frac{2dp}{p^3};$$

§ XXIV.] *THE PERPENDICULAR UPON THE TANGENT.* 171

hence
$$\left(u + \frac{d^2u}{d\theta^2}\right)du = -\frac{dp}{p^3},$$

or since
$$du = -\frac{dr}{r^2},$$

$$u + \frac{d^2u}{d\theta^2} = \frac{r^2}{p^3} \cdot \frac{dp}{dr}.$$

The Perpendicular upon an Asymptote.

173. When the point of contact P passes to infinity the tangent at P becomes an asymptote, and the subtangent OT coincides with the perpendicular upon the asymptote. Hence (θ_1 denoting a value of θ for which r is infinite) the length of this perpendicular is given by the expression $-\dfrac{d\theta}{du}\bigg]_{\theta_1}$, and like the polar subtangent is, when positive, to be laid off in the direction $\theta_1 - 90°$.

This expression for the perpendicular upon the asymptote is also easily derived by evaluating that given in Art. 146. Thus—

$$r \sin(\theta_1 - \theta)\bigg]_{\theta_1} = \frac{\sin(\theta_1 - \theta)}{u}\bigg]_{\theta_1} = -\frac{d\theta}{du}\bigg]_{\theta_1}.$$

Points of Inflexion.

174. When, as in Fig. 37, the curve lies between the tangent and the pole, it is obvious that r and p will increase and decrease together; that is, $\dfrac{dp}{dr}$ will be positive. When on the other hand the curve lies on the other side of the tangent, $\dfrac{dp}{dr}$ is negative. Hence at a point of inflexion $\dfrac{dp}{dr}$ must change sign.

Now, since p is always positive, it follows from the equation deduced in Art. 172 that the sign of this expression is the same as that of

$$u + \frac{d^2u}{d\theta^2}; \qquad \qquad (1)$$

hence *at a point of inflexion this expression must change sign.*

175. As an illustration, let us determine the point of inflexion of the curve traced in Art. 147; viz.,

$$r = \frac{a\theta^2}{\theta^2 - 1}.$$

In this case
$$u = \frac{1}{a}\left(1 - \theta^{-2}\right);$$

whence
$$\frac{du}{d\theta} = \frac{2}{a}\theta^{-3}, \quad \text{and} \quad \frac{d^2u}{d\theta^2} = -\frac{6}{a}\theta^{-4};$$

therefore
$$u + \frac{d^2u}{d\theta^2} = \frac{1}{a}\left(1 - \theta^{-2} - 6\theta^{-4}\right)$$

$$= \frac{\theta^4 - \theta^2 - 6}{a\theta^4}.$$

Putting this expression equal to zero, the real roots are

$$\theta = \pm \sqrt{3},$$

and it is evident that, as θ passes through either of these values, the expression $u + \frac{d^2u}{d\theta^2}$ changes sign. Hence the points of inflexion are determined by

$$\theta = \pm \sqrt{3} \quad \text{and} \quad r = \frac{3a}{2}.$$

Examples XXIV.

1. Prove that, in the case of the lemniscata $r^2 = a^2 \cos 2\theta$,

$$\psi = 2\theta + \tfrac{1}{2}\pi, \quad \text{and} \quad \frac{ds}{d\theta} = \frac{a^2}{r}.$$

2. Find the subtangent of the lituus $r^2 = \dfrac{a^2}{\theta}$, and prove that the perpendicular from the origin upon the tangent is

$$\frac{2a \sqrt{\theta}}{\sqrt{(1 + 4\theta^2)}}.$$

3. Find the polar subtangent of the spiral $r(\varepsilon^\theta + \varepsilon^{-\theta}) = a$.

$$-\frac{a}{\varepsilon^\theta - \varepsilon^{-\theta}}.$$

4. Find the value of p in the case of the curve $r^n = a^n \sin n\theta$.

$$p = a(\sin n\theta)^{1+\frac{1}{n}}.$$

5. In the case of the parabola referred to the focus

$$r = \frac{2a}{1 + \cos \theta}, \text{ prove that } p^2 = ar.$$

6. In the case of the equilateral hyperbola

$$r^2 \cos 2\theta = a^2, \text{ prove that } p = \frac{a^2}{r}.$$

7. In the case of the lemniscata

$$r^2 = a^2 \cos 2\theta, \text{ prove that } p = \frac{r^3}{a^2}.$$

8. In the case of the ellipse $r = \dfrac{a(1 - e^2)}{1 - e \cos \theta}$, the pole being at

the focus, determine p.

$$p = \frac{a(1-e^2)}{\sqrt{(1-2e\cos\theta+e^2)}}.$$

9. In the case of the cardioid

$$r = a(1 + \cos\theta), \text{ prove that } r^3 = 2ap^2.$$

10. Show that the curve $r\theta \sin\theta = a$ has a point of inflexion at which $r = \dfrac{2a}{\pi}$.

XXV.

Curvature.

176. If, while a point P moves along a given curve at the rate $\dfrac{ds}{dt}$, it be regarded as carrying with it the tangent and normal lines, each of these lines will rotate about the moving point P at the angular rate $\dfrac{d\phi}{dt}$, ϕ denoting the inclination of the tangent line to the axis of x.

The point P is always moving in a direction perpendicular to the normal with the velocity $\dfrac{ds}{dt}$. Let us consider the motion of a point A on the normal at a given distance k from P on the concave side of the arc. While this point is carried forward by the motion of P with the velocity $\dfrac{ds}{dt}$ in a direction perpendicular to the normal, it is at the same time carried backward, by the rotation of this line about P, with the

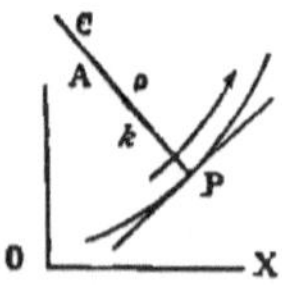

FIG. 38.

§ XXV.] CURVATURE. 175

velocity $\frac{kd\phi}{dt}$; since this is the velocity with which A would move if the point P occupied a fixed position in the plane; and the *direction* of this motion is evidently directly opposite to that of P. Hence the actual velocity of A will be

$$\frac{ds}{dt} - k\frac{d\phi}{dt},$$

in a direction parallel to the tangent at P.

Let ρ denote the value of k which reduces this expression to zero, and let C (Fig. 38) be the corresponding position of A: then,

$$\frac{ds}{dt} - \rho\frac{d\phi}{dt} = 0;$$

whence $\qquad PC = \rho = \frac{ds}{d\phi}.\qquad \ldots \ldots \ldots (1)$

177. The value of ρ determined by this equation is, in general, variable; for, if the point P move along the curve with a given linear velocity $\frac{ds}{dt}$, the angular velocity $\frac{d\phi}{dt}$ will generally be variable. If however we suppose the angular velocity $\frac{d\phi}{dt}$ to become constant, at the instant when P passes a given position on the curve, $\frac{ds}{d\phi}$, the value of ρ, will likewise become constant, and C will remain stationary. When this hypothesis is made, the curvature of the path of P becomes constant, for P describes a circle whose centre is C, and whose radius is ρ. Hence this circle is called *the circle of curvature* corresponding to the given position of P; C is accordingly called *the centre of curvature*, and ρ is called *the radius of curvature*.

The Direction of the Radius of Curvature.

178. If in Fig. 38 the arrow indicates the positive direction of s; the case represented is that in which ϕ and s increase together, and therefore the value of ρ as determined by equation (1), Art. 176, is positive. Hence it is evident that when ρ is positive its direction from P is that of PC in Fig. 38; namely, $\phi + 90°$. In other words, to a person looking along the curve in the positive direction of ds, ρ, when positive, is laid off on *the left-hand side of the curve.*

For example, let the curve be the four-cusped hypocycloid,

$$x = a \cos^3 \psi, \qquad y = a \sin^3 \psi.$$

It was shown in Art. 166 that for this curve

$$ds = 3a \sin \psi \cos \psi \, d\psi, \quad \text{and} \quad \phi = \pi - \psi;$$

hence $$d\phi = - d\psi,$$

and $$\rho = \frac{ds}{d\phi} = - 3a \sin \psi \cos \psi. \quad \ldots \ldots \quad (1)$$

When ψ is in the first quadrant ρ is negative; its direction is therefore $\phi - \tfrac{1}{2}\pi = \tfrac{1}{2}\pi - \psi$, which is in the first quadrant. When ψ is in the second quadrant ρ is positive and its direction is $\phi + \tfrac{1}{2}\pi = \tfrac{3}{2}\pi - \psi$, which is in the second quadrant.

The Radius of Curvature in Rectangular Coordinates.

179. To express ρ in terms of derivatives with reference to x, we have

$$\phi = \tan^{-1}\frac{dy}{dx}, \quad \text{and} \quad \frac{ds}{dx} = \sqrt{\left[1 + \left(\frac{dy}{dx}\right)^2\right]};$$

hence
$$\frac{d\phi}{dx} = \frac{\dfrac{d^2y}{dx^2}}{1 + \left(\dfrac{dy}{dx}\right)^2},$$

and
$$\rho = \frac{\dfrac{ds}{dx}}{\dfrac{d\phi}{dx}} = \frac{\left[1 + \left(\dfrac{dy}{dx}\right)^2\right]^{\frac{3}{2}}}{\dfrac{d^2y}{dx^2}} \quad \ldots \ldots (1)$$

Since $\dfrac{ds}{dx}$ is assumed to be positive, ϕ should be so taken as to cause x to increase with s, and it must be remembered that the direction of ρ is $\phi + 90°$ when ρ is positive, in accordance with the remark in the preceding article.

180. To illustrate the application of the above formula, we find the radius of curvature of the ellipse

$$y = \pm \frac{b}{a}\sqrt{(a^2 - x^2)}. \quad \ldots \ldots (1)$$

Differentiating, $\quad \dfrac{dy}{dx} = \mp \dfrac{bx}{a\sqrt{(a^2 - x^2)}}, \quad \ldots \ldots (2)$

and $\quad \dfrac{d^2y}{dx^2} = \mp \dfrac{ab}{(a^2 - x^2)^{\frac{3}{2}}}. \quad \ldots \ldots (3)$

Putting $b = a\sqrt{(1 - e^2)}$ we obtain

$$1 + \left(\frac{dy}{dx}\right)^2 = \frac{a^2 - e^2x^2}{a^2 - x^2};$$

whence, substituting in equation (1) of the preceding article,

$$\rho = \mp \frac{(a^2 - e^2x^2)^{\frac{3}{2}}}{a^2 \sqrt{(1-e^2)}}. \quad \ldots \ldots \quad (4)$$

Expressions for ρ in which x is not the Independent Variable.

181. To express ρ in terms of derivatives with reference to y, we have

$$\phi = \cot^{-1}\frac{dx}{dy}, \quad \text{and} \quad \frac{ds}{dy} = \sqrt{\left[1 + \left(\frac{dx}{dy}\right)^2\right]};$$

whence $\quad \dfrac{d\phi}{dy} = -\dfrac{\frac{d^2x}{dy^2}}{1 + \left(\frac{dx}{dy}\right)^2}, \quad$ and $\quad \rho = -\dfrac{\left[1 + \left(\frac{dx}{dy}\right)^2\right]^{\frac{3}{2}}}{\frac{d^2x}{dy^2}}.$

In this case ds and dy were assumed to have the same sign, hence ϕ must be taken so as to cause y to increase.

182. When x and y are expressed in terms of a third variable we employ the formula deduced below.

Differentiating

$$\phi = \tan^{-1}\frac{dy}{dx},$$

both dx and dy being regarded as variable, we have

$$d\phi = \frac{\frac{dx\, d^2y - dy\, d^2x}{dx^2}}{1 + \left(\frac{dy}{dx}\right)^2} = \frac{dx\, d^2y - dy\, d^2x}{dx^2 + dy^2};$$

whence $\quad \rho = \dfrac{ds}{d\phi} = \dfrac{(dx^2 + dy^2)^{\frac{3}{2}}}{dx\, d^2y - dy\, d^2x}. \quad \ldots \ldots \quad (1)$

Examples XXV.

1. Find the radius of curvature of the *cycloid*

$$x = a(\psi - \sin\psi), \qquad y = a(1 - \cos\psi).$$

Prove that $\phi = \frac{1}{2}(\pi - \psi)$, and use $\rho = \dfrac{ds}{d\phi}$.

$$\rho = -2\sqrt{(2ay)}.$$

2. Find the radius of curvature of the *parabola* $y^2 = 4ax$.

$$\rho = \mp 2\frac{(a+x)^{\frac{3}{2}}}{\sqrt{a}}.$$

3. Find the radius of curvature of the *catenary*

$$y = \frac{c}{2}\left(\varepsilon^{\frac{x}{c}} + \varepsilon^{-\frac{x}{c}}\right),$$

and show that its numerical value equals that of the normal at the same point. *See Art.* 165.

$$\rho = \frac{y^2}{c}.$$

4. Find the radius of curvature of the *semi-cubical parabola*

$$ay^2 = x^3.$$

$$\rho = \frac{(4a + 9x)^{\frac{3}{2}} x^{\frac{1}{2}}}{6a}.$$

5. Find the radius of curvature of the *logarithmic curve*

$$y = a\varepsilon^{\frac{x}{c}}.$$

$$\rho = \frac{[c^2 + y^2]^{\frac{3}{2}}}{cy}.$$

6. Find the radius of curvature of the *cissoid*

$$y = \frac{x^{\frac{3}{2}}}{(2a-x)^{\frac{1}{2}}}.$$

$$\rho = \frac{a\sqrt{x}\,(8a-3x)^{\frac{3}{2}}}{3\,(2a-x)^2}.$$

7. Find the radius of curvature of the *parabola*

$$\sqrt{x} + \sqrt{y} = 2\sqrt{a}.$$

$$\rho = \frac{(x+y)^{\frac{3}{2}}}{\sqrt{a}}.$$

8. Find the radius of curvature of the *cubical parabola*

$$a^2 y = x^3.$$

$$\rho = \frac{(a^4 + 9x^4)^{\frac{3}{2}}}{6a^4 x}.$$

9. Find the radius of curvature of the *prolate cycloid*

$$x = a\psi - b\sin\psi, \qquad y = a - b\cos\psi.$$

$$\rho = \frac{(a^2 + b^2 - 2ab\cos\psi)^{\frac{3}{2}}}{b\,(a\cos\psi - b)}.$$

XXVI.

Envelopes.

183. The curves determined by an equation involving x and y together with constants to which arbitrary values may be assigned are said to constitute a *system* of curves. The arbitrary constants are called *parameters*. When but one of the parameters is regarded as variable, denoting it by α, the general equation of the system of curves may be expressed thus:

$$f(x, y, \alpha) = 0. \quad \ldots \ldots \ldots \quad (1)$$

When the curves of a system mutually intersect (the intersections not being fixed points), there usually exists a curve which touches each curve of the system obtained by causing the value of α to vary.

For example, the ellipses whose axes are fixed in position, and whose semi-axes have a constant sum, constitute such a system; and, if we regard the ellipse as varying continuously from the position in which one semi-axis is zero to that in which the other is zero, it is evident that the boundary of that portion of the plane which is swept over by the perimeter of the varying ellipse is a curve to which the ellipse is tangent in all its positions. A curve having this relation to a given system of curves is called the *envelope of the system*.

Every point on an envelope may be regarded as the limiting position of the point of intersection of two members of the given system of curves, when the difference between the corresponding values of α is indefinitely diminished. For this reason, the envelope is sometimes called *the locus of the ultimate intersections* of the curves of the given system.

184. If we differentiate equation (1) of the preceding article (regarding α as a variable as well as x and y) the resulting equation will be of the general form

$$f'_x(x, y, \alpha)\, dx + f'_y(x, y, \alpha)\, dy + f'_\alpha(x, y, \alpha)\, d\alpha = 0. \quad (2)$$

In this equation each term may be separately obtained by differentiating the given equation on the supposition that the quantity indicated by the subscript is alone variable. See Art. 64.

From equation (2) we derive

$$\frac{dy}{dx} = -\frac{f'_x(x, y, \alpha)}{f'_y(x, y, \alpha)} - \frac{f'_\alpha(x, y, \alpha)}{f'_y(x, y, \alpha)}\frac{d\alpha}{dx}. \quad \cdots \quad (3)$$

In Fig. 39 let PC be the curve corresponding to a particular value of α, and let P be the point (x, y); then the expression for $\frac{dy}{dx}$ given in equation (3) determines the direction in which the point P is actually moving when x, y, and α vary simultaneously. This direction depends there-

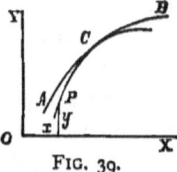

FIG. 39.

fore in part upon the arbitrary value given to the ratio $\frac{d\alpha}{dx}$.

185. Now if α were constant $d\alpha$ would vanish, and equation (3) would become

$$\frac{dy}{dx} = -\frac{f'_x(x, y, \alpha)}{f'_y(x, y, \alpha)}. \quad \cdots \quad (4)$$

This expression for $\frac{dy}{dx}$ determines the direction in which P moves when PC is a fixed curve.

Let AB be an arc of the envelope, and let C be its point of contact with PC. Now, if P be placed at the point C, it is obvious that it can move only in the direction of the common tangent at C, *whether α be fixed or variable*. It follows therefore that, at every point at which a curve belonging to the system touches the envelope, the expressions for $\frac{dy}{dx}$ given in equations (3) and (4) must be identical in value.

Assuming that $f'_x(x, y, \alpha)$ and $f'_y(x, y, \alpha)$ do not become infinite for any finite values of x and y, the above condition requires that

$$f'_\alpha(x, y, \alpha) = 0. \quad \cdots \quad (5)$$

§ XXVI.] ENVELOPES.

Hence the coordinates of every point of the envelope must satisfy simultaneously equations (5) and (1); the equation of the envelope is therefore obtained by eliminating α between these two equations.

186. Let it be required to find the *envelope of the circles having for diameters the double ordinates of the parabola*

$$y^2 = 4ax.$$

If we denote by α the abscissa of the centre of the variable circle, its radius will be the ordinate of the point on the parabola of which α is the abscissa, the equation of the circle will therefore be

$$y^2 + (x - \alpha)^2 - 4a\alpha = 0. \quad \ldots \ldots (1)$$

Differentiating with reference to the variable parameter α, we have

$$- 2(x - \alpha) - 4a = 0,$$

or
$$\alpha = 2a + x; \ldots \ldots \ldots (2)$$

substituting in (1), and reducing, we obtain

$$y^2 = 4a(a + x). \ldots \ldots \ldots (3)$$

The envelope is, therefore, a parabola equal to the given parabola and having its focus at the vertex of the given parabola.

Two Variable Parameters.

187. When the equation of the given curve contains two variable parameters connected by an equation, only one of these parameters can be regarded as arbitrary, since, by means of the equation connecting them, one of the parameters can be eliminated. Instead, however, of eliminating one of the parameters at once, it is often better to proceed as in the following example.

Required, *the envelope of a straight line of fixed length* a, *which moves with its extremities on two rectangular axes.*

Denoting the intercepts on the axes by α and β, the equation of the line is

$$\frac{x}{\alpha} + \frac{y}{\beta} = 1, \quad \ldots \ldots \ldots \quad (1)$$

α and β being two variable parameters which, by the conditions of the problem, are connected by the relation

$$\alpha^2 + \beta^2 = a^2. \quad \ldots \ldots \ldots \quad (2)$$

Differentiating (1) and (2) with respect to α and β as variables, we have

$$\frac{x\,d\alpha}{\alpha^2} + \frac{y\,d\beta}{\beta^2} = 0, \quad \ldots \ldots \ldots \quad (3)$$

and

$$\alpha\,d\alpha + \beta\,d\beta = 0. \quad \ldots \ldots \ldots \quad (4)$$

We have now four equations from which we are to eliminate α, β, and the ratio $\dfrac{d\alpha}{d\beta}$. Transposing and dividing (3) by (4), we obtain

$$\frac{x}{\alpha^3} = \frac{y}{\beta^3}.$$

Substituting in (1) the value of y derived from the last equation, we have

$$x(\alpha^2 + \beta^2) = \alpha^3;$$

whence by equation (2)

$$\alpha = x^{\frac{1}{3}} a^{\frac{2}{3}}.$$

In like manner we find

$$\beta = y^{\frac{1}{3}} a^{\frac{2}{3}}.$$

Hence, substituting in (2)

$$x^{\frac{2}{3}} + y^{\frac{2}{3}} = a^{\frac{2}{3}}.$$

The envelope is therefore a *four-cusped hypocycloid*.

Evolutes.

188. In Fig. 40 let C be the centre of curvature of the given curve: this point is so determined (see Art. 176) as to have no motion in a direction perpendicular to the normal PC, but since ρ is in general variable, it has a motion in the direction PC. Hence C describes a curve to which the normal PC is always tangent at the point C. Moreover, since P has no motion in the direction PC, if we regard P as a fixed point on this line, the rate of C along this moving line will be identical with its rate along the curve which it describes. Hence the motion of PC is the same as that of a tangent line rolling upon the curve described by C, while P, a fixed point of this tangent, describes the original curve.

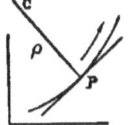

Fig. 40.

The curve described by the centre of curvature C is therefore called the *evolute* of the curve described by P, and the latter is called an *involute* of the former.

189. Since the evolute of a given curve is the curve to which all the normals to the given curve are tangent, it is evidently the envelope of these normals.

The equation of the normal at the point (x, y) of a given curve may be written in the form

$$x' - x + (y' - y)\frac{dy}{dx} = 0, \quad \ldots \quad (1)$$

(x', y') being any point of the normal. See Art. 163.

In this equation y and $\dfrac{dy}{dx}$ are functions of x determined by the equation of the given curve, and x is to be regarded as the arbitrary parameter. Hence, differentiating with reference to x, we have

$$-1 - \left(\frac{dy}{dx}\right)^2 + (y' - y)\frac{d^2y}{dx^2} = 0. \quad \ldots \quad (2)$$

The equation of the evolute is therefore the relation between x' and y' which arises from the elimination of x between equations (1) and (2).

190. As an illustration, let it be required to find the evolute of the common parabola

$$y = 2a^{\frac{1}{2}}x^{\frac{1}{2}};$$

whence $\quad \dfrac{dy}{dx} = \left(\dfrac{a}{x}\right)^{\frac{1}{2}}, \quad$ and $\quad \dfrac{d^2y}{dx^2} = -\dfrac{a^{\frac{1}{2}}}{2x^{\frac{3}{2}}}$.

Substituting, we obtain from equation (2) of the preceding article

$$x^{\frac{3}{2}} = -\tfrac{1}{2}a^{\frac{1}{2}}y';$$

whence, from equation (1) of the same article,

$$27ay'^2 = 4(x' - 2a)^3,$$

the equation of the evolute, which is, therefore, a *semi-cubical parabola* having its cusp at the point $(2a, 0)$.

191. It is frequently desirable to express the equation of the normal in terms of some parameter other than x before differentiating. Thus, let us determine *the evolute of the ellipse* by means of the equation of the normal in terms of the eccentric angle.

The equations of the ellipse are

$$x = a \cos \psi, \quad \text{and} \quad y = b \sin \psi;$$

whence $\quad dx = -a \sin \psi \, d\psi, \quad \text{and} \quad dy = b \cos \psi \, d\psi.$

Substitution in the equation of the normal,

$$(x' - x) \, dx + (y' - y) \, dy = 0,$$

gives $\quad ax' \sin \psi - by' \cos \psi - (a^2 - b^2) \sin \psi \cos \psi = 0.$

Differentiating, we have

$$ax' \cos \psi + by' \sin \psi - (a^2 - b^2)(\cos^2 \psi - \sin^2 \psi) = 0;$$

eliminating y' and x' successively, and dropping the accents,

$$ax = (a^2 - b^2) \cos^3 \psi \quad \text{and} \quad by = -(a^2 - b^2) \sin^3 \psi;$$

whence $\quad (ax)^{\frac{2}{3}} + (by)^{\frac{2}{3}} = (a^2 - b^2)^{\frac{2}{3}}.$

Examples XXVI.

1. Find the envelope of the system of parabolas represented by the equation

$$y^2 = \frac{\alpha^2}{c}(x - \alpha),$$

in which α is an arbitrary parameter and c a fixed constant.

$$y^2 = \frac{4}{27c} x^3.$$

2. Find the envelope of the circles described on the double ordinates of an ellipse as diameters.

$$\frac{x^2}{a^2+b^2}+\frac{y^2}{b^2}=1.$$

3. Find the envelope of the ellipses, the product of whose semi-axes is equal to the constant c^2.

The conjugate hyperbolas, $2xy = \pm c^2$.

4. Find the envelope of a perpendicular to the normal to the parabola, $y^2 = 4ax$, drawn through the intersection of the normal with the axis.

$$y^2 = 4a(2a - x).$$

5. Find the envelope of the ellipses whose axes are fixed in position, and whose semi-axes have a constant sum c.

The four-cusped hypocycloid, $x^{\frac{2}{3}} + y^{\frac{2}{3}} = c^{\frac{2}{3}}$.

6. Given the equation of the *catenary*

$$y = \frac{a}{2}\left(\varepsilon^{\frac{x}{a}} + \varepsilon^{-\frac{x}{a}}\right);$$

prove that

$$y' = 2y, \quad \text{and} \quad x' = x - \frac{y}{2}\left(\varepsilon^{\frac{x}{a}} - \varepsilon^{-\frac{x}{a}}\right),$$

and deduce the equation of the evolute.

$$x' = a \log \frac{y' \pm (y'^2 - 4a^2)^{\frac{1}{2}}}{2a} \mp \frac{y'}{4a}(y'^2 - 4a^2)^{\frac{1}{2}}.$$

7. Derive the equation of the evolute to the hyperbola, its equations in terms of an auxiliary angle being

$$x = a \sec \psi \quad \text{and} \quad y = b \tan \psi.$$

The equation of the normal is

$$ax \sin \psi + by = (a^2 + b^2) \tan \psi,$$

and the equation of the evolute is

$$a^{\frac{2}{3}} x^{\frac{2}{3}} - b^{\frac{2}{3}} y^{\frac{2}{3}} = (a^2 + b^2)^{\frac{2}{3}}.$$

8. Find the equation of the evolute of the cycloid.

The equation of the normal is

$$x + y \frac{\sin \psi}{1 - \cos \psi} - a\psi = 0.$$

The equations of the evolute are

$$x = a(\psi + \sin \psi) \quad \text{and} \quad y = -a(1 - \cos \psi).$$

The evolute is therefore a cycloid situated below the axis of x, having its vertex at the origin. *See equations* (3), *Art.* 158.

CHAPTER X.

FUNCTIONS OF TWO OR MORE VARIABLES

XXVII.

The Derivative Regarded as the Limit of a Ratio.

192. THE difference between two values of a variable is frequently expressed by prefixing the symbol Δ to the symbol denoting the variable, and the difference between corresponding values of any function of the variable, by prefixing Δ to the symbol denoting the function. Hence x and $x + \Delta x$ denote two values of the independent variable, and $\Delta f(x)$ denotes the difference between the corresponding values of $f(x)$; that is, if $y = f(x)$,

$$\Delta y = \Delta f(x) = f(x + \Delta x) - f(x). \quad . \quad . \quad . \quad (1)$$

If we put $\Delta x = 0$, we shall have $\Delta y = 0$;

hence the ratio $\quad \dfrac{\Delta y}{\Delta x} = \dfrac{f(x + \Delta x) - f(x)}{\Delta x} \quad . \quad . \quad . \quad . \quad (2)$

takes the indeterminate form $\dfrac{0}{0}$ when $\Delta x = 0$. The value assumed in this case is called *the limiting value of the ratio of the increments*, Δy and Δx, when the absolute values of these increments are diminished indefinitely.

193. To determine this limiting value, for a particular value a of x, we put a for x and z for Δx in the second member of equation (2), and evaluate for $z=0$, by the ordinary process (see Art. 82). Thus

$$\left.\frac{f(a+z)-f(a)}{z}\right]_0 = f'(a). \quad \ldots \quad (3)$$

Therefore when Δx is diminished indefinitely, the limiting value of $\frac{\Delta y}{\Delta x}$ corresponding to $x = a$ is $\left.\frac{dy}{dx}\right]_a$, and, since a denotes any value of x, we have in general

$$\text{limit of } \frac{\Delta y}{\Delta x} = \frac{dy}{dx}.$$

If we denote by e the difference between the values of $\frac{\Delta y}{\Delta x}$ and $\frac{dy}{dx}$, we shall have

$$\frac{\Delta y}{\Delta x} = \frac{dy}{dx} + e, \quad \ldots \quad (4)$$

and the result established in the preceding article may be expressed thus—

$$e = 0 \quad \text{when} \quad \Delta x = 0;$$

in other words, *e is a quantity that vanishes with Δx.*

Partial Derivatives.

194. Let $\quad u = f(x, y),$

in which x and y are two independent variables. The derivative of u with reference to x, y being regarded as constant, is denoted by $\frac{d}{dx}u$, and the derivative of u with reference to y, x being constant, by $\frac{d}{dy}u$. These derivatives are called *the partial* derivatives of u with reference to x and y respectively.

Adopting this notation, the result established in Art. 64 may be expressed thus:

$$du = \frac{d}{dx}u \cdot dx + \frac{d}{dy}u \cdot dy;$$

provided u *denotes a function that can be expressed by means of the elementary functions differentiated in Chapters* II *and* III.

It is now to be proved that this result is universally true.

195. Let $\Delta_x u$ denote the increment of u corresponding to Δx, y being unchanged, $\Delta_y u$ the increment corresponding to Δy, x being unchanged, and Δu the increment which u receives when x and y receive the simultaneous increments Δx and Δy. Let

$$u' = f(x + \Delta x, y),$$

and
$$u'' = f(x + \Delta x, y + \Delta y);$$

then
$$\Delta_x u = u' - u,$$

$$\Delta_y u' = u'' - u',$$

and
$$\Delta u = u'' - u;$$

hence
$$\Delta u = \Delta_x u + \Delta_y u'. \quad \dots \quad \dots \quad (1)$$

Denoting by Δt the interval of time in which x, y, and u receive the increments Δx, Δy, and Δu, we have

$$\frac{\Delta u}{\Delta t} = \frac{\Delta_x u}{\Delta t} + \frac{\Delta_y u'}{\Delta t}. \quad \dots \quad \dots \quad (2)$$

Since Δu is the actual increment of u in the interval Δt, the limit of the first member of equation (2) is, by Art. 193, $\dfrac{du}{dt}$, the rate of u. The limit of $\dfrac{\Delta_x u}{\Delta t}$ is the rate which u would have

§ XXVII.]   PARTIAL DERIVATIVES.   193

were x the only variable; and, since $\dfrac{d}{dx}u \cdot dx$ is the value which du assumes when this supposition is made, if we put

$$\frac{d}{dx}u \cdot dx = d_x u,$$

this rate will be denoted by $\dfrac{d_x u}{dt}$. Hence by equation (4), Art. 193, equation (2) becomes

$$\frac{du}{dt} + e = \frac{d_x u}{dt} + e' + \frac{d_y u'}{dt} + e'',$$

in which e, e', and e'' vanish with Δt; but when $\Delta t = 0$, $\Delta x = 0$, and therefore $u' = u$; hence, putting $\Delta t = 0$, we have

$$\frac{du}{dt} = \frac{d_x u}{dt} + \frac{d_y u}{dt}.$$

Therefore $\qquad du = d_x u + d_y u;$

that is, $\qquad du = \dfrac{d}{dx}u \cdot dx + \dfrac{d}{dy}u \cdot dy.$

196. This result is usually written in the form

$$du = \frac{du}{dx}dx + \frac{du}{dy}dy,$$

but when written in this form it must be remembered that the fractions in the second member represent *partial derivatives*, the symbol du in the numerators standing for the quantities denoted above by $d_x u$ and $d_y u$, which are sometimes called *partial differentials*. The du that appears in the first member is called the *total differential* of u when x and y are both variable.

The above result is easily extended to functions of more than two independent variables.

Examples XXVII.

1. Given $u = (x^2 + y^2)^{\frac{1}{3}}$, prove that

$$x\frac{du}{dx} + y\frac{du}{dy} = \tfrac{2}{3}u.$$

2. Given $u = \dfrac{xy}{x+y}$, prove that

$$x\frac{du}{dx} + y\frac{du}{dy} = u.$$

3. Given $u = \tan^{-1}\left(\dfrac{x-y}{x+y}\right)^{\frac{3}{2}}$, prove that

$$x\frac{du}{dx} + y\frac{du}{dy} = 0.$$

4. Given $u = \log_y x$, to find $\dfrac{du}{dx}$ and $\dfrac{du}{dy}$.

$$\frac{du}{dx} = \frac{1}{x \log y}\ ;\quad \frac{du}{dy} = \frac{-\log x}{y(\log y)^2}.$$

XXVIII.

The Second and Higher Derivatives regarded as Limits.

197. In Art. 193 it is shown that

$$\frac{\varDelta y}{\varDelta x} = \frac{dy}{dx} + e.$$

In this equation e is a function of x and likewise of $\varDelta x$; hence the derivative $\dfrac{de}{dx}$ is in general a function of x and of $\varDelta x$. It is

§ XXVIII.] *THE SECOND DERIVATIVE AS A LIMIT.* 195

also proved in the same article that e becomes zero when $\varDelta x$ vanishes; that is, *e assumes a constant value independent of the value of* x *when* $\varDelta$x *becomes zero*; hence, when $\varDelta x$ is zero, *the derivative* of e with reference to x must take the value zero, whatever be the value of x; in other words,

$$\frac{de}{dx} \text{ vanishes with } \varDelta x.$$

In a similar manner it may be shown that each of the higher derivatives of e with reference to x vanishes when $\varDelta x = 0$.

198. Since $\frac{\varDelta y}{\varDelta x}$ is a function of x, $\varDelta \frac{\varDelta y}{\varDelta x}$ will denote the increment of this function corresponding to $\varDelta x$. Employing the symbol $\frac{\varDelta}{\varDelta x}$ to denote the operation of taking this increment, and dividing the result by $\varDelta x$, we obtain, by applying to this function the principle expressed in equation (4), Art. 193,

$$\frac{\varDelta}{\varDelta x} \cdot \frac{\varDelta y}{\varDelta x} = \frac{d}{dx} \cdot \frac{\varDelta y}{\varDelta x} + e', \quad \ldots \ldots \quad (1)$$

$$= \frac{d}{dx}\left(\frac{dy}{dx} + e\right) + e'$$

$$= \frac{d^2y}{dx^2} + \frac{de}{dx} + e'.$$

In this equation both e' and $\frac{de}{dx}$ vanish with $\varDelta x$ by the preceding article; hence the sum of these quantities likewise vanishes with $\varDelta x$, and may be denoted by e. Thus we write

$$\frac{\varDelta}{\varDelta x} \cdot \frac{\varDelta y}{\varDelta x} = \frac{d^2y}{dx^2} + e. \quad \ldots \ldots \quad (2)$$

199. Since Δx is an arbitrary quantity it may be regarded as constant, whence $\Delta \dfrac{\Delta y}{\Delta x}$ is the increment of a fraction whose denominator is constant; but this is evidently equivalent to the result obtained by dividing the increment of the numerator by the denominator; that is,

$$\Delta \frac{\Delta y}{\Delta x} = \frac{\Delta \cdot \Delta y}{\Delta x}.$$

The numerator $\Delta \cdot \Delta y$ is usually denoted by the symbol $\Delta^2 y$; hence equation (2) may be written thus:

$$\frac{\Delta^2 y}{\Delta x^2} = \frac{d^2 y}{dx^2} + e, \quad \ldots \ldots \quad (3)$$

and, since e vanishes with Δx, it follows that the second derivative is the limit of the expression in the first member of equation (3).

In a similar manner it may be shown that each of the higher derivatives is the limit of the expression obtained by substituting Δ for d in the symbol denoting the derivative.

Higher Partial Derivatives.

200. The partial derivatives of u with reference to x and y are themselves functions of x and y. Their partial derivatives, viz.,

$$\frac{d}{dx}\cdot\frac{du}{dx}, \quad \frac{d}{dy}\cdot\frac{du}{dx}, \quad \frac{d}{dx}\cdot\frac{du}{dy}, \quad \text{and} \quad \frac{d}{dy}\cdot\frac{du}{dy},$$

are called partial derivatives of u of the *second order*.

It will now be shown that the second and third of these derivatives, although results of different operations, are in fact identical; that is, that

$$\frac{d}{dy}\cdot\frac{du}{dx} = \frac{d}{dx}\cdot\frac{du}{dy}.$$

§ XXVIII.] HIGHER PARTIAL DERIVATIVES.

Employing the notation introduced in Art. 195, we have

$$\Delta_x u = f(x + \Delta x, y) - f(x, y);$$

if in this equation we replace y by $y + \Delta y$, we obtain a new value of $\Delta_x u$, and, denoting this value by $\Delta'_x u$, we have

$$\Delta'_x u = f(x + \Delta x, y + \Delta y) - f(x, y + \Delta y).$$

Since this change in the value of $\Delta_x u$ results from the increment received by y, the expression for the increment received by $\Delta_x u$ will be $\Delta_y(\Delta_x u)$; hence

$$\Delta_y(\Delta_x u) = \Delta'_x u - \Delta_x u,$$

or

$$\Delta_y(\Delta_x u) = f(x+\Delta x, y+\Delta y) - f(x, y+\Delta y) - f(x+\Delta x, y) + f(x, y).$$

The value of $\Delta_x(\Delta_y u)$, obtained in a precisely similar manner, is identical with that just given; hence

$$\Delta_y(\Delta_x u) = \Delta_x(\Delta_y u). \quad \ldots \ldots \quad (1)$$

Since Δx is constant, we have, as in Art. 199,

$$\frac{\Delta_y(\Delta_x u)}{\Delta x} = \Delta_y \cdot \frac{\Delta_x u}{\Delta x}.$$

Hence, dividing both members of equation (1) by $\Delta x \cdot \Delta y$, we have

$$\frac{\Delta_y}{\Delta y} \cdot \frac{\Delta_x u}{\Delta x} = \frac{\Delta_x}{\Delta x} \cdot \frac{\Delta_y u}{\Delta y}, \quad \ldots \ldots \quad (2)$$

or, employing the symbol $\dfrac{\Delta}{\Delta x}$ as in Art. 198,

$$\frac{\Delta}{\Delta y} \cdot \frac{\Delta}{\Delta x} u = \frac{\Delta}{\Delta x} \cdot \frac{\Delta}{\Delta y} u.$$

From this result, by a course of reasoning similar to that employed in Art. 198, we obtain

$$\frac{d}{dy} \cdot \frac{du}{dx} = \frac{d}{dx} \cdot \frac{du}{dy}. \qquad (3)$$

201. The partial derivatives of the second order are usually denoted by

$$\frac{d^2u}{dx^2}, \qquad \frac{d^2u}{dx\,dy}, \qquad \frac{d^2u}{dy^2},$$

the factors dx and dy in the denominator of the second being, by virtue of formula (3), interchangeable, as in the case of an ordinary product.

The numerators of the above fractions are of course not identical. Compare Art. 196.

Formula (3) of the preceding article is readily verified in any particular case. Thus, given

$$u = y^x,$$

whence $\quad \dfrac{du}{dx} = y^x \log y, \quad$ and $\quad \dfrac{du}{dy} = xy^{x-1};$

$\therefore \quad \dfrac{d}{dy} \cdot \dfrac{du}{dx} = y^{x-1}(x \log y + 1) = \dfrac{d}{dx} \cdot \dfrac{du}{dy}.$

Examples XXVIII.

1. Given $u = \sec(y + ax) + \tan(y - ax)$, prove that

$$\frac{d^2u}{dx^2} = a^2 \frac{d^2u}{dy^2}.$$

2. Verify the theorem $\dfrac{d^2u}{dx\,dy} = \dfrac{d^2u}{dy\,dx}$ when $u = \sin(xy^2)$.

3. Verify the theorem $\dfrac{d^2u}{dx\,dy} = \dfrac{d^2u}{dy\,dx}$ when $u = \log \tan(ax + y^2)$.

4. Verify the theorem $\dfrac{d^3u}{dy^2\,dx} = \dfrac{d^3u}{dx\,dy^2}$ when $u = \tan^{-1}\dfrac{x}{y}$.

5. Verify the theorem $\dfrac{d^3u}{dy\,dx^2} = \dfrac{d^3u}{dx^2\,dy}$ when $u = y\log(1+xy)$.

6. Given $u = \sin x \cos y$, prove that

$$\frac{d^4u}{dy^2\,dx^2} = \frac{d^4u}{dx^2\,dy^2} = \frac{d^4u}{dx\,dy\,dx\,dy}.$$

7. Given $u = x^3z^4 + \varepsilon^xy^2z^3 + x^2y^3z^2$, derive

$$\frac{d^4u}{dx^2\,dy\,dz} = 6y\varepsilon^xz^2 + 8yz.$$

8. Given $u = \dfrac{1}{\sqrt{(4ab-c^2)}}$, prove that

$$\frac{d^2}{dc^2}u = \frac{d^2}{da\,db}u.$$

9. Given $u = (x+y)^2$, prove that

$$x\frac{d^2u}{dx^2} + y\frac{d^2u}{dx\,dy} = \frac{du}{dx}.$$

10. Given $u = \dfrac{1}{(x^2+y^2+z^2)^{\frac{1}{2}}}$, prove that

$$\frac{d^2u}{dx^2} + \frac{d^2u}{dy^2} + \frac{d^2u}{dz^2} = 0.$$

AN

ELEMENTARY TREATISE

ON THE

INTEGRAL CALCULUS

FOUNDED ON THE

METHOD OF RATES OR FLUXIONS

BY

WILLIAM WOOLSEY JOHNSON

PROFESSOR OF MATHEMATICS AT THE UNITED STATES NAVAL ACADEMY
ANNAPOLIS MARYLAND

FOURTH THOUSAND.

NEW YORK:
JOHN WILEY AND SONS,
53 East Tenth Street,
1893.

COPYRIGHT,
1881,
BY JOHN WILEY AND SONS.

PRESS OF J. J. LITTLE & CO.,
NOS. 10 TO 20 ASTOR PLACE, NEW YORK.

PREFACE.

THIS work, as at present issued, is designed as a shorter course in the Integral Calculus, to accompany the abridged edition of the treatise on the Differential Calculus, by Professor J. Minot Rice and the writer. It is intended hereafter to publish a volume commensurate with the full edition of the work above mentioned, of which the present shall form a part, but which shall contain a fuller treatment of many of the subjects here treated, including Definite Integrals, and the Mechanical Applications of the Calculus, as well as Elliptic Integrals, Differential Equations, and the subjects of Probabilities and Averages. The conception of Rates has been employed as the foundation of the definitions, and of the whole subject of the integration of known functions. The connection between integration, as thus defined, and the process of summation, is established in Section VII. Both of these views of an integral—namely, as a quantity generated at a given rate, and as the limit of a sum—have been freely used in expressing geometrical and physical quantities in the integral form.

The treatises of Bertrand, Frenet, Gregory, Todhunter, and Williamson, have been freely consulted. My thanks are due to Professor Rice for very many valuable suggestions in the course of the work, and for performing much the larger share of the work of revising the proof-sheets.

W. W. J.

U. S. NAVAL ACADEMY, *July*. 1881.

CONTENTS.

CHAPTER I.

ELEMENTARY METHODS OF INTEGRATION.

I.

	PAGE
Integrals	1
The differential of a curvilinear area	3
Definite and indefinite integrals	4
Elementary theorems	6
Fundamental integrals	7
Examples I	10

II.

Direct integration	14
Rational fractions	15
Denominators of the second degree	16
Denominators of degrees higher than the second	19
Denominators containing equal roots	22
Examples II	26

III.

Trigonometric integrals 33

Cases in which $\int \sin^m \theta \cos^n \theta \, d\theta$ is directly integrable 34

The integrals $\int \sin^2 \theta \, d\theta$, and $\int \cos^2 \theta \, d\theta$ 36

The integrals $\int \dfrac{d\theta}{\sin \theta \cos \theta}$, $\int \dfrac{d\theta}{\sin \theta}$, and $\int \dfrac{d\theta}{\cos \theta}$ 37

	PAGE
Miscellaneous trigonometric integrals	38
The integration of $\dfrac{d\theta}{a + b \cos\theta}$	40
Examples III	43

CHAPTER II.

METHODS OF INTEGRATION—CONTINUED.

IV.

Integration by change of independent variable	50
Transformation of trigonometric forms	51
Limits of a transformed integral	53
The reciprocal of x employed as the new independent variable	53
A power of x employed as the new independent variable	54
Examples IV	56

V.

Integrals containing radicals	59
Radicals of the form $\sqrt{(ax^2 + b)}$	61
The integration of $\dfrac{dx}{\sqrt{(x^2 \pm a^2)}}$	64
Transformation to trigonometric forms	65
Radicals of the form $\sqrt{(ax^2 + bx + c)}$	67
The integrals $\displaystyle\int \dfrac{dx}{\sqrt{[(x-\alpha)(x-\beta)]}}$ and $\displaystyle\int \dfrac{dx}{\sqrt{[(x-\alpha)(\beta-x)]}}$	68
Examples V	70

VI.

Integration by parts	77
A geometrical illustration	78
Applications	78
Formulas of reduction	81
Reduction of $\displaystyle\int \sin^m\theta\, d\theta$ and $\displaystyle\int \cos^m\theta\, d\theta$	82
Reduction of $\displaystyle\int \sin^m\theta \cos^n\theta\, d\theta$	84

	PAGE
Illustrative examples	87
Extension of the formula employed in integration by parts	89
Taylor's theorem	90
Examples VI	91

VII.

Definite integrals	97
Multiple-valued integrals	100
Formulas of reduction for definite integrals	101
Elementary theorems relating to definite integrals	104
Change of independent variable in a definite integral	105
The differentiation of an integral	106
Integration under the integral sign	109
The definite integral regarded as the limiting value of a sum	111
Additional formulas of integration	115
Examples VII	117

CHAPTER III.

GEOMETRICAL APPLICATIONS.

VIII.

Areas generated by variable lines having fixed directions	123
Application to the witch	124
Application to the parabola when referred to oblique coordinates	126
The employment of an auxiliary variable	126
Areas generated by rotating variable lines	128
The area of the lemniscata	129
The area of the cissoid	130
A transformation of the polar formulas	130
Application to the folium	131
Examples VIII	134

IX.

The volumes of solids of revolution	141
The volume of an ellipsoid	143
Solids of revolution regarded as generated by cylindrical surfaces	144
Double integration	145
Determination of the volume of a solid by double integration	149

	PAGE
The determination of volumes by triple integration	150
Elements of area and volume	152
Polar elements	154
The determination of volumes by polar formulas	155
Polar coordinates in space	157
Application to the volume generated by the revolution of a cardioid	159
Examples IX	160

X.

Rectification of plane curves	168
Rectification of the semi-cubical parabola	168
Rectification of the four-cusped hypocycloid	169
Change of sign of ds	170
Polar coordinates	170
Rectification of curves of double curvature	171
Rectification of the loxodromic curve	172
Examples X	173

XI.

Surfaces of solids of revolution	178
Quadrature of surfaces in general	179
The expression in partial derivatives for sec γ	180
The determination of surfaces by polar formulas	181
Examples XI	183

XII.

Areas generated by straight lines moving in planes	186
Applications	187
Sign of the generated area	189
Areas generated by lines whose extremities describe closed circuits	190
Amsler's Planimeter	191
Examples XII	193

XIII.

Approximate expressions for areas and volumes	195
Simpson's rules	197
Cotes' method of approximation	198

CONTENTS. ix

PAGE

Weddle's rule 199
The five-eight rule... 199
The comparative accuracy of Simpson's first and second rules 200
The application of these rules to solids 200
Woolley's rule.. 201
    Examples XIII.. 202

CHAPTER IV.

MECHANICAL APPLICATIONS.

XIV.

Definitions... 204
Statical moment 204
Centres of gravity... 206
Polar formulas 208
Centre of gravity of the lemniscata..................................... 209
Solids of revolution 209
Centre of gravity of a spherical cap 210
The properties of Pappus... 210
    Examples XIV... 212

XV.

Moments of inertia... 219
Moment of inertia of a straight line 220
Radii of gyration 220
Radius of gyration of a sphere 221
Radii of gyration about parallel axes.................................. 222
Application to the cone.. 223
Polar moments of inertia 225
    Examples XV 225

THE
INTEGRAL CALCULUS.

CHAPTER I.

ELEMENTARY METHODS OF INTEGRATION.

I.

Integrals.

1. IN an important class of problems, the required quantities are magnitudes generated in given intervals of time with rates which are either given in terms of the time t, or are readily expressed in terms of the assumed rate of some other independent variable.

For example, the velocity of a freely falling body is known to be expressed by the equation

$$v = gt, \quad \quad \quad \quad \quad (1)$$

in which t is the number of seconds which have elapsed since the instant of rest, and g is a constant which has been determined experimentally. If s denotes the distance of the body at the time t, from a fixed origin taken on the line of motion, v is the rate of s; that is,

$$v = \frac{ds}{dt};$$

hence equation (1) is equivalent to

$$ds = gt\,dt, \ldots \qquad \ldots \quad (2)$$

which expresses the differential of s in terms of t and dt. Now it is obvious that $\tfrac{1}{2}gt^2$ is a function of t having a differential equal to the value of ds in equation (2); and, moreover, since two functions which have the same differential (and hence the same rate) can differ only by a constant, the most general expression for s is

$$s = \tfrac{1}{2}gt^2 + C, \ldots \ldots \quad (3)$$

in which C denotes an undetermined constant.

2. A variable thus determined from its rate or differential is called an *integral*, and is denoted by prefixing to the given differential expression the symbol $\int$, which is called the integral sign.* Thus, from equation (2) we have

$$s = \int gt\,dt,$$

which therefore expresses that s is a variable whose differential is $gt\,dt$; and we have shown that

$$\int gt\,dt = \tfrac{1}{2}gt^2 + C.$$

The constant C is called the *constant of integration;* its occurrence in equation (3) is explained by the fact that we have not determined the origin from which s is to be measured.

---

* The origin of this symbol, which is a modification of the long s, will be explained hereafter. See Art. 100.

§ I.] THE DIFFERENTIAL OF A CURVILINEAR AREA. 3

If we take this origin at the point occupied by the body when at rest, we shall have $s = 0$ when $t = 0$, and therefore from equation (3) $C = 0$; whence the equation becomes $s = \frac{1}{2}gt^2$.

The Differential of a Curvilinear Area.

3. The area included between a curve, whose equation is given, the axis of x and two ordinates affords an instance of the second case mentioned in the first paragraph of Art. 1; namely, that in which the rate of the generated quantity, although not given in terms of t, can be readily expressed by means of the assumed rate of some other independent variable.

Let BPD in Fig. 1 be the curve whose equation is supposed to be given in the form

$$y = f(x).$$

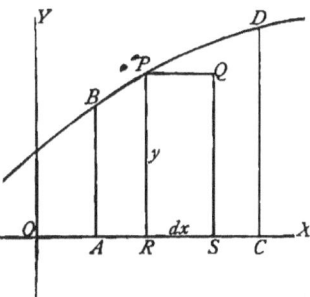

Fig. 1.

Supposing the variable ordinate PR to move from the position AB to the position CD, the required area $ABDC$ is the final value of the variable area $ABPR$, denoted by A, which is generated by the motion of the ordinate. The rate at which the area A is generated can be expressed in terms of the rate of the independent variable x. The required and the assumed rates are denoted, respectively, by $\dfrac{dA}{dt}$ and $\dfrac{dx}{dt}$; and, to express the former in terms of the latter, it is necessary to express dA in terms of dx. Since x is an independent variable, we may assume dx to be constant; the rate at which A is generated is then a variable rate, because PR or y is of variable length, while moving at a constant rate along the axis of x. Now dA is the increment which A would receive in the time

dt, were the rate of A to become constant (see Diff. Calc., Art. 17). If, now, at the instant when the ordinate passes the position PR in the figure, its length should become constant, the rate of the area would become constant, and the increment which would then be received in the time dt, namely, the rectangle $PQSR$, represents dA. Since the base RS of this rectangle is dx, we have

$$dA = y dx = f(x)dx. \qquad \cdots \qquad (1)$$

Hence, by the definition given in Art. 2, A is an integral, and is denoted by

$$A = \int f(x)dx. \qquad \cdots \cdots \cdots (2)$$

Definite Integrals.

4. Equation (2) expresses that A is a function of x, whose differential is $f(x)dx$; this function, like that considered in Art. 2, involves an undetermined constant. In fact, the expression $\int f(x)dx$ is manifestly insufficient to represent precisely the area $ABPR$, because OA, *the initial value of* x, is not indicated. The indefinite character of this expression is removed by writing this value as a subscript to the integral sign; thus, denoting the initial value by a, we write

$$A = \int_a f(x)dx, \qquad \cdots \cdots \cdots (3)$$

in which the subscript is *that value of* x *for which the integral has the value zero.*

If we denote the *final value* of x (OC in the figure) by b, the area $ABDC$, which is a particular value of A, is denoted by writing this value of x at the top of the integral sign, thus,

$$ABDC = \int_a^b f(x)dx. \quad \ldots \ldots \quad (4)$$

This last expression is called a *definite integral*, and a and b are called its *limits*. In contradistinction, the expression $\int f(x)dx$ is called an *indefinite integral*.

5. As an application of the general expressions given in the last two articles, let the given curve be the parabola

$$y = x^2.$$

Equation (2) becomes in this case

$$A = \int x^2 dx.$$

Now, since $\tfrac{1}{3}x^3$ is a function whose differential is $x^2 dx$, this equation gives

$$A = \int x^2 dx = \tfrac{1}{3}x^3 + C, \quad \ldots \ldots \quad (1)$$

in which C is undetermined.

Now let us suppose the limiting ordinates of the required area to be those corresponding to $x = 1$ and $x = 3$. The variable area of which we require a special value is now represented by $\int_1 x^2 dx$, which denotes that value of the indefinite integral which vanishes when $x = 1$. If we put $x = 1$ in the general expression in equation (1), namely $\tfrac{1}{3}x^3 + C$, we have $\tfrac{1}{3} + C$; hence if we subtract this quantity from the general expression, we shall have an expression which becomes zero when $x = 1$. We thus obtain

$$A = \int_1 x^2 dx = \tfrac{1}{3}x^3 - \tfrac{1}{3}.$$

Finally, putting, in this expression for the variable area, $x = 3$, we have for the required area

$$\int_1^3 x^2 dx = \tfrac{1}{3}3^3 - \tfrac{1}{3} = 8\tfrac{2}{3}.$$

6. It is evident that the definite integral obtained by this process is simply *the difference between the values of the indefinite integral at the upper and lower limits.* This difference may be expressed by attaching the limits to the symbol] affixed to the value of the indefinite integral. Thus the process given in the preceding article is written thus,

$$\int_1^3 x^2 dx = \tfrac{1}{3}x^3 + C \Big]_1^3 = 9 - \tfrac{1}{3} = 8\tfrac{2}{3}.$$

The essential part of this process is the determination of the indefinite integral or function whose differential is equal to the given expression. This is called the *integration* of the given differential expression.

Elementary Theorems.

7. *A constant factor may be transferred from one side of the integral sign to the other.* In other words, if m is a constant and u a function of x,

$$\int mu\,dx = m\int u\,dx.$$

Since each member of this equation involves an arbitrary constant, the equation only implies that the two members have the same differential. The differential of an integral is by definition the quantity under the integral sign. Now the second member is the product of a constant by a variable factor; hence its differential is $m\,d\Big[\int u\,dx\Big]$, that is, $m\,u\,dx$, which is also the differential of the first member.

8. This theorem is useful not only in removing constant factors from under the integral sign, but also in introducing such factors when desired. Thus, given the integral

$$\int x^n \, dx\,;$$

recollecting that

$$d(x^{n+1}) = (n + 1)x^n \, dx,$$

we introduce the constant factor $n + 1$ under the integral sign; thus,

$$\int x^n \, dx = \frac{1}{n+1} \int (n+1) x^n \, dx = \frac{1}{n+1} x^{n+1} + C.$$

9. *If a differential expression be separated into parts, its integral is the sum of the integrals of the several parts.* That is, if $u, v, w, \cdots$ are functions of x,

$$\int (u + v + w + \cdots) dx = \int u \, dx + \int v \, dx + \int w \, dx + \cdots$$

For, since the differential of a sum is the sum of the differentials of the several parts, the differential of the second member is identical with that of the first member, and each member involves an arbitrary constant

Thus, for example,

$$\int (2 - \sqrt{x}) \, dx = \int 2 \, dx - \int x^{\frac{1}{2}} \, dx = 2x - \tfrac{2}{3} x^{\frac{3}{2}} + C,$$

the last term being integrated by means of the formula deduced in Art. 8.

Fundamental Integrals.

10. The integrals whose values are given below are called the *fundamental integrals*. The constants of integration are generally omitted for convenience.

Formula (*a*) is given in two forms, the first of which is derived in Art. 8, while the second is simply the result of putting $n = -m$. It is to be noticed that this formula gives an indeterminate result when $n = -1$; but in this case, formula (*b*) may be employed.*

The remaining formulas are derived directly from the formulas for differentiation; except that (j'), (k'), (l'), and (m') are derived from (j), (k), (l), and (m) by substituting $\dfrac{x}{a}$ for x.

$$\int x^n dx = \frac{x^{n+1}}{n+1} + C \qquad \int \frac{dx}{x^m} = -\frac{1}{(m-1)x^{m-1}} + C \quad (a)$$

$$\int \frac{dx}{x} = \log(\pm x) \dagger + C \quad \ldots \ldots \ldots \ldots (b)$$

$$\int a^x dx = \frac{a^x}{\log a} + C \qquad \int \varepsilon^x dx = \varepsilon^x + C \quad \ldots \ldots (c)$$

$$\int \cos\theta \, d\theta = \sin\theta + C \quad \ldots \ldots \ldots \ldots (d)$$

$$\int \sin\theta \, d\theta = -\cos\theta + C \quad \ldots \ldots \ldots \ldots (e)$$

---

* Applying formula (*a*) to the definite integral $\int_a^b x^n dx$, we have

$$\int_a^b x^n dx = \frac{b^{n+1} - a^{n+1}}{n+1},$$

which takes the form $\dfrac{0}{0}$ when $n = -1$; but, evaluating in the usual manner,

$$\left.\frac{b^{n+1} - a^{n+1}}{n+1}\right]_{n=-1} = \left.\frac{b^{n+1}\log b - a^{n+1}\log a}{1}\right]_{n=-1} = \log b - \log a\,;$$

a result identical with that obtained by employing formula (*b*).

† That sign is to be employed which makes the logarithm real. See Diff. Calc., Art. 43.

$$\int \frac{d\theta}{\cos^2\theta} = \int \sec^2\theta\, d\theta = \tan\theta + C \quad \ldots \ldots \quad (f)$$

$$\int \frac{d\theta}{\sin^2\theta} = \int \csc^2\theta\, d\theta = -\cot\theta + C \quad \ldots \ldots \quad (g)$$

$$\int \frac{\sin\theta\, d\theta}{\cos^2\theta} = \int \sec\theta \tan\theta\, d\theta = \sec\theta + C \quad \ldots \quad (h)$$

$$\int \frac{\cos\theta\, d\theta}{\sin^2\theta} = \int \csc\theta \cot\theta\, d\theta = -\csc\theta + C \quad \ldots \quad (i)$$

$$\int \frac{dx}{\sqrt{(1-x^2)}} = \sin^{-1} x + C = -\cos^{-1} x + C' \quad \ldots \quad (j)$$

$$\int \frac{dx}{\sqrt{(a^2-x^2)}} = \sin^{-1}\frac{x}{a} + C = -\cos^{-1}\frac{x}{a} + C' \ldots \quad (j')$$

$$\int \frac{dx}{1+x^2} = \tan^{-1} x + C = -\cot^{-1} x + C' \ldots \ldots \quad (k)$$

$$\int \frac{dx}{a^2+x^2} = \frac{1}{a}\tan^{-1}\frac{x}{a} + C = -\frac{1}{a}\cot^{-1}\frac{x}{a} + C' \ldots \quad (k')$$

$$\int \frac{dx}{x\sqrt{(x^2-1)}} = \sec^{-1} x + C = -\csc^{-1} x + C' \ldots \quad (l)$$

$$\int \frac{dx}{x\sqrt{(x^2-a^2)}} = \frac{1}{a}\sec^{-1}\frac{x}{a} + C = -\frac{1}{a}\csc^{-1}\frac{x}{a} + C' \ldots \quad (l')$$

$$\int \frac{dx}{\sqrt{(2x-x^2)}} = \mathrm{vers}^{-1} x + C \quad \ldots \ldots \quad (m)$$

$$\int \frac{dx}{\sqrt{(2ax-x^2)}} = \mathrm{vers}^{-1}\frac{x}{a} + C \quad \ldots \ldots \quad (m')$$

Examples I.

Find the values of the following integrals:

1. $\int \dfrac{dx}{\sqrt{x}}$, $\qquad\qquad 2\sqrt{x}$.

2. $\int \dfrac{dx}{x^2}$, $\qquad\qquad -\dfrac{1}{x}$.

3. $\int_1 \dfrac{dx}{x^{\frac{3}{2}}}$, $\qquad\qquad 2 - \dfrac{2}{\sqrt{x}}$.

4. $\int_1^\infty \dfrac{dx}{x^5}$, $\qquad\qquad \tfrac{1}{4}$.

5. $\int_0 \sqrt{x}\, dx$, $\qquad\qquad \tfrac{2}{3} x^{\frac{3}{2}}$.

6. $\int_1 (x-1)^2 dx$, $\qquad\qquad \dfrac{x^3}{3} - x^2 + x - \tfrac{1}{3}$.

7. $\int_0^{\frac{a}{b}} (a-bx)^2 dx$, $\qquad a^2 x - abx + \dfrac{b^2 x^3}{3}\bigg]_0^{\frac{a}{b}} = \dfrac{a^3}{3b}$.

8. $\int_{-a}^a (a+x)^3 dx$, $\qquad a^3 x + \dfrac{3a^2 x^2}{2} + ax^3 + \dfrac{x^4}{4}\bigg]_{-a}^a = 4a^4$.

9. $\int_1^{a^2} \dfrac{dx}{x}$, $\qquad\qquad 2 \log a$.

10. $\int_{-1}^{-2} \dfrac{dx}{x}$, $\qquad \log(-x)\bigg]_{-1}^{-2} = \log 2$.

EXAMPLES.

11. $\int_{a}^{4a} \frac{(a+x)^2}{\sqrt{x}} dx,$ $\quad 2\sqrt{x}(a^2 + \frac{2}{3}ax + \frac{1}{5}x^2)\Big]_{a}^{4a} = 23\frac{11}{15} \cdot a^{\frac{5}{2}}.$

12. $\int_{0}^{y} \varepsilon^x dx,$ $\qquad \varepsilon^y - 1.$

13. $\int_{0}^{\theta} \sin\theta\, d\theta,$ $\qquad 1 - \cos\theta.$

14. $\int_{0}^{\pi} \cos x\, dx,$ $\qquad \sin x\Big]_{0}^{\pi} = 0.$

15. $\int_{0}^{\frac{1}{4}\pi} \frac{d\theta}{\cos^2\theta},$ $\qquad \tan\theta\Big]_{0}^{\frac{1}{4}\pi} = 1.$

16. $\int_{0}^{\frac{1}{2}} \frac{dx}{\sqrt{(a^2 - x^2)}},$ $\qquad \sin^{-1}\frac{x}{a}\Big]_{0}^{\frac{1}{2}} = \frac{\pi}{6}.$

17. $\int_{-\infty}^{\infty} \frac{dx}{a^2 + x^2},$ $\qquad \frac{\pi}{a}.$

18. $\int_{1}^{\infty} \frac{dx}{x\sqrt{(x^2 - 1)}},$ $\qquad \frac{\pi}{2}.$

19. If a body is projected vertically upward, its velocity after t units of time is expressed by

$$v = a - gt,$$

a denoting the initial velocity; find the space s_1 described in the time t_1 and the greatest height to which the body will rise.

$$s_1 = \int_{0}^{t_1} v\, dt = at_1 - \tfrac{1}{2}gt_1^2,$$

$$\text{when } v = 0, t = \frac{a}{g}, s = \frac{a^2}{2g}.$$

20. If the velocity of a pendulum is expressed by

$$v = a \cos \frac{\pi t}{2\tau},$$

the position corresponding to $t = 0$ being taken as origin, find an expression for its position s at the time t, and the extreme positive and negative values of s.

$$s = \frac{2\tau a}{\pi} \sin \frac{\pi t}{2\tau}.$$

$$s = \pm \frac{2\tau a}{\pi} \text{ when } t = \tau, 3\tau, 5\tau, \text{ etc.}$$

21. Find the area included between the axis of x and a branch of the curve

$$y = \sin x.$$
      2.

22. Show that the area between the axis of x, the parabola

$$y^2 = 4ax,$$

and any ordinate is two thirds of the rectangle whose sides are the ordinate and the corresponding abscissa.

23. Find (α) the area included by the axes, the curve

$$y = \varepsilon^x,$$

and the ordinate corresponding to $x = 1$, and (β) the whole area between the curve and axes on the left of the axis of y.

$$(\alpha)\ \varepsilon - 1,\ (\beta)\ 1.$$

24. Find the area between the parabola of the nth degree,

$$a^{n-1} y = x^n,$$

and the coordinates of the point (a, a).
$$\frac{a^2}{n+1}.$$

§ I.]   EXAMPLES.   13

25. Show that the area between the axis of x, the rectangular hyperbola

$$xy = 1,$$

the ordinate corresponding to $x = 1$, and any other ordinate is equivalent to the Napierian logarithm of the abscissa of the latter ordinate.

For this reason Napierian logarithms are often called *hyperbolic logarithms*.

26. Find the whole area between the axes, the curve

$$y^n x^m = a^{n+m},$$

and the ordinate for $x = a$, m and n being positive.

If $n > m$, $\qquad \dfrac{na^2}{n-m}$;

if $n \leq m$, $\qquad \infty$.

27. If the ordinate BR of any point B on the circle

$$x^2 + y^2 = a^2$$

be produced so that $BR \cdot RP = a^2$, prove that the whole area between the locus of P and its asymptotes is double the area of the circle.

28. Find the whole area between the axis of x and the curve

$$y(a^2 + x^2) = a^3.$$

$\qquad \pi a^2.$

29. Find the area between the axis of x and one branch of the *companion to the cycloid*, the equations of which are

$$x = a\psi \qquad y = a(1 - \cos\psi).$$

$\qquad 2\pi a^2.$

II.

Direct Integration.

11. In any one of the formulas of Art. 10, we may of course substitute for x and dx any function of x and its differential. For instance, if in formula (*b*) we put $x - a$ in place of x, we have

$$\int \frac{dx}{x - a} = \log(x - a) \quad \text{or} \quad \log(a - x),$$

according as x is greater or less than a.

When a given integral is obviously the result of such a substitution in one of the fundamental integrals, or can be made to take this form by the introduction of a constant factor, it is said to be *directly integrable*. Thus, $\int \sin mx \, dx$ is directly integrable by formula (*c*); for, if in this formula we put mx for θ, we have

$$\int \sin mx \cdot m \, dx = -\cos mx,$$

hence

$$\int \sin mx \, dx = \frac{1}{m} \int \sin mx \cdot m \, dx = -\frac{1}{m} \cos mx.$$

So also in $\int \sqrt{(a + bx^2)} \, x \, dx$,

the quantity $x \, dx$ becomes the differential of the binomial $(a + bx^2)$ when we introduce the constant factor $2b$, hence this integral can be converted into the result obtained by putting $(a + bx^2)$ in place of x in $\int \sqrt{x} \, dx$, which is a case of formula (*a*). Thus

$$\int \sqrt{(a + bx^2)} \, x \, dx = \frac{1}{2b} \int (a + bx^2)^{\frac{1}{2}} 2bx \, dx = \frac{1}{3b} (a + bx^2)^{\frac{3}{2}}.$$

§ II.] DIRECT INTEGRATION.

12. A simple algebraic or trigonometric transformation sometimes suffices to render an expression directly integrable, or to separate it into directly integrable parts. Thus, since $-\sin x\, dx$ is the differential of $\cos x$, we have by formula (b)

$$\int \tan x\, dx = \int \frac{\sin x\, dx}{\cos x} = -\log \cos x.$$

So also, by formula (f),

$$\int \tan^2 \theta\, d\theta = \int (\sec^2 \theta - 1)\, d\theta = \tan \theta - \theta;$$

by (e) and (a),

$$\int \sin^3 \theta\, d\theta = \int (1 - \cos^2 \theta) \sin \theta\, d\theta = -\cos \theta + \tfrac{1}{3} \cos^3 \theta;$$

by (j) and (a),

$$\int \sqrt{\left(\frac{1+x}{1-x}\right)} dx = \int \frac{1+x}{\sqrt{(1-x^2)}} dx$$

$$= \int \frac{dx}{\sqrt{(1-x^2)}} - \frac{1}{2}\int (1-x^2)^{-\frac{1}{2}}(-2x\, dx) = \sin^{-1} x - \sqrt{(1-x^2)}.$$

Rational Fractions.

13. When the coefficient of dx in an integral is a fraction whose terms are rational functions of x, the integral may generally be separated into parts directly integrable. If the denominator is of the first degree, we proceed as in the following example.

Given the integral $\int \frac{x^2 - x + 3}{2x + 1} dx;$

by division,

$$\frac{x^2 - x + 3}{2x + 1} = \frac{x}{2} - \frac{3}{4} + \frac{15}{4} \frac{1}{2x + 1},$$

hence

$$\int \frac{x^2 - x + 3}{2x + 1} dx = \frac{1}{2}\int x\, dx - \frac{3}{4}\int dx + \frac{15}{4}\int \frac{dx}{2x+1}$$

$$= \frac{x^2}{4} - \frac{3x}{4} + \frac{15}{8} \log(2x+1).$$

When the denominator is of higher degree, it is evident that we may, by division, make the integration depend upon that of a fraction in which the degree of the numerator is lower than that of the denominator by at least a unit. We shall consider therefore fractions of this form only.

Denominators of the Second Degree.

14. If the denominator is of the second degree, it will (after removing a constant, if necessary) either be the square of an expression of the first degree, or else such a square increased or diminished by a constant. As an example of the first case, let us take

$$\int \frac{x+1}{(x-1)^2} dx.$$

The fraction may be decomposed thus:

$$\frac{x+1}{(x-1)^2} = \frac{x-1+2}{(x-1)^2} = \frac{1}{x-1} + \frac{2}{(x-1)^2};$$

hence

$$\int \frac{x+1}{(x-1)^2} dx = \int \frac{dx}{x-1} + 2\int \frac{dx}{(x-1)^2}$$

$$= \log(x-1) - \frac{2}{x-1}.$$

15. The integral $\displaystyle\int \frac{x+3}{x^2 + 2x + 6} dx$

affords an example of the second case, for the denominator may be written in the form

$$x^2 + 2x + 6 = (x + 1)^2 + 5.$$

Decomposing the fraction as in the preceding article,

$$\frac{x+3}{(x+1)^2+5} = \frac{x+1}{(x+1)^2+5} + \frac{2}{(x+1)^2+5};$$

whence

$$\int \frac{x+3}{x^2+2x+6}dx = \int \frac{(x+1)\,dx}{(x+1)^2+5} + 2\int \frac{dx}{(x+1)^2+5}.$$

The first of the integrals in the second member is directly integrable by formula (*b*), since the differential of the denominator is $2(x+1)\,dx$, and the second is a case of formula (*k'*). Therefore

$$\int \frac{x+3}{x^2+2x+6}dx = \tfrac{1}{2}\log(x^2+2x+6) + \frac{2}{\sqrt{5}}\tan^{-1}\frac{x+1}{\sqrt{5}}.$$

16. To illustrate the third case, let us take

$$\int \frac{2x+1}{x^2-x-6}dx,$$

in which the denominator is equivalent to $(x - \tfrac{1}{2})^2 - 6\tfrac{1}{4}$, and can therefore be resolved into real factors of the first degree. We can then decompose the fraction into fractions having these factors for denominators. Thus, in the present example, assume

$$\frac{2x+1}{x^2-x-6} = \frac{A}{x-3} + \frac{B}{x+2}, \quad \cdots \quad (1)$$

in which A and B are numerical quantities to be determined. Multiplying by $(x-3)(x+2)$,

$$2x+1 = A(x+2) + B(x-3). \quad \cdots \quad (2)$$

Since equation (2) is an algebraic identity, we may in it assign any value we choose to x. Putting $x = 3$, we find

$$7 = 5A, \quad \text{whence} \quad A = \tfrac{7}{5},$$

putting $x = -2$,

$$-3 = -5B, \quad \text{whence} \quad B = \tfrac{3}{5}.$$

Substituting these values in (1),

$$\frac{2x+1}{x^2 - x - 6} = \frac{7}{5(x-3)} + \frac{3}{5(x+2)},$$

whence

$$\int \frac{2x+1}{x^2 - x - 6} dx = \tfrac{7}{5} \log (x-3) + \tfrac{3}{5} \log (x+2).$$

17. If the denominator, in a case of the kind last considered, is denoted by $(x-a)(x-b)$, a and b are evidently the roots of the equation formed by putting this denominator equal to zero. The cases considered in Art. 14 and Art. 15 are respectively those in which the roots of this equation are equal, and those in which the roots are imaginary. When the roots are real and unequal, if the numerator does not contain x, the integral can be reduced to the form

$$\int \frac{dx}{(x-a)(x-b)},$$

and by the method given in the preceding article we find

$$\int \frac{dx}{(x-a)(x-b)} = \frac{1}{a-b} \Big[\log(x-a) - \log(x-b) \Big]$$

$$= \frac{1}{a-b} \log \frac{x-a}{x-b}, \quad \cdots \cdots \quad (A)^*$$

---

* The formulas of this series are collected together at the end of Chapter II., for convenience of reference. See Art. 101.

in which, when $x < a$, $\log(a - x)$ should be written in place of $\log(x - a)$. [See note on formula (*b*), Art. 10.]

If $b = -a$, this formula becomes

$$\int \frac{dx}{x^2 - a^2} = \frac{1}{2a} \log \frac{x-a}{x+a} \quad \ldots \quad (A')$$

Integrals of the special forms given in (*A*) and (*A'*) may be evaluated by the direct application of these formulas. Thus, given the integral

$$\int \frac{dx}{2x^2 + 3x - 2};$$

if we place the denominator equal to zero, we have the roots $a = \tfrac{1}{2}$, $b = -2$; whence by formula (*A*),

$$\int \frac{dx}{2x^2 + 3x - 2} = \tfrac{1}{2}\int \frac{dx}{(x - \tfrac{1}{2})(x + 2)} = \frac{1}{2} \cdot \frac{1}{2\tfrac{1}{2}} \log \frac{x - \tfrac{1}{2}}{x + 2};$$

or, since $\log(2x - 1)$ differs from $\log(x - \tfrac{1}{2})$ only by a constant, we may write

$$\int \frac{dx}{2x^2 + 3x - 2} = \frac{1}{5} \log \frac{2x - 1}{x + 2}.$$

Denominators of Higher Degree.

18. When the denominator is of a degree higher than the second, we may in like manner suppose it resolved into factors corresponding to the roots of the equation formed by placing it equal to zero. The fraction (of which we suppose the numerator to be lower in degree than the denominator) may now be decomposed into partial fractions. If the roots are all real and unequal, we assume these partial fractions as in Art. 16; there being one assumed fraction for each factor.

If, however, a pair of imaginary roots occurs, the factor corresponding to the pair is of the form $(x-\alpha)^2 + \beta^2$, and the partial fraction must be assumed in the form

$$\frac{Ax+B}{(x-\alpha)^2 + \beta^2};$$

for we are only entitled to assume that the numerator of each partial fraction is lower in degree than its denominator (otherwise the given fraction which is the sum of the partial fractions would not have this property).

19. For example, given

$$\int \frac{x+3}{(x^2+1)(x-1)} dx.$$

Assume

$$\frac{x+3}{(x^2+1)(x-1)} = \frac{Ax+B}{x^2+1} + \frac{C}{x-1}, \quad \ldots \quad (1)$$

whence

$$x + 3 = (x-1)(Ax+B) + (x^2+1)C.$$

Putting $x = 1$,

$$4 = 2C, \quad \text{whence} \quad C = 2;$$

putting $x = 0$,

$$3 = -B + C, \quad \text{whence} \quad B = -1.$$

To determine A, any convenient third value may be given to x; for example, if we put $x = -1$, we have

$$2 = -2(-A+B) + 2C \quad \therefore \quad A = -2.$$

Substituting in (1),

$$\frac{x+3}{(x^2+1)(x-1)} = \frac{2}{x-1} - \frac{2x+1}{x^2+1},$$

therefore

$$\int \frac{x+3}{(x^2+1)(x-1)}dx = 2\int \frac{dx}{x-1} - \int \frac{2x\,dx}{x^2+1} - \int \frac{dx}{x^2+1}$$

$$= 2\log(x-1) - \log(x^2+1) - \tan^{-1}x.$$

20. If the denominator admits of factors which are functions of x^2, and the numerator is also a function of x^2, we may with advantage first decompose into fractions having these factors for denominators. Thus, given

$$\int \frac{x^2 dx}{x^4 - a^4}.$$

Putting y for x^2 in the fraction, we first find

$$\frac{y}{y^2 - a^4} = \frac{1}{2(y+a^2)} + \frac{1}{2(y-a^2)},$$

hence

$$\int \frac{x^2 dx}{x^4 - a^4} = \tfrac{1}{2}\int \frac{dx}{x^2 - a^2} + \tfrac{1}{2}\int \frac{dx}{x^2 + a^2},$$

therefore [see equation (A'), Art. 17],

$$\int \frac{x^2 dx}{x^4 - a^4} = \frac{1}{4a}\log\frac{x-a}{x+a} + \frac{1}{2a}\tan^{-1}\frac{x}{a}.$$

This method may sometimes be employed when the numerator is not a function of x^2; thus, since

$$\frac{1}{x^4 - a^4} = \frac{1}{2a^2(x^2 - a^2)} - \frac{1}{2a^2(x^2 + a^2)},$$

we have

$$\frac{x}{x^4 - a^4} = \frac{x}{2a^2(x^2 - a^2)} - \frac{x}{2a^2(x^2 + a^2)},$$

hence
$$\int \frac{x\,dx}{x^4 - a^4} = \frac{1}{4a^2} \log \frac{x^2 - a^2}{x^2 + a^2}.$$

21. The fraction corresponding to a pair of equal roots, that is, to a factor in the denominator of the form $(x - a)^2$, is (see Art. 14) equivalent to a pair of fractions of the form

$$\frac{A}{x - a} + \frac{B}{(x - a)^2},$$

we may, therefore, at once assume the partial fractions in this form. We proceed in like manner when a higher power of a linear factor occurs. For example, given

$$\int \frac{x + 2}{(x - 1)^3 (x + 1)} dx;$$

we assume

$$\frac{x + 2}{(x - 1)^3 (x + 1)} = \frac{A}{(x - 1)^3} + \frac{B}{(x - 1)^2} + \frac{C}{x - 1} + \frac{D}{x + 1}.$$

whence

$$x + 2 = [A + B(x - 1) + C(x - 1)^2](x + 1) + D(x - 1)^3. \quad (1)$$

Putting $x = 1$, we have

$$3 = 2A \quad \therefore \quad A = \tfrac{3}{2}.$$

The values of B and C may be determined as follows: if we substitute the value just determined for A, equation (1), is identically satisfied by $x = 1$, hence it may be divided by $x - 1$. We thus obtain

$$-\tfrac{1}{2} = [B + C(x - 1)](x + 1) + D(x - 1)^2 \quad (2)$$

§ II.] MULTIPLE ROOTS.

in which we may again put $x = 1$, whence $B = -\frac{1}{4}$. In like manner from (2), we obtain

$$\tfrac{1}{4} = C(x + 1) + D(x - 1),$$

from which $C = \frac{1}{8}$, and $D = -\frac{1}{8}$. Therefore

$$\int \frac{x+2}{(x-1)^3(x+1)} dx = \frac{3}{2}\int\frac{dx}{(x-1)^3} - \frac{1}{4}\int\frac{dx}{(x-1)^2} + \frac{1}{8}\int\frac{dx}{x-1} - \frac{1}{8}\int\frac{dx}{x+1}$$

$$= -\frac{3}{4(x-1)^2} + \frac{1}{4(x-1)} + \frac{1}{8}\log\frac{x-1}{x+1}.$$

22. In this example, after obtaining the values of A and D from equation (1) by putting $x = 1$, and $x = -1$, two equations from which B and C might be obtained by elimination could have been derived by giving to x any two other values. Convenient equations for determining B and C may also be obtained by putting $x = 1$ in two equations successively derived by differentiation from the identical equation (1). In the first differentiation we may reject all terms containing $(x - 1)^2$; since these terms, and also those derived from them by the second differentiation, will vanish when $x = 1$. Thus, from equation (1), Art. 21, we obtain

$$1 = A + 2Bx + 2C(x^2 - 1) + \text{terms containing } (x - 1)^2.$$

Putting $x = 1$, and $A = \frac{3}{2}$, we have $B = -\frac{1}{4}$. Differentiating again and substituting the value of B,

$$0 = -\frac{1}{2} + 4Cx + \text{terms containing } (x - 1),$$

and, putting $x = 1$ in this last equation, $C = \frac{1}{8}$.

23. When the method of differentiation is applied to a case in which more than one multiple root occurs, it is best to proceed with each root separately. Thus given,

$$\int \frac{x+1}{(x-1)^2(x+2)^2} dx,$$

$$\frac{x+1}{(x-1)^2(x+2)^2} = \frac{A}{(x-1)^2} + \frac{B}{x-1} + \frac{C}{(x+2)^2} + \frac{D}{x+2}$$

whence

$$x+1 = [A+B(x-1)](x+2)^2 + [C+D(x+2)](x-1)^2 \quad . . . (1)$$

Putting $x = 1$, and $x = -2$, we derive

$$A = \frac{2}{9}, \qquad C = -\frac{1}{9}.$$

Differentiating (1), we have

$$1 = 2A(x+2) + B(x+2)^2 + \text{terms containing } (x-1),$$

whence, putting $x = 1$, and $A = \frac{2}{9}$, we have $B = -\frac{1}{27}$.

Again, differentiating (1), we have

$$1 = 2C(x-1) + D(x-1)^2 + \text{terms containing } (x+2),$$

whence, putting $x = -2$, and $C = -\frac{1}{9}$, we have $D = \frac{1}{27}$.

Therefore

$$\int \frac{x+1}{(x-1)^2(x+2)^2} dx = -\frac{2}{9(x-1)} + \frac{1}{9(x+2)} + \frac{1}{27} \log \frac{x+2}{x-1}.$$

24. Instead of assuming the partial fractions with undetermined numerators, it is sometimes possible to proceed more expeditiously as in the following examples:

Given
$$\int \frac{1}{x^3(1+x^2)}dx;$$
putting the numerator in the form $1 + x^2 - x^2$, we have

$$\int \frac{1}{x^3(1+x^2)}dx = \int \frac{1+x^2}{x^3(1+x^2)}dx - \int \frac{x^2}{x^3(1+x^2)}dx$$
$$= \int \frac{dx}{x^3} - \int \frac{1}{x(1+x^2)}dx.$$

Treating the last integral in like manner,

$$\int \frac{1}{x^3(1+x^2)}dx = \int \frac{dx}{x^3} - \int \frac{dx}{x} + \int \frac{x\,dx}{1+x^2}$$
$$= -\frac{1}{2x^2} - \log x + \tfrac{1}{2}\log(1+x^2) = -\frac{1}{2x^2} + \log \frac{\sqrt{(1+x^2)}}{x}.$$

Again, given
$$\int \frac{1}{x^2(1+x)^2}dx;$$
putting the numerator in the form $(1+x)^2 - 2x - x^2$, we have

$$\int \frac{1}{x^2(1+x)^2}dx = \int \frac{dx}{x^2} - \int \frac{2+x}{x(1+x)^2}dx$$
$$= \int \frac{dx}{x^2} - 2\int \frac{dx}{x(1+x)} + \int \frac{dx}{(1+x)^2}.$$

Hence by equation (A), Art. 17,

$$\int \frac{dx}{x^2(1+x)^2} = -\frac{1}{x} - 2\log \frac{x}{1+x} - \frac{1}{1+x}.$$

Examples II.

1. $\int \dfrac{dx}{a-x}$, $\qquad -\log(a-x)$.

2. $\int \dfrac{dx}{(a-x)^2}$, $\qquad \dfrac{1}{a-x}$.

3. $\int \dfrac{x\,dx}{a^2+x^2}$, $\qquad \dfrac{1}{2}\log(a^2+x^2)$.

4. $\int \sqrt{(a^2-x^2)}\, x\, dx$, $\qquad \dfrac{a^3-(a^2-x^2)^{3/2}}{3}$.

5. $\int \dfrac{x^2\,dx}{a^3-x^3}$, $\qquad \dfrac{1}{3}\log\dfrac{a^3}{a^3-x^3}$.

6. $\int \dfrac{x\,dx}{\sqrt{(a^2-x^2)}}$, $\qquad a-\sqrt{(a^2-x^2)}$.

7. $\int (a^2+3x^2)^3 x\, dx$, $\qquad \dfrac{(a^2+3x^2)^4}{24}$.

8. $\int (a+mx)^2\, dx$, $\qquad \dfrac{(a+mx)^3-a^3}{3m}$.

9. $\int \dfrac{dx}{\sin^2 2x}$, $\qquad -\dfrac{\cot 2x}{2}$.

10. $\int \cos^3 x \sin x\, dx$, $\qquad \dfrac{1-\cos^4 x}{4}$.

11. $\int \dfrac{\cos\theta\, d\theta}{\sin^3\theta}$, $\qquad -\dfrac{1}{2}\csc^2\theta$.

12. $\int \sec^2 3x \tan 3x\, dx$, $\qquad \dfrac{\sec^2 3x - 1}{9}$.

§ II.] EXAMPLES. 27

13. $\int a^{mx} dx$, $\qquad\qquad\qquad\dfrac{a^{mx}}{m \log a}$.

14. $\int (\varepsilon^x - 1)^3 dx$, $\qquad\qquad \frac{1}{3}\varepsilon^{3x} - \frac{3}{2}\varepsilon^{2x} + 3\varepsilon^x - x$.

15. $\int (1 + 3\sin^2 x)^2 \sin x \cos x \, dx$, $\qquad\qquad \dfrac{(1 + 3\sin^2 x)^3}{18}$.

16. $\int_0^{2a} \dfrac{(a - x) \, dx}{\sqrt{(2ax - x^2)}}$, $\qquad\qquad \sqrt{(2ax - x^2)}\Big]_0^{2a} = 0$.

17. $\int_0^{\frac{\pi}{2}} \cos^3 \theta \, d\theta$, $\qquad\qquad \dfrac{2}{3}$.

18. $\int \sec^4 \theta \, d\theta$, $\qquad\qquad \tan \theta + \dfrac{1}{3} \tan^3 \theta$.

19. $\int \tan^3 x \, dx$, $\qquad\qquad \dfrac{1}{2}\tan^2 x + \log \cos x$.

20. $\int_0^{\frac{\pi}{4}} \sec^4 x \tan x \, dx$, $\qquad\qquad \dfrac{1}{4}\sec^4 x\Big]_0^{\frac{\pi}{4}} = \dfrac{3}{4}$.

21. $\int \sqrt{\dfrac{a - x}{a + x}} \, dx$, $\qquad\qquad a \sin^{-1}\dfrac{x}{a} + \sqrt{(a^2 - x^2)}$.

22. $\int_{\frac{\pi}{4}}^{\frac{\pi}{2}} \cot^3 \theta \, d\theta$, $\qquad\qquad \dfrac{1 - \log 2}{2}$.

23. $\int \sqrt{\dfrac{2a - x}{x}} \, dx$, $\qquad\qquad \sqrt{(2ax - x^2)} + a \operatorname{vers}^{-1}\dfrac{x}{a}$.

24. $\int \sin (\alpha - 2\theta) \, d\theta$, $\qquad\qquad \dfrac{\cos (\alpha - 2\theta)}{2}$.

25. $\int \dfrac{\cos x\, dx}{a - b \sin x}$, $\qquad -\dfrac{1}{b} \log(a - b \sin x)$.

26. $\int_{\frac{\pi}{6}}^{\frac{\pi}{4}} \dfrac{dx}{\tan x}$, $\qquad \tfrac{1}{2} \log 2$.

27. $\int_{\frac{\pi}{4}}^{\frac{\pi}{2}} \dfrac{dx}{\tan x}$, $\qquad \tfrac{1}{2} \log 2$.

28. $\int_{\frac{1}{4}}^{\frac{1}{2}} \dfrac{dx}{x \log x}$, $\qquad \log(-\log x) \Big]_{\frac{1}{4}}^{\frac{1}{2}} = -\log 2$.

29. $\int \dfrac{dx}{e^x + e^{-x}}$, $\qquad \tan^{-1} e^x$.

30. $\int \dfrac{x^2\, dx}{x^6 + 1}$, $\qquad \dfrac{1}{3} \tan^{-1} x^3$.

31. $\int \dfrac{x\, dx}{\sqrt{(a^4 - x^4)}}$, $\qquad \dfrac{1}{2} \sin^{-1} \dfrac{x^2}{a^2}$.

32. $\int \dfrac{dx}{\sqrt{(5 - 3x^2)}}$, $\qquad \dfrac{1}{\sqrt{3}} \sin^{-1} \dfrac{x\sqrt{3}}{\sqrt{5}}$.

33. $\int \dfrac{dx}{2 + 5x^2}$, $\qquad \dfrac{1}{\sqrt{10}} \tan^{-1} \dfrac{x\sqrt{5}}{\sqrt{2}}$.

34. $\int_{1}^{\infty} \dfrac{dx}{x\sqrt{(2x^2 - 1)}}$, $\qquad \dfrac{\pi}{4}$.

35. $\int_{0}^{1} \dfrac{dx}{x^2 + x + 1}$, $\qquad \dfrac{2}{\sqrt{3}} \tan^{-1} \dfrac{2x + 1}{\sqrt{3}} \Big]_{0}^{1} = \dfrac{\pi}{3\sqrt{3}}$.

36. $\int_0^1 \dfrac{dx}{\sqrt{(5-4x-x^2)}},$ $\cos^{-1}\frac{2}{3}.$

37. $\int \dfrac{\sqrt{(x^2-a^2)}}{x}\,dx\left[=\int\dfrac{x^2-a^2}{x\sqrt{(x^2-a^2)}}\,dx\right],$

$\sqrt{(x^2-a^2)} - a\sec^{-1}\dfrac{x}{a}.$

38. $\int_0^{\frac{1}{2}a} \dfrac{x^2}{a-x}\,dx,$ $a^2(\log 2 - \tfrac{3}{8}).$

39. $\int \dfrac{4x^2-x+3}{x^2+1}\,dx,$ $4x - \tfrac{1}{2}\log(x^2+1) - \tan^{-1}x.$

40. $\int \dfrac{x^2+x+1}{x^2-x+1}\,dx,$ $x+\log(x^2-x+1)+\dfrac{2}{\sqrt{3}}\tan^{-1}\dfrac{2x-1}{\sqrt{3}}.$

41. $\int \dfrac{x^3-1}{x^2-4}\,dx,$ $x+\dfrac{3}{4}\log\dfrac{x-2}{x+2}.$

42. $\int \dfrac{(1+x)^2}{x-x^2}\,dx,$ $\log\dfrac{x}{(1-x)^4} - x.$

43. $\int \dfrac{(2x+1)^2\,dx}{2x+3},$ $x^2-x+2\log(2x+3).$

44. $\int \dfrac{2x+3}{(2x+1)^2}\,dx,$ $\tfrac{1}{2}\log(2x+1) - \dfrac{1}{2x+1}.$

45. $\int \dfrac{x^2-3x+3}{x^2-3x+2}\,dx,$ $x+\log\dfrac{x-2}{x-1}.$

46. $\int_0^a \dfrac{dx}{x^2-2ax\cos\alpha+a^2},$

$\dfrac{1}{a\sin\alpha}\tan^{-1}\dfrac{x-a\cos\alpha}{a\sin\alpha}\bigg]_0^a = \dfrac{\pi-\alpha}{2a\sin\alpha}.$

47. $\int \dfrac{dx}{x^2 - 2ax \sec \alpha + a^2}$, $\quad \dfrac{1}{2a \tan \alpha} \log \dfrac{x - a \sec \alpha - a \tan \alpha}{x - a \sec \alpha + a \tan \alpha}$.

48. $\int \dfrac{dx}{2x^2 - 4x - 7}$, $\quad \dfrac{\sqrt{2}}{12} \log \dfrac{2x - 2 - 3\sqrt{2}}{2x - 2 + 3\sqrt{2}}$.

49. $\int \dfrac{x^2\, dx}{1 - x^6}$, $\quad \dfrac{1}{6} \log \dfrac{1 + x^3}{1 - x^3}$.

50. $\int \dfrac{3x - 1}{x^3 - x^2 - 2x}\, dx$, $\quad \dfrac{1}{6} \log \dfrac{x^3 (x - 2)^5}{(x + 1)^8}$.

51. $\int \dfrac{x\, dx}{(x + 2)(x + 3)^2}$, $\quad 2 \log \dfrac{x + 3}{x + 2} - \dfrac{3}{x + 3}$.

52. $\int \dfrac{x\, dx}{x^3 + x^2 + x + 1}$, $\quad \dfrac{1}{2}\left[\tan^{-1} x + \log \dfrac{\sqrt{(x^2 + 1)}}{x + 1} \right]$.

53. $\int \dfrac{x^2\, dx}{x^4 + x^2 - 2}$, $\quad \dfrac{1}{6} \log \dfrac{x - 1}{x + 1} + \dfrac{\sqrt{2}}{3} \tan^{-1} \dfrac{x}{\sqrt{2}}$.

54. $\int \dfrac{x^2 - x + 2}{x^4 - 5x^2 + 4}\, dx$, $\quad \dfrac{2}{3} \log \dfrac{x + 1}{x + 2} + \dfrac{1}{3} \log \dfrac{x - 2}{x - 1}$.

55. $\int \dfrac{dx}{x^3 - x^2 - x + 1}$, $\quad \dfrac{1}{4} \log \dfrac{x + 1}{x - 1} - \dfrac{1}{2(x - 1)}$.

56. $\int \dfrac{3x + 1}{x^4 - 1}\, dx$, $\quad \log \dfrac{(x - 1)\sqrt{(x + 1)}}{(x^2 + 1)^{\frac{3}{4}}} - \dfrac{1}{2} \tan^{-1} x$.

57. $\int \dfrac{dx}{1 + x^3}$, $\quad \dfrac{1}{6} \log \dfrac{(x + 1)^2}{x^2 - x + 1} + \dfrac{1}{\sqrt{3}} \tan^{-1} \dfrac{2x - 1}{\sqrt{3}}$.

58. $\int \dfrac{x^2\, dx}{(x - 1)^2 (x^2 + 1)}$, $\quad \dfrac{1}{2} \log (x - 1) - \dfrac{1}{4} \log (x^2 + 1) - \dfrac{1}{2(x - 1)}$.

§ II.]   EXAMPLES.   31

59. $\int \dfrac{dx}{x(1 + x + x^2 + x^3)}$,   Partial Fractions

$\log x - \dfrac{1}{2}\log(1 + x) - \dfrac{1}{4}\log(1 + x^2) - \dfrac{1}{2}\tan^{-1}x.$

60. $\int \dfrac{x^2 - 1}{x^4 + x^2 + 1} dx$, $\dfrac{1}{2}\log\dfrac{x^2 - x + 1}{x^2 + x + 1}$.

61. $\int \dfrac{x^2 + x - 1}{x^3 + x^2 - 6x} dx$, $\dfrac{1}{6}\log x + \dfrac{1}{2}\log(x - 2) + \dfrac{1}{3}\log(x + 3)$.

62. $\int \dfrac{x^2 dx}{x^4 - x^2 - 12}$, $\dfrac{1}{7}\log\dfrac{x - 2}{x + 2} + \dfrac{\sqrt{3}}{7}\tan^{-1}\dfrac{x}{\sqrt{3}}$.

63. $\int \dfrac{x^2 dx}{(x^2 - 1)^2}$, $\dfrac{1}{4}\log\dfrac{x - 1}{x + 1} - \dfrac{x}{2(x^2 - 1)}$.

64. $\int \dfrac{2x^2 - 3a^2}{x^4 - a^4} dx$, $\dfrac{5}{2a}\tan^{-1}\dfrac{x}{a} - \dfrac{1}{4a}\log\dfrac{x - a}{x + a}$.

65. $\int \dfrac{x \, dx}{x^4 - x^2 - 2}$, $\dfrac{1}{6}\log\dfrac{x^2 - 2}{x^2 + 1}$.

66. $\int \dfrac{dx}{(x^2 + a^2)(x + b)}$, $\dfrac{1}{b^2 + a^2}\left[\log\dfrac{x + b}{\sqrt{(x^2 + a^2)}} + \dfrac{b}{a}\tan^{-1}\dfrac{x}{a}\right]$.

67. $\int_0^\infty \dfrac{dx}{(x^2 + a^2)(x^2 + b^2)}$, $\dfrac{\pi}{2ab(a + b)}$.

68. $\int_a^\infty \dfrac{a^2 dx}{x^2(a^2 + x^2)}$, $\dfrac{4 - \pi}{4a}$.

69. $\int \dfrac{x + 1}{x(1 + x^2)} dx$, $\tan^{-1}x + \log\dfrac{x}{\sqrt{(1 + x^2)}}$.

70. $\int \dfrac{dx}{x^4(x^2 + 1)}$, $\tan^{-1}x + \dfrac{1}{x} - \dfrac{1}{3x^3}$.

32   ELEMENTARY METHODS OF INTEGRATION.   [Ex. II.

71. $\int \frac{dx}{x(1+x)^2}$, $\log \frac{x}{1+x} + \frac{1}{1+x}$.

72. $\int \frac{dx}{x(a+bx^3)}$, $\frac{1}{3a} \log \frac{x^3}{a+bx^3}$.

73. $\int \frac{dx}{x^4(a+bx^3)}$, $-\frac{1}{3ax^3} + \frac{b}{3a^2} \log \frac{a+bx^3}{x^3}$.

74. Find the whole area enclosed by both loops of the curve
$$y^2 = x^2(1 - x^2).$$
$\frac{4}{3}$.

75. Find the area enclosed between the asymptote corresponding to $x = a$, and the curve
$$x^2 y^2 + a^2 x^2 = a^2 y^2.$$
$2a^2$.

76. Find the whole area enclosed by the curve
$$a^2 y^4 = x^4 (a^2 - x^2).$$
$\tfrac{8}{15} a^2$.

77. Find the area enclosed by the catenary
$$y = \frac{c}{2}\left[\varepsilon^{\frac{x}{c}} + \varepsilon^{-\frac{x}{c}}\right],$$
the axes and any ordinate.

$\frac{c^2}{2}\left[\varepsilon^{\frac{x}{c}} - \varepsilon^{-\frac{x}{c}}\right]$.

78. Find the whole area between the witch
$$xy^2 = 4a^2(2a - x)$$
and its asymptote. *See Ex.* 23.

$4\pi a^2$.

III.

Trigonometric Integrals.

25. The transformation, $\tan^2 \theta = \sec^2 \theta - 1$, suffices to separate all integrals of the form

$$\int \tan^n \theta \, d\theta, \quad \ldots \ldots \ldots \quad (1)$$

in which n is an integer, into directly integrable parts. Thus, for example,

$$\int \tan^5 \theta \, d\theta = \int \tan^3 \theta \, (\sec^2 \theta - 1) \, d\theta$$

$$= \frac{\tan^4 \theta}{4} - \int \tan^3 \theta \, d\theta.$$

Transforming the last integral in like manner, we have

$$\int \tan^5 \theta \, d\theta = \frac{\tan^4 \theta}{4} - \frac{\tan^2 \theta}{2} + \int \tan \theta \, d\theta;$$

hence (see Art. 12)

$$\int \tan^5 \theta \, d\theta = \frac{\tan^4 \theta}{4} - \frac{\tan^2 \theta}{2} - \log \cos \theta.$$

When the value of n in (1) is even, the value of the final integral will be θ. When n is negative, the integral takes the form

$$\int \cot^n \theta \, d\theta,$$

which may be treated in a similar manner.

26. Integrals of the form

$$\int \sec^n \theta \, d\theta \quad \cdots \cdots \cdots \quad (2)$$

are readily evaluated *when* n *is an even number*, thus

$$\int \sec^6 \theta \, d\theta = \int (\tan^2 + 1)^2 \sec^2 \theta \, d\theta$$

$$= \int \tan^4 \theta \sec^2 \theta \, d\theta + 2 \int \tan^2 \theta \sec^2 \theta \, d\theta + \int \sec^2 \theta \, d\theta$$

$$= \frac{\tan^5 \theta}{5} + \frac{2 \tan^3 \theta}{3} + \tan \theta.$$

If n in expression (2) is odd, the method to be explained in Section VI is required.

Integrals of the form $\int \operatorname{cosec}^n \theta \, d\theta$ are treated in like manner.

Cases in which $\sin^m \theta \cos^n \theta \, d\theta$ *is directly integrable.*

27. If n is a *positive odd number*, an integral of the form

$$\int \sin^m \theta \cos^n \theta \, d\theta \quad \cdots \cdots \cdots \quad (3)$$

is directly integrable in terms of sin θ. Thus,

$$\int \sin^2 \theta \cos^5 \theta \, d\theta = \int \sin^2 \theta \, (1 - \sin^2 \theta)^2 \cos \theta \, d\theta$$

$$= \frac{\sin^3 \theta}{3} - \frac{2 \sin^5 \theta}{5} + \frac{\sin^7 \theta}{7}.$$

This method is evidently applicable even when m is fractional or negative. Thus, putting y for sin θ,

§ III.]   TRIGONOMETRIC INTEGRALS.   35

$$\int \frac{\cos^3 \theta}{\sin^{\frac{3}{2}} \theta} d\theta = \int \frac{(1-y^2) dy}{y^{\frac{3}{2}}} = \int y^{-\frac{3}{2}} dy - \int y^{\frac{1}{2}} dy;$$

hence

$$\int \frac{\cos^3 \theta}{\sin^{\frac{3}{2}} \theta} d\theta = -2y^{-\frac{1}{2}} - \frac{2}{3} y^{\frac{3}{2}} = -\frac{2}{3} \cdot \frac{3 + \sin^2 \theta}{\sqrt{(\sin \theta)}}.$$

When m in expression (3) is a positive odd number, the integral is evaluated in a similar manner.

28. An integral of the form (3) is also directly integrable *when* m + n *is an even negative integer*, in other words, when it can be written in the form

$$\int \frac{\sin^m \theta \, d\theta}{\cos^{m+2q} \theta} = \int \tan^m \theta \sec^{2q} \theta \, d\theta,$$

in which q *is positive.*

For example,

$$\int \frac{d\theta}{\sin^{\frac{3}{2}} \theta \cos^{\frac{5}{2}} \theta} = \int (\tan \theta)^{-\frac{3}{2}} \sec^4 \theta \, d\theta$$

$$= \int (\tan \theta)^{-\frac{3}{2}} (\tan^2 \theta + 1) \sec^2 \theta \, d\theta;$$

hence

$$\int \frac{d\theta}{\sin^{\frac{3}{2}} \theta \cos^{\frac{5}{2}} \theta} = \frac{2}{3} \tan^{\frac{3}{2}} \theta - \frac{2}{\tan^{\frac{1}{2}} \theta}.$$

It may be more convenient to express the integral in terms of cot θ and cosec θ, thus

$$\int \frac{\cos^4 \theta \, d\theta}{\sin^8 \theta} = \int \cot^4 \theta (\cot^2 \theta + 1) \operatorname{cosec}^2 \theta \, d\theta$$

$$= -\frac{\cot^7 \theta}{7} - \frac{\cot^5 \theta}{5}.$$

Integrals of the forms treated in Art. 25 and Art. 26 are included in the general form (3), Art. 27. Except in the cases already considered, and in the special cases given below, the method of reduction given in Section VI is required in the evaluation of integrals of this form.

The Integrals $\int \sin^2\theta\, d\theta$, and $\int \cos^2\theta\, d\theta$.

29. These integrals are readily evaluated by means of the transformations

$$\sin^2\theta = \tfrac{1}{2}(1 - \cos 2\theta), \quad \text{and} \quad \cos^2\theta = \tfrac{1}{2}(1 + \cos 2\theta).$$

Thus

$$\int \sin^2\theta\, d\theta = \tfrac{1}{2}\int d\theta - \tfrac{1}{2}\int \cos 2\theta\, d\theta = \tfrac{1}{2}\theta - \tfrac{1}{4}\sin 2\theta,$$

or, since $\sin 2\theta = 2\sin\theta\cos\theta$,

$$\int \sin^2\theta\, d\theta = \tfrac{1}{2}(\theta - \sin\theta\cos\theta). \quad \ldots \ldots (B)$$

In like manner

$$\int \cos^2\theta\, d\theta = \tfrac{1}{2}(\theta + \sin\theta\cos\theta). \quad \ldots \ldots (C)$$

Since $\sin^2\theta + \cos^2\theta = 1$, the sum of these integrals is $\int d\theta$; accordingly we find the sum of their values to be θ.

In the applications of the Integral Calculus, these integrals frequently occur with the limits 0 and $\tfrac{1}{2}\pi$; from (B) and (C) we derive

$$\int_0^{\frac{\pi}{2}} \sin^2\theta\, d\theta = \int_0^{\frac{\pi}{2}} \cos^2\theta\, d\theta = \tfrac{1}{4}\pi.$$

The Integrals $\int \dfrac{d\theta}{\sin\theta\cos\theta}$, $\int \dfrac{d\theta}{\sin\theta}$, *and* $\int \dfrac{d\theta}{\cos\theta}$.

30. We have

$$\int \frac{d\theta}{\sin\theta\cos\theta} = \int \frac{\sec^2\theta\, d\theta}{\tan\theta} = \log\tan\theta. \quad \ldots \quad (D)$$

Again, using the transformation,

$$\sin\theta = 2\sin\tfrac{1}{2}\theta\cos\tfrac{1}{2}\theta,$$

we have

$$\int \frac{d\theta}{\sin\theta} = \int \frac{\tfrac{1}{2}d\theta}{\sin\tfrac{1}{2}\theta\cos\tfrac{1}{2}\theta} = \int \frac{\sec^2\tfrac{1}{2}\theta\, \tfrac{1}{2}d\theta}{\tan\tfrac{1}{2}\theta};$$

hence

$$\int \frac{d\theta}{\sin\theta} = \log\tan\tfrac{1}{2}\theta. \quad \ldots \ldots \quad (E)$$

This integral may also be evaluated thus,

$$\int \frac{d\theta}{\sin\theta} = \int \frac{\sin\theta\, d\theta}{\sin^2\theta} = \int \frac{\sin\theta\, d\theta}{1-\cos^2\theta}.$$

Since $\sin\theta\, d\theta = -d(\cos\theta)$, the value of the last integral is, by formula (A'), Art. 17,

$$\frac{1}{2}\log\frac{1-\cos\theta}{1+\cos\theta} = \log\sqrt{\frac{1-\cos\theta}{1+\cos\theta}};$$

and, multiplying both terms of the fraction by $1-\cos\theta$, we have

$$\int \frac{d\theta}{\sin\theta} = \log\frac{1-\cos\theta}{\sin\theta}. \quad \ldots \ldots \quad (E')$$

31. Since $\cos\theta = \sin(\tfrac{1}{2}\pi + \theta)$, we derive from formula (E),

$$\int \frac{d\theta}{\cos\theta} = \int \frac{d\theta}{\sin(\tfrac{1}{2}\pi+\theta)} = \log\tan\left[\frac{\pi}{4}+\frac{\theta}{2}\right]. \quad \ldots \quad (F)$$

By employing a process similar to that used in deriving formula (E'), we have also

$$\int \frac{d\theta}{\cos\theta} = \log\frac{1+\sin\theta}{\cos\theta}. \quad \ldots \ldots \quad (F')$$

Miscellaneous Trigonometric Integrals.

32. A trigonometric integral may sometimes be reduced, by means of the formulas for trigonometric transformation, to one of the forms integrated in the preceding articles. For example, let us take the integral

$$\int \frac{d\theta}{a\sin\theta + b\cos\theta}.$$

Putting
$$a = k\cos\alpha, \qquad b = k\sin\alpha, \quad \ldots \quad (1)$$
we have

$$\int \frac{d\theta}{a\sin\theta + b\cos\theta} = \frac{1}{k}\int \frac{d\theta}{\sin(\theta+\alpha)}.$$

Hence by formula (E)

$$\int \frac{d\theta}{a\sin\theta + b\cos\theta} = \frac{1}{k}\log\tan\frac{1}{2}(\theta+\alpha);$$

or, since equations (1) give

$$k = \sqrt{(a^2+b^2)}, \qquad \tan\alpha = \frac{b}{a},$$

$$\int \frac{d\theta}{a\sin\theta + b\cos\theta} = \frac{1}{\sqrt{(a^2+b^2)}}\log\tan\frac{1}{2}\left[\theta + \tan^{-1}\frac{b}{a}\right].$$

§ III.] *MISCELLANEOUS TRIGONOMETRIC INTEGRALS.* 39

33. The expression $\sin m\theta \sin n\theta \, d\theta$ may be integrated by means of the formula

$$\cos(m-n)\theta - \cos(m+n)\theta = 2\sin m\theta \sin n\theta;$$

whence

$$\int \sin m\theta \sin n\theta \, d\theta = \frac{\sin(m-n)\theta}{2(m-n)} - \frac{\sin(m+n)\theta}{2(m+n)} \quad . \quad (1)$$

In like manner, from

$$\cos(m-n)\theta + \cos(m+n)\theta = 2\cos m\theta \cos n\theta,$$

we derive

$$\int \cos m\theta \cos n\theta \, d\theta = \frac{\sin(m-n)\theta}{2(m-n)} + \frac{\sin(m+n)\theta}{2(m+n)} \quad . \quad (2)$$

When $m = n$, the first term of the second member of each of these equations takes an indeterminate form. Evaluating this term, we have

$$\int \sin^2 n\theta \, d\theta = \frac{\theta}{2} - \frac{\sin 2n\theta}{4n}, \quad \ldots \quad (3)$$

and

$$\int \cos^2 n\theta \, d\theta = \frac{\theta}{2} + \frac{\sin 2n\theta}{4n}. \quad \ldots \quad (4)$$

Using the limits 0 and π we have, from (1) and (2), *when* m *and* n *are unequal integers*,

$$\int_0^\pi \sin m\theta \sin n\theta \, d\theta = \int_0^\pi \cos m\theta \cos n\theta \, d\theta = 0; \quad . \quad (5)$$

but, when m and n are *equal integers*, we have from (3) and (4)

$$\int_0^\pi \sin^2 n\theta \, d\theta = \int_0^\pi \cos^2 n\theta \, d\theta = \frac{\pi}{2} \quad \ldots \quad (6)$$

34. To integrate $\sqrt{(1 + \cos\theta)} \, d\theta$; we use the formula

$$2\cos^2 \tfrac{1}{2}\theta = 1 + \cos\theta,$$

whence
$$\sqrt{(1+\cos\theta)} = \pm\sqrt{2}\cos\tfrac{1}{2}\theta,$$
in which the positive sign is to be taken, provided the value of θ is between 0 and π. Supposing this to be the case, we have

$$\int \sqrt{(1+\cos\theta)}\,d\theta = \sqrt{2}\int\cos\tfrac{1}{2}\theta\,d\theta$$
$$= 2\sqrt{2}\sin\tfrac{1}{2}\theta.$$

For example, we have the definite integral

$$\int_0^{\frac{\pi}{2}} \sqrt{(1+\cos\theta)}\,d\theta = 2\sqrt{2}\sin\frac{\pi}{4} = 2.$$

Integration of $\dfrac{d\theta}{a+b\cos\theta}$.

35. By means of the formulas

$$1 = \cos^2\tfrac{1}{2}\theta + \sin^2\tfrac{1}{2}\theta \quad\text{and}\quad \cos\theta = \cos^2\tfrac{1}{2}\theta - \sin^2\tfrac{1}{2}\theta,$$

we have

$$\int \frac{d\theta}{a+b\cos\theta} = \int \frac{d\theta}{(a+b)\cos^2\tfrac{1}{2}\theta + (a-b)\sin^2\tfrac{1}{2}\theta}.$$

Multiplying numerator and denominator by $\sec^2\tfrac{1}{2}\theta$, this becomes

$$\int \frac{\sec^2\tfrac{1}{2}\theta\,d\theta}{a+b+(a-b)\tan^2\tfrac{1}{2}\theta},$$

and, putting for abbreviation

$$\tan\tfrac{1}{2}\theta = y,$$

we have, since $\tfrac{1}{2}\sec^2\tfrac{1}{2}\theta\,d\theta = dy$,

$$\int \frac{d\theta}{a+b\cos\theta} = 2\int \frac{dy}{a+b+(a-b)y^2}.$$

§ III.] MISCELLANEOUS TRIGONOMETRIC INTEGRALS.

The form of this integral depends upon the relative values of a and b. Assuming a to be positive, if b, which may be either positive or negative, is numerically less than a, we may put

$$\frac{a+b}{a-b} = c^2.$$

The integral may then be written in the form

$$\frac{2}{a-b}\int \frac{dy}{c^2 + y^2},$$

the value of which is, by formula (k'),

$$\frac{2}{c(a-b)} \tan^{-1} \frac{y}{c}.$$

Hence, substituting their values for y and c, we have, in this case,

$$\int \frac{d\theta}{a+b\cos\theta} = \frac{2}{\sqrt{(a^2-b^2)}} \tan^{-1}\left[\sqrt{\frac{a-b}{a+b}} \tan\tfrac{1}{2}\theta\right]. \quad . \;(G)$$

If, on the other hand, b is numerically greater than a, this expression for the integral involves imaginary quantities; but putting

$$\frac{b+a}{b-a} = c^2,$$

the integral becomes

$$\frac{2}{b-a}\int \frac{dy}{c^2 - y^2},$$

the value of which is, by formula (A'), Art. 17,

$$\frac{1}{c(b-a)} \log \frac{c+y}{c-y}.$$

Therefore, in this case,

$$\int \frac{d\theta}{a + b \cos \theta} = \frac{1}{\sqrt{(b^2 - a^2)}} \log \frac{\sqrt{(b+a)} + \sqrt{(b-a)} \tan \tfrac{1}{2}\theta}{\sqrt{(b+a)} - \sqrt{(b-a)} \tan \tfrac{1}{2}\theta}. \quad . \ (G)$$

36. If $e < 1$, formula (G) of the preceding article gives

$$\int \frac{d\theta}{1 + e \cos \theta} = \frac{2}{\sqrt{(1 - e^2)}} \tan^{-1}\left[\sqrt{\frac{1-e}{1+e}} \tan \tfrac{1}{2}\theta\right]. \ . \ (1)$$

Putting

$$\sqrt{\frac{1-e}{1+e}} \cdot \tan \tfrac{1}{2}\theta = \tan \tfrac{1}{2}\phi, \quad \ldots \ldots \quad (2)$$

and noticing that $\phi = 0$ when $\theta = 0$, we may write

$$\int_0 \frac{d\theta}{1 + e \cos \theta} = \frac{\phi}{\sqrt{(1 - e^2)}}. \quad \ldots \ldots \quad (3)$$

Now, if in equation (1) we put ϕ for θ and change the sign of e, we obtain

$$\int_0 \frac{d\phi}{1 - e \cos \phi} = \frac{2}{\sqrt{(1 - e^2)}} \tan^{-1}\left[\sqrt{\frac{1+e}{1-e}} \tan \tfrac{1}{2}\phi\right];$$

hence, by equation (2),

$$\int_0 \frac{d\phi}{1 - e \cos \phi} = \frac{\theta}{\sqrt{(1 - e^2)}}. \quad \ldots \ldots \quad (4)$$

Equations (3) and (4) are equivalent to

$$\frac{d\theta}{1 + e \cos \theta} = \frac{d\phi}{\sqrt{(1 - e^2)}}, \quad \ldots \ldots \quad (5)$$

and

$$\frac{d\phi}{1 - e \cos \phi} = \frac{d\theta}{\sqrt{(1 - e^2)}}, \quad \ldots \ldots \quad (6)$$

§ III.] TRIGONOMETRIC INTEGRALS. 43

the product of which gives

$$(1 + e \cos \theta)(1 - e \cos \phi) = 1 - e^2 \quad . \quad . \quad . \quad . \quad (7)$$

By means of these relations any expression of the form

$$\int \frac{d\theta}{(1 + e \cos \theta)^n},$$

where n is a positive integer, may be reduced to an integrable form. For

$$\int \frac{d\theta}{(1 + e \cos \theta)^n} = \int \frac{d\theta}{1 + e \cos \theta} \frac{1}{(1 + e \cos \theta)^{n-1}};$$

hence, by equations (5) and (7),

$$\int_0 \frac{d\theta}{(1 + e \cos \theta)^n} = \frac{1}{(1 - e^2)^{n - \frac{1}{2}}} \int_0 (1 - e \cos \phi)^{n-1} d\phi.$$

By expanding $(1 - e \cos \phi)^{n-1}$, the last expression is reduced to a series of integrals involving powers of $\cos \phi$; these may be evaluated by the methods given in this section and Section VI, and the results expressed in terms of θ by means of equation (2) or of equation (7).

Examples III.

1. $\int \tan^4 mx \, dx,$ $\qquad \dfrac{\tan^3 mx}{3m} - \dfrac{\tan mx}{m} + x.$

2. $\int_0^{\frac{\pi}{4}} \tan^7 x \, dx,$ $\qquad \dfrac{5}{12} - \frac{1}{2} \log 2.$

3. $\int \sec^4 (\theta + \alpha) \, d\theta,$ $\qquad \dfrac{\tan^3(\theta + \alpha)}{3} + \tan (\theta + \alpha).$

44 ELEMENTARY METHODS OF INTEGRATION. [Ex. III.

4. $\int_0^{\frac{\pi}{m}} \sin^3 mx\, dx$, $\qquad\qquad\qquad \dfrac{4}{3m}$.

5. $\int \sin^2 \theta \cos^3 \theta\, d\theta$, $\qquad\qquad \dfrac{\sin^3 \theta}{3} - \dfrac{\sin^5 \theta}{5}$.

6. $\int \sqrt{(\sin \theta)} \cos^5 \theta\, d\theta$, $\qquad \dfrac{2}{3}\sin^{\frac{3}{2}}\theta - \dfrac{4}{7}\sin^{\frac{7}{2}}\theta + \dfrac{2}{11}\sin^{\frac{11}{2}}\theta$.

7. $\int_0^{\frac{\pi}{2}} \cos^4 \theta \sin^3 \theta\, d\theta$, $\qquad\qquad \dfrac{2}{35}$.

8. $\int \dfrac{\sin^3 \theta\, d\theta}{\sqrt{(\cos \theta)}}$, $\qquad\qquad \tfrac{2}{5} \cos^{\frac{5}{2}}\theta - 2 \cos^{\frac{1}{2}}\theta$.

9. $\int \dfrac{d\theta}{\sin^2 \theta \cos^2 \theta}$, $\quad$ Multiply by $\sin^2\theta + \cos^2\theta$. $\quad \tan\theta - \cot\theta$.

10. $\int \dfrac{\sin^3 x}{\cos^5 x}\, dx$, $\qquad\qquad$ See Art. 28. $\qquad\qquad \dfrac{\tan^4 x}{4}$.

11. $\int \dfrac{d\theta}{\sin^3 \theta \cos^3 \theta}$, $\qquad\qquad \tfrac{1}{2}(\tan^2 \theta - \cot^2 \theta) + 2 \log \tan \theta$.

12. $\int \dfrac{\sqrt{(\sin \theta)}\, d\theta}{\cos^{\frac{5}{2}} \theta}$, $\qquad\qquad \tfrac{2}{3} \tan^{\frac{3}{2}}\theta$.

13. $\int \dfrac{\sin^3 x\, dx}{\cos^6 x}$, $\qquad\qquad \dfrac{1}{5 \cos^5 x} - \dfrac{1}{3 \cos^3 x}$.

14. $\int \dfrac{\sin^3 x\, dx}{\cos^8 x}$, $\qquad\qquad \dfrac{\tan^5 x}{5} + \dfrac{\tan^3 x}{3}$.

15. $\int \sin^2 \theta \cos^2 \theta\, d\theta$, $\qquad\qquad \tfrac{1}{16}[2\theta - \sin 2\theta \cos 2\theta]$.

§ III.]   EXAMPLES.   45

16. $\int_0^{\frac{\pi}{m}} \sin^2 mx\, dx,$ $\qquad\qquad \dfrac{\pi}{2m}.$

17. $\int \dfrac{\sin^2\theta\, d\theta}{\cos\theta},$ $\qquad\qquad \log\tan\left[\dfrac{\pi}{4}+\dfrac{\theta}{2}\right] - \sin\theta.$

18. $\int_{\frac{\pi}{3}}^{\frac{\pi}{2}} \dfrac{\cos^2\theta\, d\theta}{\sin\theta},$ $\qquad\qquad \tfrac{1}{2}(\log 3 - 1).$

19. $\int \dfrac{d\theta}{\sin\theta + \cos\theta},$ $\qquad\qquad \dfrac{1}{\sqrt{2}}\log\tan\left[\dfrac{\theta}{2}+\dfrac{\pi}{8}\right].$

20. $\int \dfrac{dx}{1 + \cos x},$ $\qquad\qquad \tan\tfrac{1}{2}x.$

21. $\int_{\frac{\pi}{2}} \dfrac{dx}{1 - \cos x},$ $\qquad\qquad 1 - \cot\tfrac{1}{2}x.$

22. $\int \dfrac{dx}{1 \pm \sin x},$

Multiply both terms of the fraction by $1 \mp \sin x.$ $\qquad \tan x \pm \sec x.$

23. $\int \dfrac{d\theta}{\sec\theta \pm \tan\theta},$ $\qquad\qquad \log\tan\left[\dfrac{\pi}{4}+\dfrac{\theta}{2}\right] \pm \log\cos\theta.$

24. $\int \cos\theta \cos 3\theta\, d\theta.$ *See Art. 33.* $\qquad \tfrac{1}{8}\sin 4\theta + \tfrac{1}{4}\sin 2\theta.$

25. $\int_0^{\frac{\pi}{2}} \cos\theta \cos 2\theta\, d\theta,$ $\qquad\qquad \dfrac{1}{3}.$

26. $\int_0^{\frac{\pi}{4}} \sin^2\theta \sin 2\theta\, d\theta,$ $\qquad\qquad \tfrac{1}{2}\sin^4\theta\Big]_0^{\frac{\pi}{4}} = \tfrac{1}{8}.$

27. $\int_0^{\frac{\pi}{2}} \sin 3\theta \sin 2\theta\, d\theta,$ $\qquad\qquad \dfrac{2}{5}.$

28. $\int_0^\theta \sin m\theta \cos n\theta\, d\theta$,

$$\frac{1-\cos(m+n)\theta}{2(m+n)} + \frac{1-\cos(m-n)\theta}{2(m-n)}.$$

29. $\int \cos x \cos 2x \cos 3x\, dx$,

Reduce products to sums by means of equation (2), Art. 33.

$$\frac{1}{4}\left[\frac{\sin 6x}{6} + \frac{\sin 4x}{4} + \frac{\sin 2x}{2} + x\right].$$

30. $\int_0^\pi \sqrt{(1-\cos x)}\, dx$, $\qquad 2\sqrt{2}$.

31. $\int \dfrac{dx}{a^2 \cos^2 x + b^2 \sin^2 x}$, $\qquad \dfrac{1}{ab}\tan^{-1}\left[\dfrac{b}{a}\tan x\right]$.

32. $\int \dfrac{dx}{1+\cos^2 x}$, $\qquad \dfrac{1}{\sqrt{2}}\tan^{-1}\dfrac{\tan x}{\sqrt{2}}$.

33. $\int \dfrac{dx}{a^2 \cos^2 x - b^2 \sin^2 x}$, $\qquad \dfrac{1}{2ab}\log\dfrac{a+b\tan\theta}{a-b\tan\theta}$.

34. $\int \dfrac{\sin x\, dx}{\sqrt{(3\cos^2 x + 4\sin^2 x)}}$, $\qquad \cos^{-1}\{\tfrac{1}{2}\cos x\}$.

35. $\int \dfrac{\sin x \cos^2 x\, dx}{1 + a^2 \cos^2 x}$,

Putting y for $\cos x$, the integral becomes $-\int \dfrac{y^2\, dy}{1+a^2 y^2}$.

$$-\dfrac{\cos x}{a^2} + \dfrac{\tan^{-1}(a\cos x)}{a^3}.$$

§ III.] EXAMPLES. 47

36. $\int \dfrac{d\theta}{a + b \sin \theta}$.

Put $\sin \theta = \cos (\theta - \tfrac{1}{2}\pi)$, *and use formulas* (G) *and* (G').

If $a > b$, $\dfrac{2}{\sqrt{(a^2 - b^2)}} \tan^{-1} \left[\sqrt{\dfrac{a-b}{a+b}} \tan \dfrac{2\theta - \pi}{4} \right]$.

If $a < b$, $\dfrac{1}{\sqrt{(b^2 - a^2)}} \log \dfrac{\sqrt{(b+a)} + \sqrt{(b-a)} \tan(\tfrac{1}{2}\theta - \tfrac{1}{4}\pi)}{\sqrt{(b+a)} - \sqrt{(b-a)} \tan(\tfrac{1}{2}\theta - \tfrac{1}{4}\pi)}$.

37. $\int \dfrac{d\theta}{3 + 5 \cos \theta}$, $\tfrac{1}{4} \log \dfrac{2 + \tan \tfrac{1}{2}\theta}{2 - \tan \tfrac{1}{2}\theta}$.

38. $\int \dfrac{d\theta}{5 + 3 \cos \theta}$, $\tfrac{1}{2} \tan^{-1}[\tfrac{1}{2} \tan \tfrac{1}{2}\theta]$.

39. $\int \dfrac{d\theta}{5 - 4 \cos \theta}$, $\tfrac{2}{3} \tan^{-1}\{3 \tan \tfrac{1}{2}\theta\}$.

40. $\int \dfrac{d\theta}{2 \cos \theta - 1}$, $\dfrac{1}{\sqrt{3}} \log \dfrac{1 - \sqrt{3} \tan \tfrac{1}{2}\theta}{1 + \sqrt{3} \tan \tfrac{1}{2}\theta}$.

41. $\int_0^{\tfrac{\pi}{2}} \dfrac{d\theta}{3 - \cos \theta}$, $\dfrac{\tan^{-1}\sqrt{2}}{\sqrt{2}}$.

42. $\int_0^{\tfrac{\pi}{3}} \dfrac{d\theta}{2 - \cos \theta}$, $\dfrac{\pi}{2\sqrt{3}}$.

43. $\int \dfrac{d\theta}{(1 + e \cos \theta)^2}$, *See Art.* 36.

$\dfrac{1}{(1 - e^2)^{\tfrac{3}{2}}} \cos^{-1} \dfrac{e + \cos \theta}{1 + e \cos \theta} - \dfrac{e}{1 - e^2} \dfrac{\sin \theta}{1 + e \cos \theta}$.

44. $\int_0^{\pi} \dfrac{d\theta}{(1 + e \cos \theta)^3}$, $\dfrac{(2 + e^2) \pi}{2(1 - e^2)^{\tfrac{5}{2}}}$.

45. $\int \dfrac{p \cos x + q \sin x}{a \cos x + b \sin x} dx,$

Solution :—

By adding and subtracting an undetermined constant, the fraction may be written in the form

$$\dfrac{p \cos x + q \sin x + A (a \cos x + b \sin x)}{a \cos x + b \sin x} - A,$$

we may now assume

$$p \cos x + q \sin x + A (a \cos x + b \sin x) = k (b \cos x - a \sin x);$$

the expression is then readily integrated, and A and k so determined as to make the equation last written an identity. The result is

$$\int \dfrac{p \cos x + q \sin x}{a \cos x + b \sin x} dx = \dfrac{ap + bq}{a^2 + b^2} x + \dfrac{bp - aq}{a^2 + b^2} \log (a \cos x + b \sin x).$$

46. $\int \dfrac{dx}{a + b \tan x},$ *See Ex.* 45.

$$\dfrac{ax}{a^2 + b^2} + \dfrac{b}{a^2 + b^2} \log (a \cos x + b \sin x).$$

47. Find the area of the ellipse

$$x = a \cos \phi \qquad y = b \sin \phi.$$

$$- 4ab \int_{\frac{1}{2}\pi}^{0} \sin^2 \phi\, d\phi = \pi ab.$$

48. Find the area of the cycloid

$$x = a (\psi - \sin \psi) \qquad y = a (1 - \cos \psi).$$

$$a^2 \int_0^{2\pi} (1 - \cos \psi)^2 d\psi = 3a^2 \pi.$$

49. Find the area of the trochoid $\quad (b < a)$

$$x = a\psi - b \sin \psi \qquad y = a - b \cos \psi.$$

$$(2a^2 + b^2)\pi.$$

50. Find the area of the loop, and also the area between the curve and the asymptote, in the case of the strophoid whose polar equation is

$$r = a (\sec \theta \pm \tan \theta).$$

Solution :—

Using θ as an auxiliary variable, we have

$$x = a(1 \pm \sin \theta) \qquad y = a\left[\tan \theta \pm \frac{\sin^2 \theta}{\cos \theta}\right],$$

the upper sign corresponding to the infinite branch, and the lower to the loop. Hence, for the half areas we obtain

$$+ a^2 \int_0^{\frac{1}{2}\pi} \sin \theta \, d\theta + a^2 \int_0^{\frac{1}{2}\pi} \sin^2 \theta \, d\theta = a^2 \left[1 + \frac{\pi}{4}\right]$$

and $\quad -a^2 \int_{\frac{1}{2}\pi}^0 \sin \theta \, d\theta + a^2 \int_{\frac{1}{2}\pi}^0 \sin^2 \theta \, d\theta = a^2 \left[1 - \frac{\pi}{4}\right].$

CHAPTER II.

Methods of Integration—Continued.

IV.

Integration by Change of Independent Variable.

37. If x is the independent variable used in expressing an integral, and y is any function of x, the integral may be expressed in terms of y, by substituting for x and dx their values in terms of y and dy. By properly assuming the function y, the integral may frequently be made to take a directly integrable form. For example, the integral

$$\int \frac{x\,dx}{(ax+b)^2}$$

will obviously be simplified by assuming

$$y = ax + b$$

for the new independent variable. This assumption gives

$$x = \frac{y-b}{a}, \qquad \text{whence} \qquad dx = \frac{dy}{a};$$

substituting, we have

$$\int \frac{x\,dx}{(ax+b)^2} = \frac{1}{a^2} \int \frac{(y-b)\,dy}{y^2}$$

$$= \frac{1}{a^2} \log y + \frac{b}{a^2 y};$$

§ IV.]   *CHANGE OF INDEPENDENT VARIABLE.*   51

or replacing y by x in the result,

$$\int \frac{x\, dx}{(ax+b)^2} = \frac{1}{a^2}\log(ax+b) + \frac{b}{a^2(ax+b)}.$$

38. Again, if in the integral

$$\int \frac{dx}{\epsilon^x - 1}$$

we put $y = \epsilon^x$, whence

$$x = \log y, \qquad \text{and} \qquad dx = \frac{dy}{y},$$

we have

$$\int \frac{dx}{\epsilon^x - 1} = \int \frac{dy}{y(y-1)}.$$

Hence, by formula (A), Art. 17,

$$\int \frac{dx}{\epsilon^x - 1} = \log \frac{y-1}{y} = \log(\epsilon^x - 1) - x.$$

It is easily seen that, by this change of independent variable, any integral in which the coefficient of dx is a rational function of ϵ^x, may be transformed into one in which the coefficient of dy is a rational function of y.

Transformation of Trigonometric Forms.

39. When in a trigonometric integral the coefficient of $d\theta$ is a rational function of $\tan\theta$, the integral will take a rational algebraic form if we put

$$\tan\theta = x, \qquad \text{whence} \qquad d\theta = \frac{dx}{1+x^2}.$$

For example, by this transformation, we have

$$\int \frac{d\theta}{1 + \tan\theta} = \int \frac{dx}{(1 + x^2)(1 + x)}.$$

Decomposing the fraction in the latter integral, we have

$$\int \frac{d\theta}{1 + \tan\theta} = \frac{1}{2}\int \frac{dx}{1 + x^2} - \frac{1}{2}\int \frac{x\,dx}{1 + x^2} + \frac{1}{2}\int \frac{dx}{1 + x}$$

$$= \tfrac{1}{2}\tan^{-1}x - \tfrac{1}{4}\log(1 + x^2) + \tfrac{1}{2}\log(1 + x)$$

$$= \frac{1}{2}\left[\theta + \log\frac{1 + \tan\theta}{\sec\theta}\right],$$

or $\int \dfrac{d\theta}{1 + \tan\theta} = \tfrac{1}{2}[\theta + \log(\cos\theta + \sin\theta)].$

40. The method given in the preceding article may be employed when the coefficient of $d\theta$ is a *homogeneous rational function of* $\sin\theta$ *and* $\cos\theta$, *of a degree indicated by an even integer;* for such a function is a rational function of $\tan\theta$. It may also be noticed that, when the coefficient of $d\theta$ is *any rational function* of $\sin\theta$ and $\cos\theta$, the integral becomes rational and algebraic if we put

$$z = \tan\frac{\theta}{2};$$

for this gives

$$\sin\theta = \frac{2z}{1 + z^2}, \qquad \cos\theta = \frac{1 - z^2}{1 + z^2}, \qquad d\theta = \frac{2\,dz}{1 + z^2}.$$

This transformation has in fact been already employed in the integration of $\dfrac{d\theta}{a + b\cos\theta}$. See Art. 35.

Limits of the Transformed Integral.

41. When a definite integral is transformed by a change of independent variable, it is necessary to make a corresponding change in the limits. If, for example, in the integral

$$\int_a^\infty \frac{dx}{(a^2+x^2)^2}$$

we put $\quad x = a \tan \theta, \quad$ whence $\quad dx = a \sec^2\theta \, d\theta,$

we must at the same time replace the limits a and ∞, which are values of x, by $\tfrac{1}{4}\pi$ and $\tfrac{1}{2}\pi$, the corresponding values of θ. Thus

$$\int_a^\infty \frac{dx}{(a^2+x^2)^2} = \frac{1}{a^3}\int_{\frac{\pi}{4}}^{\frac{\pi}{2}} \cos^2\theta \, d\theta$$

$$= \frac{1}{2a^3}\Big[\theta + \sin\theta\cos\theta\Big]_{\frac{\pi}{4}}^{\frac{\pi}{2}} = \frac{\pi-2}{8a^3}.$$

The Reciprocal of x taken as the New Independent Variable.

42. In the case of fractional integrals, it is sometimes useful to take the reciprocal of x as the new independent variable. For example, let the given integral be

$$\int \frac{dx}{x^3(x+1)^2}.$$

Putting $\quad x = \dfrac{1}{y}, \quad$ whence $\quad dx = -\dfrac{dy}{y^2},$

we have

$$\int \frac{dx}{x^3 (x+1)^2} = -\int \frac{y^3\, dy}{y^2\left(1+\dfrac{1}{y}\right)^2} = -\int \frac{y^3\, dy}{(y+1)^2}.$$

Transforming again by putting $z = y + 1$, the integral becomes

$$-\int \frac{(z-1)^3}{z^2}\, dz = -\int z\, dz + 3\int dz - 3\int \frac{dz}{z} + \int \frac{dz}{z^2}$$

$$= -\frac{z^2}{2} + 3z - 3\log z - \frac{1}{z}.$$

Therefore, since $z = y + 1 = \dfrac{1}{x} + 1 = \dfrac{x+1}{x}$,

$$\int \frac{dx}{x^3 (x+1)^2} = -\frac{(x+1)^2}{2x^2} + \frac{3(x+1)}{x} - \frac{x}{x+1} - 3\log \frac{x+1}{x}.$$

A Power of x taken as the New Independent Variable.

43. The transformation of an integral by the assumption,

$$y = x^n. \qquad \qquad (1)$$

is not generally useful, since the substitution

$$x = y^{\frac{1}{n}}, \qquad \text{whence} \qquad dx = \frac{1}{n} y^{\frac{1}{n}-1}\, dy,$$

will usually introduce radicals. Exceptional cases, however, occur. For, since logarithmic differentiation of equation (1) gives

$$\frac{dx}{x} = \frac{dy}{ny}, \qquad \dots \dots \dots (2)$$

it is evident that, *if the expression to be integrated is the product of $\frac{dx}{x}$ and a function of x^n, the transformed expression will be the product of $\frac{dy}{ny}$ and the like function of y.*

For example, the expression

$$\frac{(x^4 - 1)\,dx}{x(x^4 + 1)},$$

which is the product of $\frac{dx}{x}$ and a rational function of x^4, becomes

$$\frac{y - 1}{4y(y + 1)}\,dy,$$

a rational function of y. Hence, decomposing the fraction in the latter expression, we have

$$\int \frac{(x^4 - 1)\,dx}{x(x^4 + 1)} = \frac{1}{4} \int \frac{y - 1}{y(y + 1)}\,dy = \frac{1}{4} \log \frac{(y + 1)^2}{y}$$

$$= \log \frac{\sqrt{(x^4 + 1)}}{x}.$$

44. When this method is applied to an integral whose form at the same time suggests the employment of the reciprocal, as in Art. 42, we may at once assume $y = x^{-n}$. Thus, given the integral

$$\int_1^\infty \frac{dx}{x^4(2 + x^3)};$$

we obtain

$$\int_1^\infty \frac{dx}{x^4(2+x^3)} = -\frac{1}{3}\int_1^0 \frac{y\,dy}{2y+1}$$

$$= \left[-\frac{y}{6} + \frac{\log(2y+1)}{12}\right]_1^0 = \frac{2-\log 3}{12}.$$

45. The same mode of transforming may be employed to simplify the coefficient of $\frac{dx}{x}$, when this coefficient is not a rational function of x^n. Thus, the integral

$$\int \frac{dx}{x\sqrt{(x^3-a^3)}}$$

will take the form of the fundamental integral (l'), if we put

$$x^3 = y^2, \qquad \text{whence} \qquad \frac{dx}{x} = \frac{2}{3}\frac{dy}{y}.$$

Making the substitutions, we have

$$\int \frac{dx}{x\sqrt{(x^3-a^3)}} = \frac{2}{3}\int \frac{dy}{y\sqrt{(y^2-a^3)}} = \frac{2}{3a^{\frac{3}{2}}}\sec^{-1}\frac{y}{a^{\frac{3}{2}}} = \frac{2}{3a^{\frac{3}{2}}}\sec^{-1}\left(\frac{x}{a}\right)^{\frac{3}{2}}.$$

Examples IV.

√ 1. $\int \frac{1+x}{(2+x)^2}\,dx$, $\qquad\qquad \log(2+x) + \frac{1}{2+x}.$

§ IV.]   EXAMPLES.   57

2. $\displaystyle\int \frac{x\,dx}{(1-x)^3}$, $\displaystyle\frac{2x-1}{2(1-x)^2}$.

3. $\displaystyle\int \frac{x^2-x+1}{(2x+1)^2}\,dx$, $\displaystyle\frac{2x+1}{8} - \frac{\log(2x+1)}{2} - \frac{7}{8(2x+1)}$.

4. $\displaystyle\int_{-1}^{0} \frac{x^2\,dx}{(x+2)^3}$, $\displaystyle\log y + \frac{4y-2}{y^2}\bigg]_1^2 = \log 2 - \frac{1}{2}$.

5. $\displaystyle\int \frac{dx}{1+\varepsilon^x}$, $x - (\log 1 + \varepsilon^x)$.

6. $\displaystyle\int \frac{dx}{\varepsilon^x - \varepsilon^{-x}}$, $\displaystyle\frac{1}{2}\log\frac{\varepsilon^x-1}{\varepsilon^x+1}$.

7. $\displaystyle\int_{-\infty}^{0} \frac{\varepsilon^{2x}\,dx}{\varepsilon^x+1}$, $1 - \log 2$.

8. $\displaystyle\int \frac{\varepsilon^x+1}{1-\varepsilon^{-x}}\,dx$, $\varepsilon^x + 2\log(\varepsilon^x - 1)$.

9. $\displaystyle\int \frac{2+\tan\theta}{3-\tan\theta}\,d\theta$, $\displaystyle\frac{\theta - \log(3\cos\theta - \sin\theta)}{2}$.

10. $\displaystyle\int \frac{d\theta}{\tan^2\theta - 1}$, $\displaystyle\frac{1}{4}\log\frac{\tan\theta-1}{\tan\theta+1} - \frac{\theta}{2}$.

11. $\displaystyle\int \frac{\tan^2\theta\,d\theta}{\tan^2\theta - 1}$, $\displaystyle\frac{1}{4}\log\frac{\tan\theta-1}{\tan\theta+1} + \frac{\theta}{2}$.

12. $\displaystyle\int \frac{\cos\theta\,d\theta}{a\cos\theta - b\sin\theta}$, $\displaystyle\frac{a\theta - b\log(a\cos\theta - b\sin\theta)}{a^2+b^2}$.

13. $\int \dfrac{\cos\theta\, d\theta}{\cos(\alpha+\theta)}$, Put $\theta' = \alpha + \theta$.

$$(\theta+\alpha)\cos\alpha - \sin\alpha \log\cos(\theta+\alpha).$$

14. $\int \dfrac{\sin(\theta+\alpha)}{\sin(\theta+\beta)}\, d\theta$,

$$(\theta+\beta)\cos(\alpha-\beta) + \sin(\alpha-\beta)\log\sin(\theta+\beta).$$

15. $\int \tan(\theta+\alpha)\cos\theta\, d\theta$, $-\cos\theta + \sin\alpha \log\tan\dfrac{2\theta+2\alpha+\pi}{4}$

16. $\int_0^a \dfrac{\cos\theta\, d\theta}{\sin(\alpha+\theta)}$, $\cos\alpha \log(2\cos\alpha) + \alpha\sin\alpha.$

17. $\int_0^{\frac{\pi}{3}} \dfrac{\cos\frac{1}{2}\theta}{\cos\theta}\, d\theta$, $\dfrac{1}{\sqrt{2}}\log\dfrac{\sqrt{2}+2\sin\theta'}{\sqrt{2}-2\sin\theta'}\Big]_0^{\frac{\pi}{6}} = \dfrac{\log(3+2\sqrt{2})}{\sqrt{2}}$.

18. $\int \dfrac{\sin\frac{1}{2}\theta\, d\theta}{\sin\theta}$, $\log\tan\dfrac{\pi+\theta}{4}$.

19. $\int \dfrac{x^3\, dx}{(a^2+x^2)^2}$, $\frac{1}{2}\log(a^2+x^2) + \dfrac{a^2}{2(a^2+x^2)}$.

20. $\int \dfrac{dx}{x^3(1+x^2)}$, $\log\dfrac{\sqrt{(1+x^2)}}{x} - \dfrac{1}{2x^2}$.

21. $\int_0^\infty \dfrac{x^2\, dx}{(1+x^2)^3}$, $\dfrac{1}{4}\int_0^{\frac{\pi}{2}} \sin^2 2\theta\, d\theta = \dfrac{\pi}{16}$.

22. $\int_1^\infty \dfrac{dx}{x^3(1+x^2)^2}$, $\frac{3}{4} - \log 2.$

23. $\int \dfrac{dx}{x^3(x+1)}$, $\qquad -\dfrac{1}{2x^2} + \dfrac{1}{x} - \log \dfrac{x+1}{x}$.

24. $\int \dfrac{dx}{(1-x)^3 x}$, $\qquad \dfrac{1}{2(1-x)^2} + \dfrac{1}{1-x} + \log \dfrac{x}{1-x}$.

25. $\int_1^\infty \dfrac{dx}{x^3(x^2+2)}$, $\qquad \left[-\dfrac{y^2}{4} + \tfrac{1}{4}\log(2y^2+1)\right]_1^0 = \dfrac{2-\log 3}{8}$.

26. $\int \dfrac{dx}{x(a+bx^4)}$, $\qquad \dfrac{1}{4a} \log \dfrac{x^4}{a+bx^4}$.

27. $\int \dfrac{dx}{x(x^n+a^n)}$, $\qquad \dfrac{1}{na^n} \log \dfrac{x^n}{x^n+a^n}$.

28. $\int \dfrac{(x^2+1)\,dx}{x(x^3-1)}$, $\qquad \tfrac{2}{3} \log(x^3-1) - \log x$.

29. $\int \dfrac{dx}{x\sqrt{x^n-a^n}}$, $\qquad \dfrac{2}{na^{\frac{n}{2}}} \sec^{-1} \left(\dfrac{x}{a}\right)^{\frac{n}{2}}$.

V.

Integrals Containing Radicals.

46. An integral containing a single radical, in which the expression under the radical sign is of the first degree, is *rationalized*, that is, transformed into a rational integral, by taking the radical as the value of the new independent variable. Thus, given the integral

$$\int \dfrac{dx}{1+\sqrt{x+1}},$$

putting
$$y = \sqrt{(x+1)},$$
whence $x = y^2 - 1$, and $dx = 2y\,dy$,
we have

$$\int \frac{dx}{1+\sqrt{(x+1)}} = 2\int \frac{y\,dy}{1+y} = 2\int dy - 2\int \frac{dy}{1+y}$$

$$= 2y - 2\log(1+y)$$

$$= 2\sqrt{(x+1)} - 2\log[1+\sqrt{(x+1)}].$$

47. The same method evidently applies whenever all the radicals which occur in the integral are powers of a single radical, in which the expression under the radical sign is linear. Thus, in the integral

$$\int_1^2 \frac{dx}{(x-1)^{\frac{2}{3}} + (x-1)^{\frac{1}{2}}},$$

the radicals are powers of $(x-1)^{\frac{1}{6}}$; hence we put $y = (x-1)^{\frac{1}{6}}$, and obtain

$$\int_1^2 \frac{dx}{(x-1)^{\frac{2}{3}} + (x-1)^{\frac{1}{2}}} = 6\int_0^1 \frac{y^8\,dy}{y^4 + y^3}$$

$$= 6\int_0^1 (y-1)\,dy + 6\int_0^1 \frac{dy}{y+1} = -3 + 6\log 2.$$

48. An integral in which a binomial expression occurs under the radical sign can sometimes be reduced to the form considered above by the method of Art. 43. For example, since

$$\int \frac{dx}{x(x^3+1)^{\frac{1}{2}}}$$

fulfils the condition given in Art. 43, when $n = 3$, the quantity under the radical sign may be reduced to the first degree. Hence, in accordance with Art. 46, we may take the radical as the value of the new independent variable. Thus, putting

$$z = (x^3 + 1)^{\frac{1}{4}},$$

whence $\quad x^3 = z^4 - 1,\quad$ and $\quad \dfrac{dx}{x} = \dfrac{4z^3\,dz}{3(z^4-1)},$

we have

$$\int \frac{dx}{x(x^3+1)^{\frac{1}{4}}} = \frac{4}{3}\int \frac{z^2\,dz}{z^4-1}.$$

Decomposing the fraction in the latter integral as in Art. 20, we have finally

$$\int \frac{dx}{x(x^3+1)^{\frac{1}{4}}} = \frac{2}{3}\tan^{-1}\left[(x^3+1)^{\frac{1}{4}}\right] + \frac{1}{3}\log\frac{(x^3+1)^{\frac{1}{4}}-1}{(x^3+1)^{\frac{1}{4}}+1}.$$

Radicals of the Form $\sqrt{(ax^2 + b)}$.

49. It is evident that the method given in the preceding article is applicable to all integrals of the general form

$$\int x^{2m+1}(ax^2+b)^{n+\frac{1}{2}}dx, \quad \ldots \ldots \quad (1)$$

in which m and n are positive or negative integers. These integrals are therefore rationalized by putting

$$y = \sqrt{(ax^2 + b)}.$$

Putting $m = 0$, the form (1) includes the directly integrable case

$$\int (ax^2 + b)^{n+\frac{1}{2}}\, x\, dx.$$

50. As an illustration let us take the integral

$$\int \frac{dx}{x\sqrt{(x^2 + a^2)}};$$

putting
$$y = \sqrt{(x^2 + a^2)},$$

whence $\quad x^2 = y^2 - a^2,\quad$ and $\quad \dfrac{dx}{x} = \dfrac{y\, dy}{y^2 - a^2},$

we have

$$\int \frac{dx}{x\sqrt{(x^2 + a^2)}} = \int \frac{dy}{y^2 - a^2}.$$

Hence, by equation (A') Art. 17,

$$\int \frac{dx}{x\sqrt{(x^2 + a^2)}} = \frac{1}{2a}\log \frac{y - a}{y + a} = \frac{1}{2a}\log \frac{\sqrt{(x^2 + a^2)} - a}{\sqrt{(x^2 + a^2)} + a}.$$

Rationalizing the denominator of the fraction in this result, we have

$$\frac{\sqrt{(x^2 + a^2)} - a}{\sqrt{(x^2 + a^2)} + a} = \frac{[\sqrt{(x^2 + a^2)} - a]^2}{x^2}.$$

Therefore

$$\int \frac{dx}{x\sqrt{(x^2 + a^2)}} = \frac{1}{a}\log \frac{\sqrt{(x^2 + a^2)} - a}{x} \quad \ldots \ldots (H)$$

In a similar manner we may prove that

$$\int \frac{dx}{x\sqrt{(a^2-x^2)}} = \frac{1}{a}\log\frac{a-\sqrt{(a^2-x^2)}}{x} . \quad \ldots \quad (I)$$

51. Integrals of the form

$$\int x^{2m}(ax^2+b)^{n+\frac{1}{2}}dx \quad \ldots \quad \ldots \quad (2)$$

are reducible to the form (1) Art. 49, by first putting $y = \dfrac{1}{x}$. For example:

$$\int \frac{dx}{(ax^2+b)^{\frac{3}{2}}}$$

is of the form (2); but, putting $x = \dfrac{1}{y}$, whence

$$\sqrt{(ax^2+b)} = \frac{\sqrt{(a+by^2)}}{y} \qquad \text{and} \qquad dx = -\frac{dy}{y^2},$$

we obtain

$$\int \frac{dx}{(ax^2+b)^{\frac{3}{2}}} = -\int \frac{y\,dy}{(a+by^2)^{\frac{3}{2}}}.$$

The resulting expression is in this case directly integrable. Thus

$$\int \frac{dx}{(ax^2+b)^{\frac{3}{2}}} = \frac{1}{b\sqrt{(a+by^2)}} = \frac{x}{b\sqrt{(ax^2+b)}} . \quad \ldots \quad (J)$$

Integration of $\dfrac{dx}{\sqrt{(x^2 \pm a^2)}}$.

52. If we assume a new variable z connected with x by the relation

$$z - x = \sqrt{(x^2 \pm a^2)}, \quad \ldots \ldots \quad (1)$$

we have, by squaring,

$$z^2 - 2zx = \pm a^2, \quad \ldots \ldots \quad (2)$$

and, by differentiating this equation,

$$2(z - x)\,dz - 2z\,dx = 0;$$

whence

$$\frac{dx}{z-x} = \frac{dz}{z},$$

or by equation (1),

$$\frac{dx}{\sqrt{(x^2 \pm a^2)}} = \frac{dz}{z}. \quad \ldots \ldots \quad (3)$$

Integrating equation (3), we obtain

$$\int \frac{dx}{\sqrt{(x^2 \pm a^2)}} = \log z = \log\left[x + \sqrt{(x^2 \pm a^2)}\right]. \quad \ldots \quad (K)$$

53. Since the value of x in terms of z, derived from equation (2) of the preceding article, is rational, it is obvious that this transformation may be employed to rationalize any expression which consists of the product of $\dfrac{dx}{\sqrt{(x^2 \pm a^2)}}$ and a rational function of x. For example, let us find the value of

$$\int \sqrt{(x^2 \pm a^2)}\,dx,$$

which may be written in the form

$$\int (x^2 \pm a^2) \frac{dx}{\sqrt{(x^2 \pm a^2)}}.$$

By equation (2)

$$x = \frac{z^2 \mp a^2}{2z}, \quad \ldots \ldots \quad (4)$$

whence

$$x^2 \pm a^2 = \frac{(z^2 \pm a^2)^2}{4z^2} \ldots \ldots \quad (5)$$

Therefore, by equations (3) and (5),

$$\int \sqrt{(x^2 \pm a^2)}\, dx = \frac{1}{4} \int \frac{(z^2 \pm a^2)^2}{z^3}\, dz$$

$$= \frac{1}{4}\int z\, dz \pm \frac{a^2}{2}\int \frac{dz}{z} + \frac{a^4}{4}\int \frac{dz}{z^3}$$

$$= \frac{z^4 - a^4}{8z^2} \pm \frac{a^2}{2} \log z.$$

By equations (4) and (5), the first term of the last member is equal to $\frac{1}{2} x \sqrt{(x^2 \pm a^2)}$. Hence

$$\int \sqrt{(x^2 \pm a^2)}\, dx = \frac{x \sqrt{(x^2 \pm a^2)}}{2} \pm \frac{a^2}{2} \log [x + \sqrt{(x^2 \pm a^2)}] \quad . \quad (L)$$

Transformation to Trigonometric Forms.

54. Integrals involving either of the radicals

$$\sqrt{(a^2 - x^2)}, \qquad \sqrt{(a^2 + x^2)}, \qquad \text{or} \qquad \sqrt{(x^2 - a^2)}$$

can be transformed into rational trigonometric integrals. The transformation is effected in the first case by putting

$$x = a \sin \theta, \quad \text{whence} \quad \sqrt{(a^2 - x^2)} = a \cos \theta;$$

in the second case, by putting

$$x = a \tan \theta, \quad \text{whence} \quad \sqrt{(a^2 + x^2)} = a \sec \theta;$$

and in the third case, by putting

$$x = a \sec \theta, \quad \text{whence} \quad \sqrt{(x^2 - a^2)} = a \tan \theta.$$

55. As an illustration, let us take the integral

$$\int \sqrt{(a^2 - x^2)}\, dx\,;$$

putting $x = a \sin \theta$, we have $\sqrt{(a^2 - x^2)} = a \cos \theta$, $dx = a \cos \theta\, d\theta$; hence

$$\int \sqrt{(a^2 - x^2)}\, dx = a^2 \int \cos^2 \theta\, d\theta$$

$$= \frac{a^2 \theta}{2} + \frac{a^2 \sin \theta \cos \theta}{2},$$

by formula (*C*) Art. 29. Replacing θ by x in the result,

$$\int \sqrt{(a^2 - x^2)}\, dx = \frac{a^2}{2} \sin^{-1}\frac{x}{a} + \frac{x \sqrt{(a^2 - x^2)}}{2}. \quad . \quad . \quad (M)$$

Regarding the radical as a positive quantity, the value of θ may be restricted to the *primary* value of the symbol $\sin^{-1}\frac{x}{a}$ (see Diff. Calc., Art. 54); that is, as x passes from $-a$ to $+a$, θ passes from $-\frac{1}{2}\pi$ to $+\frac{1}{2}\pi$.

§ V.]   *INTEGRALS CONTAINING RADICALS.*   67

Radicals of the Form $\sqrt{(ax^2 + bx + c)}$.

56. When a radical of the form $\sqrt{(ax^2 + bx + c)}$ occurs in an integral, a simple change of independent variable will cause the radical to assume one of the forms considered in the preceding articles. Thus, if the coefficient of x^2 is positive,

$$\sqrt{(ax^2 + bx + c)} = \sqrt{a}\sqrt{\left[\left(x + \frac{b}{2a}\right)^2 + \frac{4ac - b^2}{4a^2}\right]},$$

in which, if we put $x + \dfrac{b}{2a} = y$, the radical takes the form $\sqrt{(y^2 + a^2)}$ or $\sqrt{(y^2 - a^2)}$, according as $4ac - b^2$ is positive or negative. If a is negative, the radical can in like manner be reduced to the form $\sqrt{(a^2 - y^2)}$ or $\sqrt{(-a^2 - y^2)}$; but the latter will never occur, since it is imaginary for all values of y, and therefore imaginary for all values of x.

For example, by this transformation, the integral

$$\int \frac{dx}{(ax^2 + bx + c)^{\frac{3}{2}}}$$

can be reduced at once to the form $(\mathcal{J})$, Art. 51. Thus

$$\int \frac{dx}{(ax^2 + bx + c)^{\frac{3}{2}}} = \int \frac{dx}{\left[a\left(x + \dfrac{b}{2a}\right)^2 + \dfrac{4ac - b^2}{4a}\right]^{\frac{3}{2}}}$$

$$= \frac{x + \dfrac{b}{2a}}{\dfrac{4ac - b^2}{4a}\sqrt{(ax^2 + bx + c)}} = \frac{4ax + 2b}{(4ac - b^2)\sqrt{(ax^2 + bx + c)}}.$$

57. When the form of the integral suggests a further change of independent variable, we may at once assume the expression for the new variable in the required form. For example, given the integral

$$\int \sqrt{(2ax - x^2)}\, x\, dx;$$

we have $\quad \sqrt{(2ax - x^2)} = \sqrt{[a^2 - (x - a)^2]}$

hence (see Art. 54), if we put $x - a = a \sin \theta$, we have

$$\sqrt{(2ax - x^2)} = a \cos \theta,$$

$$x = a(1 + \sin \theta), \qquad dx = a \cos \theta\, d\theta;$$

$$\therefore \int \sqrt{(2ax - x^2)}\, x\, dx = a^3 \int \cos^2 \theta (1 + \sin \theta)\, d\theta$$

$$= \frac{a^3}{2}(\theta + \sin \theta \cos \theta) - \frac{a^3}{3} \cos^3 \theta$$

$$= \frac{a^3}{2} \sin^{-1} \frac{x-a}{a} + \frac{a}{2}(x-a)\sqrt{(2ax - x^2)} - \frac{1}{3}(2ax - x^2)^{\frac{3}{2}}$$

$$= \frac{a^3}{2} \sin^{-1} \frac{x-a}{a} + \frac{1}{6}\sqrt{(2ax - x^2)}\,[2x^2 - ax - 3a^2].$$

The Integrals

$$\int \frac{dx}{\sqrt{[(x-\alpha)(x-\beta)]}} \quad \text{and} \quad \int \frac{dx}{\sqrt{[(x-\alpha)(\beta-x)]}}.$$

58. An integral of the form $\int \dfrac{dx}{\sqrt{(ax^2 + bx + c)}}$ may by the method of Art. 56, be reduced to the form (K), Art. 52, or to the form (j'), Art. 10, according as a is positive or negative.

But when the quantity under the radical sign can be resolved into linear factors, the formulas deduced below give the value of the integral in forms which are sometimes more convenient.

If α and β are the roots of the equation

$$ax^2 + bx + c = 0,$$

the integral may be put in the form

$$\frac{1}{\sqrt{a}} \int \frac{dx}{\sqrt{[(x-\alpha)(x-\beta)]}} \quad \text{or} \quad \frac{1}{\sqrt{(-a)}} \int \frac{dx}{\sqrt{[(x-\alpha)(\beta-x)]}},$$

according as a is positive or negative. Assuming

$$\sqrt{(x-\alpha)} = z, \quad \text{whence} \quad x = z^2 + \alpha \quad \text{and} \quad dx = 2z\,dz,$$

we have

$$\int \frac{dx}{\sqrt{[(x-\alpha)(x-\beta)]}} = 2 \int \frac{dz}{\sqrt{(z^2+\alpha-\beta)}} = 2\log[z+\sqrt{(z^2+\alpha-\beta)}],$$

by formula (K), Art. 52; hence

$$\int \frac{dx}{\sqrt{[(x-\alpha)(x-\beta)]}} = 2\log\left[\sqrt{(x-\alpha)} + \sqrt{(x-\beta)}\right] \quad . \quad . \quad . \quad (N)$$

In like manner we have

$$\int \frac{dx}{\sqrt{[(x-\alpha)(\beta-x)]}} = 2 \int \frac{dz}{\sqrt{(\beta-\alpha-z^2)}} = 2\sin^{-1}\frac{z}{\sqrt{(\beta-\alpha)}},$$

by formula (j'); hence

$$\int \frac{dx}{\sqrt{[(x-\alpha)(\beta-x)]}} = 2\sin^{-1}\sqrt{\frac{x-\alpha}{\beta-\alpha}} \quad . \quad . \quad . \quad (O)$$

It can be shown that the values given in formulas (N) and (O) differ only by constants from the results derived by employing the process given in Art. 56.

Examples V.

1. $\int \sqrt{(a-x)} \cdot x\, dx$, $\qquad -\dfrac{2}{15}(a-x)^{\frac{3}{2}}(3x+2a)$.

2. $\int \sqrt{(x+a)} \cdot x^2\, dx$, $\dfrac{2}{7}(a+x)^{\frac{7}{2}} - \dfrac{4a}{5}(a+x)^{\frac{5}{2}} + \dfrac{2a^2}{3}(a+x)^{\frac{3}{2}}$.

3. $\int \dfrac{x\, dx}{1+\sqrt{x}}$, $\qquad \dfrac{2}{3}x^{\frac{3}{2}} - x + 2\sqrt{x} - 2\log(1+\sqrt{x})$.

4. $\int \dfrac{x\, dx}{\sqrt{(x+a)}}$, $\qquad \dfrac{2}{3}(x-2a)\sqrt{(x+a)}$.

5. $\int \dfrac{dx}{\sqrt{x}-1}$, $\qquad 2\sqrt{x} + 2\log(1-\sqrt{x})$.

6. $\int_{-a}^{0} (a+x)^{\frac{1}{2}} x\, dx$, $\qquad \left[\dfrac{3v^7}{7} - \dfrac{3ay^5}{4}\right]_0^{a^{\frac{1}{2}}} = -\dfrac{9a^{\frac{7}{2}}}{28}$.

7. $\int \dfrac{dx}{x\sqrt{(2ax-a^2)}}$, $\qquad \dfrac{2}{a}\tan^{-1}\sqrt{\dfrac{2x-a}{a}}$.

8. $\int_0^a (a-x)^{\frac{3}{2}} x^2\, dx$, $\qquad \left[-\dfrac{2y^9}{9} + \dfrac{4ay^7}{7} - \dfrac{2a^2y^5}{5}\right]_{\sqrt{a}}^{0} = \dfrac{16a^{\frac{9}{2}}}{315}$.

9. $\int \dfrac{dx}{2x^{\frac{1}{2}} - x^{\frac{1}{3}}}$, $\qquad x^{\frac{1}{2}} + \dfrac{3x^{\frac{1}{3}}}{4} + \dfrac{3x^{\frac{1}{6}}}{4} + \dfrac{3\log(2x^{\frac{1}{6}}-1)}{8}$.

§ V.]   EXAMPLES.   71

10. $\int_0^1 (x+1)^{\frac{2}{3}} x\, dx,$ $\qquad \left[\dfrac{3y^8}{8} - \dfrac{3y^5}{5}\right]_1^{\sqrt[3]{2}} = \dfrac{3\sqrt[3]{4}}{10} + \dfrac{9}{40}.$

11. $\int \dfrac{1-3x}{\sqrt{(1-x)}}\, dx,$ $\qquad 2(1+x)\sqrt{(1-x)}.$

12. $\int \dfrac{x\, dx}{x - \sqrt{(x^2 - a^2)}},$ Rationalize the denominator.

$\qquad \dfrac{x^4 + (x^2-a^2)^{\frac{3}{2}}}{3a^2}.$

13. $\int \dfrac{dx}{\sqrt{(x+a)} + \sqrt{(x+b)}},$ $\qquad \dfrac{2(x+a)^{\frac{3}{2}} - 2(x+b)^{\frac{3}{2}}}{3(a-b)}$

14. $\int \dfrac{dx}{x\sqrt{(x^4+1)}},$ $\qquad \dfrac{1}{4}\log \dfrac{\sqrt{(x^4+1)} - 1}{\sqrt{(x^4+1)} + 1}.$

15. $\int \dfrac{\sqrt{(x^4+1)}\, dx}{x},$ $\qquad \dfrac{\sqrt{(x^4+1)}}{2} + \dfrac{1}{4}\log \dfrac{\sqrt{(x^4+1)} - 1}{\sqrt{(x^4+1)} + 1}.$

16. $\int \dfrac{(x^n+1)(x^n-1)^{\frac{3}{2}}}{x}\, dx,$

$\dfrac{2}{n}\left[\dfrac{(x^n-1)^{\frac{5}{2}}}{5} + \dfrac{(x^n-1)^{\frac{3}{2}}}{3} - (x^n-1)^{\frac{1}{2}} + \tan^{-1}\sqrt{(x^n-1)}\right].$

17. $\int_0^a \dfrac{x^5\, dx}{(x^2+a^2)^{\frac{3}{2}}},$ $\qquad \left[\dfrac{y^3}{3} - 2a^2 y - \dfrac{a^4}{y}\right]_a^{a\sqrt{2}} = \left[\dfrac{8}{3} - \dfrac{11\sqrt{2}}{6}\right]a^3.$

18. $\int \dfrac{\sqrt{(x^2-a^2)}}{x}\, dx \left[= \int \dfrac{x^2-a^2}{x\sqrt{(x^2-a^2)}}\, dx\right],$

$\qquad \sqrt{(x^2-a^2)} - a\sec^{-1}\dfrac{x}{a}.$

19. $\int \dfrac{x^2\,dx}{\sqrt{(x^2+a^2)}},$

$$\left[= \int \dfrac{x^2+a^2-a^2}{\sqrt{(x^2+a^2)}}\,dx. \quad \text{See formulas } (L) \text{ and } (K). \right]$$

$$\dfrac{1}{2} x \sqrt{(x^2+a^2)} - \dfrac{1}{2} a^2 \log\,[x + \sqrt{(x^2+a^2)}].$$

20. $\int \dfrac{\sqrt{(a^2-x^2)}}{x}\,dx, \qquad a \log \dfrac{a - \sqrt{(a^2-x^2)}}{x} + \sqrt{(a^2-x^2)}.$

✓ 21. $\int \dfrac{dx}{x + \sqrt{(x^2+a^2)}},$ Rationalize denominator

$$\dfrac{x}{2a^2} \sqrt{(x^2+a^2)} + \dfrac{1}{2} \log\,[x + \sqrt{(x^2+a^2)}] - \dfrac{x^2}{2a^2}.$$

22. $\int \dfrac{\sqrt{(x^2+a^2)}}{x^2}\,dx, \qquad$ See Art. 51.

$$\log\,[\sqrt{(x^2+a^2)} + x] - \dfrac{\sqrt{(x^2+a^2)}}{x}.$$

23. $\int \dfrac{dx}{\sqrt{(x^2+a^2)} - a},$

$$\log\,[\sqrt{(x^2+a^2)} + x] - \dfrac{\sqrt{(x^2+a^2)}}{x} - \dfrac{a}{x}.$$

✓ 24. $\int \dfrac{x\,dx}{\sqrt{(1+x^4)}}. \qquad$ See Formula (K). $\quad \dfrac{1}{2} \log\,[x^2 + \sqrt{(1+x^4)}].$

§ V.]   EXAMPLES.   73

25. $\int \sqrt{(ax^2 + b)}\, dx$, $[a > 0]$ Put $\sqrt{(ax^2 + b)} = z - x\sqrt{a}$.

$$\frac{b}{2\sqrt{a}} \log[x\sqrt{a} + \sqrt{(ax^2 + b)}] + \frac{1}{2} x\sqrt{(ax^2 + b)}.$$

26. $\int \dfrac{dx}{(a + x)\sqrt{(x^2 + b^2)}}$,

$$\frac{1}{\sqrt{(a^2 + b^2)}} \log \frac{x + \sqrt{(x^2 + b^2)} + a - \sqrt{(a^2 + b^2)}}{x + \sqrt{(x^2 + b^2)} + a + \sqrt{(a^2 + b^2)}}.$$

27. $\int \dfrac{dx}{x^2\sqrt{(1 + x^2)}}$, $-\dfrac{\sqrt{(1 + x^2)}}{x}$.

28. $\int_{\frac{1}{2}}^{1} \dfrac{dx}{x^2\sqrt{(1 - x^2)}}$, $-\cot\theta\Big]_{\frac{\pi}{6}}^{\frac{\pi}{2}} = \sqrt{3}$.

29. $\int \dfrac{x^3\, dx}{(x^2 - a^2)^{\frac{3}{2}}}$, $\dfrac{x^2 - 2a^2}{\sqrt{(x^2 - a^2)}}$.

30. $\int \dfrac{dx}{(p + qx)\sqrt{(x^2 + 1)}}$,

$$\frac{1}{\sqrt{(p^2 + q^2)}} \log \tan \frac{1}{2}\left[\tan^{-1} x + \tan^{-1}\frac{p}{q}\right].$$

31. $\int \dfrac{dx}{x^3\sqrt{(x^2 - 1)}}$, $-\dfrac{\sqrt{(x^2 - 1)}}{2x^2} + \dfrac{1}{2}\sec^{-1} x$.

32. $\int_{0}^{a} \dfrac{x^2\, dx}{(x^2 + a^2)^{\frac{3}{2}}}$ $\log \tan \dfrac{3\pi}{8} - \dfrac{\sqrt{2}}{2}$.

33. $\int \frac{dx}{x^4 \sqrt{(x^2-1)}}$, $x = \sec\theta$ $\qquad (2x^2+1)\frac{\sqrt{(x^2-1)}}{3x^3}$.

34. $\int \frac{dx}{(a-x)\sqrt{(a^2-x^2)}}$, $\qquad \frac{1}{a}\sqrt{\frac{a+x}{a-x}}$.

35. $\int \frac{dx}{(1+x^2)\sqrt{(1-x^2)}}$, $\qquad \frac{1}{\sqrt{2}}\tan^{-1}\frac{x\sqrt{2}}{\sqrt{(1-x^2)}}$.

36. $\int_0^a \frac{x^{\frac{5}{2}} dx}{\sqrt{(a-x)}} \left[= \int_0^a \frac{x^3 dx}{\sqrt{(ax-x^2)}} \right]$

$\qquad \left(\frac{1}{2}a\right)^3 \int_{-\frac{\pi}{2}}^{\frac{\pi}{2}} (1+\sin\theta)^3 d\theta = \frac{5\pi a^3}{16}$

37. $\int \frac{dx}{x\sqrt{(2ax-x^2)}}$, $\qquad -\frac{1}{a}\sqrt{\frac{2a-x}{x}}$.

38. $\int \frac{dx}{x^3\sqrt{(x^4-1)}}$, $\qquad$ Put $x^2 = z = \sec\theta$. $\qquad \frac{\sqrt{(x^4-1)}}{2x^2}$.

39. $\int_0^a \sqrt{(2ax-x^2)} \cdot dx$, $\qquad \frac{a^2\pi}{4}$.

40. $\int_0^a \sqrt{(2ax-x^2)} \cdot x\, dx$,

$\qquad a^3 \int_{-\frac{\pi}{2}}^{0} \cos^2\theta (1+\sin\theta) d\theta = a^3 \left[\frac{\pi}{4}-\frac{1}{3}\right]$.

41. $\int_0^a \sqrt{(2ax-x^2)} \cdot x^2 dx$,

$\qquad a^4 \int_{-\frac{\pi}{2}}^{0} \cos^2\theta (1+\sin\theta)^2 d\theta = a^4 \left[\frac{5\pi}{16}-\frac{2}{3}\right]$.

§ V.]   EXAMPLES.   75

42. $\int \dfrac{dx}{\sqrt{(2ax + x^2)}}$,

   by Art. 56, $\log [x + a + \sqrt{(2ax + x^2)}] + C$;

   by Art. 58, $\log [\sqrt{x} + \sqrt{(2a + x)}] + C'$.

43. $\int \dfrac{x\,dx}{\sqrt{(2ax + x^2)}}$, $\sqrt{(2ax + x^2)} - a \log [x + a + \sqrt{(2ax + x^2)}]$.

44. $\int \sqrt{\dfrac{x}{2a - x}}\, dx \left[= \int \dfrac{x\,dx}{\sqrt{(2ax - x^2)}} \right]$,

       $a \sin^{-1} \dfrac{x - a}{a} - \sqrt{(2ax - x^2)}$.

45. $\int \dfrac{dx}{\sqrt{(5 + 4x - x^2)}}$, by Art. 56, $\sin^{-1} \dfrac{x - 2}{3} + C$;

   by Art. 58, $2 \sin^{-1} \sqrt{\dfrac{x + 1}{6}} + C'$.

46. $\int_0^a \dfrac{dx}{\sqrt{(ax - x^2)}}$, $2 \sin^{-1} \sqrt{\dfrac{x}{a}} \Big]_0^a = \pi$.

47. $\int \dfrac{x^2\,dx}{\sqrt{(3 + 2x - x^2)}}$, $3 \sin^{-1} \dfrac{x - 1}{2} - \dfrac{(x + 3)\sqrt{(3 + 2x - x^2)}}{2}$.

48. $\int_{-2}^{-\frac{1}{2}} \dfrac{dx}{\sqrt{(2 - x - x^2)}}$, $\dfrac{\pi}{2}$.

49. $\int_a^{2a} \dfrac{dx}{\sqrt{(x^2 - ax)}}$, $\log (3 + 2\sqrt{2})$.

50. Find the area included by the rectangular hyperbola

$$y^2 = 2ax + x^2,$$

and the double ordinate of the point for which $x = 2a$.

$$a^2[6\sqrt{2} - \log(3 + 2\sqrt{2})].$$

51. Find the area included between the cissoid

$$x(x^2 + y^2) = 2ay^2$$

and the coördinates of the point (a, a); also the whole area between the curve and its asymptote.

$$\left(\frac{3}{4}\pi - 2\right)a^2, \quad \text{and} \quad 3\pi a^2.$$

52. Find the area of the loop of the strophoid

$$x(x^2 + y^2) + a(x^2 - y^2) = 0;$$

also the area between the curve and its asymptote.

$$2a^2\left(1 - \frac{\pi}{4}\right), \quad \text{and} \quad 2a^2\left(1 + \frac{\pi}{4}\right).$$

For the loop put $y = -x\dfrac{a+x}{\sqrt{(a^2-x^2)}}$, since x is negative between the limits $-a$ and 0.

53. Show that the area of the segment of an ellipse between the minor axis and any double ordinate is $ab \sin^{-1}\dfrac{x}{a} + xy$.

VI.

Integration by Parts.

59. Let u and v be any two functions of x; then since

$$d(uv) = u\,dv + v\,du,$$

$$uv = \int u\,dv + \int v\,du,$$

whence
$$\int u\,dv = uv - \int v\,du \quad \ldots \quad \ldots \quad (1)$$

By means of this formula, the integration of an expression of the form $u\,dv$, in which dv is the differential of a known function v, may be made to depend upon the integration of the expression $v\,du$. For example, if

$$u = \cos^{-1} x \quad \text{and} \quad dv = dx,$$

we have
$$du = -\frac{dx}{\sqrt{(1-x^2)}};$$

hence, by equation (1),

$$\int \cos^{-1} x \cdot dx = x \cos^{-1} x + \int \frac{x\,dx}{\sqrt{(1-x^2)}},$$

in which the new integral is directly integrable; therefore

$$\int \cos^{-1} x \cdot dx = x \cos^{-1} x - \sqrt{(1-x^2)}.$$

The employment of this formula is called *integration by parts*.

Geometrical Illustration.

60. The formula for integration by parts may be geometrically illustrated as follows. Assuming rectangular axes, let the curve be constructed in which the abscissa and ordinate of each point are corresponding values of v and u, and let this curve cut one of the axes in B. From any point P of this curve draw PR and PS, perpendicular to the axes. Now the area $PBOR$ is a value of the indefinite integral $\int u\,dv$, and in like manner the area PBS is a value of $\int v\,du$; and we have

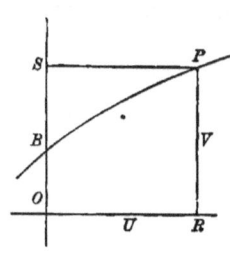

Fig. 2.

$$\text{Area } PBOR = \text{Rectangle } PSOR - \text{Area } PBS;$$

therefore

$$\int u\,dv = uv - \int v\,du.$$

Applications.

61. In general there will be more than one possible method of selecting the factors u and dv. The latter of course includes the factor dx, but it will generally be advisable to include in it any other factors which permit the direct integration of dv. After selecting the factors, it will be found convenient at once to write the product $u \cdot v$, separating the factors by a period; this will serve as a guide in forming the product

$v\,du$, which is to be written under the integral sign. Thus, let the given integral be

$$\int x^2 \log x\, dx.$$

Taking $x^2 dx$ as the value of dv, since we can integrate this expression directly, we have

$$\int x^2 \log x\, dx = \log x \cdot \frac{1}{3} x^3 - \frac{1}{3}\int x^3 \frac{dx}{x}$$

$$= \frac{1}{3} x^3 \log x - \frac{1}{3}\int x^2\, dx$$

$$= \frac{x^3}{9}(3 \log x - 1).$$

62. The form of the new integral may be such that a second application of the formula is required before a directly integrable form is produced. For example, let the given integral be

$$\int x^2 \cos x\, dx.$$

In this case we take $\cos x\, dx = dv$; so that having $x^2 = u$, the new integral will contain a lower power of x: thus

$$\int x^2 \cos x\, dx = x^2 \cdot \sin x - 2\int x \sin x\, dx.$$

Making a second application of the formula, we have

$$\int x^2 \cos x\, dx = x^2 \sin x - 2\left[x(-\cos x) + \int \cos x\, dx\right]$$

$$= x^2 \sin x + 2x \cos x - 2 \sin x.$$

63. The method of integration by parts is sometimes employed with advantage, even when the new integral is no simpler than the given one; for, in the process of successive applications of the formula, the original integral may be reproduced, as in the following example:

$$\int \varepsilon^{mx} \sin(nx + \alpha)\, dx$$

$$= \varepsilon^{mx} \cdot \frac{-\cos(nx + \alpha)}{n} + \frac{m}{n} \int \varepsilon^{mx} \cos(nx + \alpha)\, dx$$

$$= -\frac{\varepsilon^{mx} \cos(nx+\alpha)}{n} + \frac{m}{n} \varepsilon^{mx} \frac{\sin(nx+\alpha)}{n} - \frac{m^2}{n^2} \int \varepsilon^{mx} \sin(nx+\alpha)\, dx,$$

in which the integral in the second member is identical with the given integral; hence, transposing and dividing,

$$\int \varepsilon^{mx} \sin(nx + \alpha)\, dx = \frac{\varepsilon^{mx}}{m^2 + n^2}[m \sin(nx + \alpha) - n \cos(nx + \alpha)].$$

64. In some cases it is necessary to employ some other mode of transformation, in connection with the method of parts. For example, given the integral

$$\int \sec^3 \theta\, d\theta;$$

taking $dv = \sec^2 \theta\, d\theta$, we have

$$\int \sec^3 \theta\, d\theta = \sec \theta \cdot \tan \theta - \int \sec \theta \tan^2 \theta\, d\theta. \quad . \quad . \quad (1)$$

If now we apply the method of parts to the new integral, by putting

$$\sec\theta\tan\theta\,d\theta = dv,$$

the original integral will indeed be reproduced in the second member; but it will disappear from the equation, the result being an identity. If, however, in equation (1), we transform the final integral by means of the equation $\tan^2\theta = \sec^2\theta - 1$, we have

$$\int\sec^3\theta\,d\theta = \sec\theta\tan\theta - \int\sec^3\theta\,d\theta + \int\sec\theta\,d\theta.$$

Transposing,

$$2\int\sec^3\theta\,d\theta = \frac{\sin\theta}{\cos^2\theta} + \int\frac{d\theta}{\cos\theta};$$

hence, by formula (F), Art. 31,

$$\int\sec^3\theta\,d\theta = \frac{\sin\theta}{2\cos^2\theta} + \frac{1}{2}\log\tan\left[\frac{\pi}{4} + \frac{\theta}{2}\right].$$

Formulas of Reduction.

65. It frequently happens that the new integral introduced by applying the method of parts differs from the given integral only in the values of certain constants. If these constants are expressed algebraically, the formula expressing the first transformation is adapted to the successive transformations of the new integrals introduced, and is called *a formula of reduction*.

For example, applying the method of parts to the integral

$$\int x^n \varepsilon^{ax}\, dx,$$

we have

$$\int x^n \varepsilon^{ax}\, dx = x^n \cdot \frac{\varepsilon^{ax}}{a} - \frac{n}{a}\int x^{n-1} \varepsilon^{ax}\, dx, \quad \ldots \quad (1)$$

in which the new integral is of the same form as the given one, the exponent of x being decreased by unity. Equation (1) is therefore a formula of reduction for this function. Supposing n to be a positive integer, we shall finally arrive at the integral $\int \varepsilon^{ax}\, dx$, whose value is $\dfrac{\varepsilon^{ax}}{a}$. Thus, by successive application of equation (1) we have

$$\int x^n \varepsilon^{ax}\, dx = \frac{\varepsilon^{ax}}{a}\left[x^n - \frac{n}{a} x^{n-1} \ldots + (-1)^n \frac{n(n-1)\cdots 1}{a^n}\right].$$

Reduction of $\int \sin^m \theta\, d\theta$ and $\int \cos^m \theta\, d\theta$.

66. To obtain a formula of reduction, it is sometimes necessary to make a further transformation of the equation obtained by the method of parts. Thus, for the integral

$$\int \sin^m \theta\, d\theta,$$

the method of parts gives

$$\int \sin^m \theta\, d\theta = \sin^{m-1}\theta\,(-\cos\theta) + (m-1)\int \sin^{m-2}\theta \cos^2\theta\, d\theta.$$

§ VI.] REDUCTION OF TRIGONOMETRIC INTEGRALS. 83

Substituting in the latter integral $1 - \sin^2\theta$ for $\cos^2\theta$,

$$\int \sin^m \theta\, d\theta = -\sin^{m-1}\theta \cos\theta$$

$$+ (m-1)\int \sin^{m-2}\theta\, d\theta - (m-1)\int \sin^m \theta\, d\theta;$$

transposing and dividing, we have

$$\int \sin^m \theta\, d\theta = -\frac{\sin^{m-1}\theta \cos\theta}{m} + \frac{m-1}{m}\int \sin^{m-2}\theta\, d\theta, \quad \ldots \quad (1)$$

a formula of reduction in which the exponent of $\sin\theta$ is diminished two units. By successive application of this formula, we have, for example:

$$\int \sin^6 \theta\, d\theta = -\frac{\sin^5\theta \cos\theta}{6} + \frac{5}{6}\int \sin^4\theta\, d\theta$$

$$= -\frac{\sin^5\theta \cos\theta}{6} - \frac{5}{6}\cdot\frac{\sin^3\theta \cos\theta}{4} + \frac{5\cdot 3}{6\cdot 4}\int \sin^2\theta\, d\theta$$

$$= -\frac{\sin^5\theta \cos\theta}{6} - \frac{5\sin^3\theta \cos\theta}{6\cdot 4} - \frac{5\cdot 3 \sin\theta \cos\theta}{6\cdot 4\cdot 2} + \frac{5\cdot 3\cdot 1}{6\cdot 4\cdot 2}\theta.$$

67 By a process similar to that employed in deriving equation (1), or simply by putting $\theta = \frac{1}{2}\pi - \theta$ in that equation, we find

$$\int \cos^m \theta\, d\theta = \frac{\cos^{m-1}\theta \sin\theta}{m} + \frac{m-1}{m}\int \cos^{m-2}\theta\, d\theta, \quad \ldots \quad (2)$$

a formula of reduction, when m is positive.

68. It should be noticed that, when m is negative, equation (1) Art. 66 is not a formula of reduction, because the exponent in the new integral is in that case numerically greater than the exponent in the given integral. But, if we now regard the integral in the second member as the given one, the equation is readily converted into a formula of reduction. Thus, putting $-n$ for the negative exponent $m-2$, whence

$$m = -n + 2,$$

transposing and dividing, equation (1) becomes

$$\int \frac{d\theta}{\sin^n \theta} = -\frac{\cos \theta}{(n-1)\sin^{n-1}\theta} + \frac{n-2}{n-1}\int \frac{d\theta}{\sin^{n-2}\theta}. \quad \ldots \quad (3)$$

Again, putting $\theta = \tfrac{1}{2}\pi - \theta'$ in this equation, we obtain

$$\int \frac{d\theta}{\cos^n \theta} = \frac{\sin \theta}{(n-1)\cos^{n-1}\theta} + \frac{n-2}{n-1}\int \frac{d\theta}{\cos^{n-2}\theta} \quad \ldots \quad (4)$$

Reduction of $\int \sin^m \theta \cos^n \theta \, d\theta$.

69. If we put $dv = \sin^m \theta \cos \theta \, d\theta$, we have

$$\int \sin^m \theta \cos^n \theta \, d\theta = \frac{\cos^{n-1}\theta \sin^{m+1}\theta}{m+1}$$

$$+ \frac{n-1}{m+1}\int \sin^{m+2}\theta \cos^{n-2}\theta \, d\theta; \quad \ldots \quad (1)$$

but, if in the same integral we put $dv = \cos^n \theta \sin \theta \, d\theta$, we have

$$\int \sin^m \theta \cos^n \theta \, d\theta = -\frac{\sin^{m-1}\theta \cos^{n+1}\theta}{n+1}$$

$$+ \frac{m-1}{n+1}\int \sin^{m-2}\theta \cos^{n+2}\theta \, d\theta. \quad \ldots \quad (2)$$

§ VI.] REDUCTION OF TRIGONOMETRIC INTEGRALS.

When m and n are both positive, equation (1) is not a formula of reduction, since in the new integral the exponent of $\sin \theta$ is increased, while that of $\cos \theta$ is diminished. We therefore substitute in this integral

$$\sin^{m+2}\theta = \sin^m \theta (1 - \cos^2 \theta),$$

so that the last term of the equation becomes

$$\frac{n-1}{m+1}\int \sin^m \theta \cos^{n-2}\theta \, d\theta - \frac{n-1}{m+1}\int \sin^m \theta \cos^n \theta \, d\theta.$$

Hence, by this transformation, the original integral is reproduced, and equation (1) becomes

$$\left[1 + \frac{n-1}{m+1}\right]\int \sin^m \theta \cos^n \theta \, d\theta = \frac{\sin^{m+1}\theta \cos^{n-1}\theta}{m+1}$$

$$+ \frac{n-1}{m+1}\int \sin^m \theta \cos^{n-2}\theta \, d\theta.$$

Dividing by $1 + \dfrac{n-1}{m+1} = \dfrac{m+n}{m+1}$, we have

$$\int \sin^m \theta \cos^n \theta \, d\theta = \frac{\sin^{m+1}\theta \cos^{n-1}\theta}{m+n}$$

$$+ \frac{n-1}{m+n}\int \sin^m \theta \cos^{n-2}\theta \, d\theta, \quad \cdot \quad \cdot \quad (3)$$

a formula of reduction by which the exponent of $\cos \theta$ is diminished two units.

By a similar process, from equation (2), or simply by putting $\theta = \tfrac{1}{2}\pi - \theta'$ in equation (3), and interchanging m and n, we obtain

$$\int \sin^m \theta \cos^n \theta \, d\theta = -\frac{\sin^{m-1} \theta \cos^{n+1} \theta}{m+n}$$

$$+ \frac{m-1}{m+n} \int \sin^{m-2} \theta \cos^n \theta \, d\theta, \quad \ldots \quad (4)$$

a formula by which the exponent of $\sin \theta$ is diminished two units.

70. When n is positive and m negative, equation (1) of the preceding article is itself a formula of reduction, for both exponents are in that case numerically diminished. Putting $-m$ in place of m, the equation becomes

$$\int \frac{\cos^n \theta}{\sin^m \theta} d\theta = -\frac{\cos^{n-1} \theta}{(m-1)\sin^{m-1} \theta} - \frac{n-1}{m-1} \int \frac{\cos^{n-2}}{\sin^{m-2}} d\theta. \quad \ldots \quad (5)$$

Similarly, when m is positive and n negative, equation (2) gives

$$\int \frac{\sin^m \theta}{\cos^n \theta} d\theta = \frac{\sin^{m-1} \theta}{(n-1)\cos^{n-1} \theta} - \frac{m-1}{n-1} \int \frac{\sin^{m-2} \theta}{\cos^{n-2} \theta} d\theta. \quad \ldots \quad (6)$$

71. When m and n are both negative, putting $-m$ and $-n$ in place of m and n, equation (3) Art. 69 becomes

$$\int \frac{d\theta}{\sin^m \theta \cos^n \theta} = -\frac{1}{(m+n)\sin^{m-1} \theta \cos^{n+1} \theta}$$

$$+ \frac{n+1}{m+n} \int \frac{d\theta}{\sin^m \theta \cos^{n+2} \theta},$$

in which the exponent of $\cos \theta$ is numerically increased. We may therefore regard the integral in the second member as the integral to be reduced. Thus, putting n in place of $n+2$, we derive

$$\int \frac{d\theta}{\sin^m \theta \cos^n \theta} = \frac{1}{(n-1)\sin^{m-1}\theta \cos^{n-1}\theta}$$

$$+ \frac{m+n-2}{n-1} \int \frac{d\theta}{\sin^m \theta \cos^{n-2}\theta}. \quad \ldots \quad (7)$$

Putting $\theta = \tfrac{1}{2}\pi - \theta'$, and interchanging m and n, we have

$$\int \frac{d\theta}{\sin^m \theta \cos^n \theta} = - \frac{1}{(m-1)\sin^{m-1}\theta \cos^{n-1}\theta}$$

$$+ \frac{m+n-2}{m-1} \int \frac{d\theta}{\sin^{m-2}\theta \cos^n \theta}. \quad \ldots \quad (8)$$

72. The application of the formulas derived in the preceding articles to definite integrals will be given in the next section. When the value of the indefinite integral is required, it should first be ascertained whether the given integral belongs to one of the directly integrable cases mentioned in Arts. 27 and 28. If it does not, the formulas of reduction must be used, and if m and n are integers, we shall finally arrive at a directly integrable form.

As an illustration, let us take the integral

$$\int \sin^2 \theta \cos^4 \theta \, d\theta.$$

Employing formula (4) Art. 69, by which the exponent of $\sin \theta$ is diminished, we have

$$\int \sin^2 \theta \cos^4 \theta \, d\theta = -\frac{\sin \theta \cos^5 \theta}{6} + \frac{1}{6}\int \cos^4 \theta \, d\theta.$$

The last integral can be reduced by means of formula (2) Art. 67, which, when $m = 4$, gives

$$\int \cos^4 \theta \, d\theta = \frac{\cos^3 \theta \sin \theta}{4} + \frac{3}{4} \int \cos^2 \theta \, d\theta;$$

therefore

$$\int \sin^2 \theta \cos^4 \theta \, d\theta = \frac{\sin \theta \cos^5 \theta}{6} + \frac{\cos^3 \theta \sin \theta}{24} + \frac{\sin \theta \cos \theta}{16} + \frac{\theta}{16}.$$

73. Again, let the given integral be

$$\int \frac{\cos^6 \theta \, d\theta}{\sin^3 \theta}.$$

By equation (5), Art. 70, we have

$$\int \frac{\cos^6 \theta \, d\theta}{\sin^3 \theta} = -\frac{\cos^5 \theta}{2 \sin^2 \theta} - \frac{5}{2} \int \frac{\cos^4 \theta \, d\theta}{\sin \theta}.$$

We cannot apply the same formula to the new integral, since the denominator $m - 1$ vanishes; but putting $n = 4$ and $m = -1$, in equation (3) Art. 69, we have

$$\int \frac{\cos^4 \theta \, d\theta}{\sin \theta} = \frac{\cos^3 \theta}{3} + \int \frac{\cos^2 \theta \, d\theta}{\sin \theta}$$

$$= \frac{\cos^3 \theta}{3} + \int \frac{d\theta}{\sin \theta} - \int \sin \theta \, d\theta$$

$$= \frac{\cos^3 \theta}{3} + \log \tan \frac{1}{2} \theta + \cos \theta.$$

Hence

$$\int \frac{\cos^6 \theta \, d\theta}{\sin^3 \theta} = -\frac{\cos^5 \theta}{2 \sin^2 \theta} - \frac{5 \cos^3 \theta}{6} - \frac{5}{2} \log \tan \frac{1}{2} \theta - \frac{5}{2} \cos \theta.$$

Extension of the Formula.

74. Let
$$\int \phi(x)\,dx = \phi_{,}(x),$$

$$\int \phi_{,}(x)\,dx = \phi_{,,}(x),$$

etc., etc.;

then, if the functions $\phi_{,}(x), \phi_{,,}(x), \ldots \phi_{n}(x)$, which may be called the *successive integrals* of $\phi(x)$, are known, and also the successive derivatives of $f(x)$, we shall have

$$\int f(x)\phi(x)\,dx = f(x)\phi_{,}(x) - \int f'(x)\phi_{,}(x)\,dx$$

$$= f(x)\phi_{,}(x) - f'(x)\phi_{,,}(x) + \int f''(x)\phi_{,,}(x)\,dx.$$

Continuing this process, and writing for shortness $f, \phi_{,}, \ldots$ for $f(x), \phi_{,}(x) \ldots$ we have

$$\int f(x)\phi(x)\,dx = f\cdot\phi_{,} - f'\cdot\phi_{,,} + \cdots\cdots + (-1)^{n-1} f^{n-1}\cdot\phi_{n}$$

$$+ (-1)^{n} \int f^{n}\cdot\phi_{n}\,dx.$$

The application of this formula is equivalent to the use of a formula of reduction. Thus the value of $\int x^{n} \epsilon^{ax}$ given in Art. 65, may be derived immediately from it.

Taylor's Theorem.

75. If, in the formula of the preceding article, we put

$$f(x) = F'(x_0 + h - x), \qquad \text{and} \qquad \phi(x) = 1,$$

x_0 and h being constants,

$$f'(x) = -F''(x_0 + h - x), \qquad f''(x) = F'''(x_0 + h - x), \text{ etc.};$$

and $\quad \phi_{\prime}(x) = x, \quad \phi_{\prime\prime}(x) = \dfrac{x^2}{1\cdot 2}, \quad \phi_{\prime\prime\prime}(x) = \dfrac{x^3}{1\cdot 2\cdot 3}$, etc.

Hence

$$\int F'(x_0 + h - x)\, dx = F'(x_0 + h - x)\cdot x + F''(x_0 + h - x)\dfrac{x^2}{1\cdot 2}$$

$$+ \cdots + \int F^{n+1}(x_0 + h - x)\dfrac{x^n}{1\cdot 2\cdots n}\, dx.$$

Now $\quad \int F'(x_0 + h - x)\, dx = -F(x_0 + h - x);$

hence, applying the limits 0 and h, we have

$$F(x_0 + h) = F(x_0) + F'(x_0)\, h + F''(x_0)\dfrac{h^2}{1\cdot 2} + \cdots$$

$$+ \int_0^h F^{n+1}(x_0 + h - x)\dfrac{x^n\, dx}{1\cdot 2\cdots n}.$$

This formula is Taylor's Theorem, with the remainder expressed in the form of a definite integral.

Examples VI.

1. $\int_0^1 \sin^{-1} x\, dx,\qquad x\sin^{-1}x + \sqrt{(1-x^2)}\Big]_0^1 = \dfrac{\pi}{2} - 1.$

2. $\int \sec^{-1} x\, dx,\qquad x\sec^{-1}x - \log[x + \sqrt{(x^2-1)}].$

3. $\int_0^1 \tan^{-1} x\, dx,\qquad \dfrac{\pi}{4} - \dfrac{\log 2}{2}.$

4. $\int x^n \log x\, dx,\qquad \dfrac{x^{n+1}}{n+1}\left[\log x - \dfrac{1}{n+1}\right].$

5. $\int_0^{\frac{\pi}{2}} \theta \sin\theta\, d\theta,\qquad 1.$

6. $\int_0^{\frac{\pi}{2}} \theta \cos m\theta\, d\theta,\qquad \dfrac{\pi}{2m} - \dfrac{1}{m^2}.$

7. $\int x \tan^{-1} x\, dx,\qquad \dfrac{1+x^2}{2}\tan^{-1}x - \dfrac{x}{2}.$

8. $\int_0^{} x^2 \varepsilon^x\, dx,\qquad x^2\varepsilon^x - 2x\varepsilon^x + 2\varepsilon^x - 2.$

9. $\int x \sec^{-1} x\, dx,\qquad \tfrac{1}{2}[x^2\sec^{-1}x - \sqrt{(x^2-1)}].$

10. $\int_0^{\frac{\pi}{2}} \theta \sin\left[\dfrac{\pi}{4} + \theta\right]d\theta,\quad -\theta\cos\left(\dfrac{\pi}{4}+\theta\right) + \sin\left(\dfrac{\pi}{4}+\theta\right)\Big]_0^{\frac{\pi}{2}} = \dfrac{\pi\sqrt{2}}{4}.$

11. $\int x \sec^2 x \, dx$, $\qquad x \tan x + \log \cos x$.

12. $\int x \tan^2 x \, dx \left[= \int x (\sec^2 x - 1) \, dx \right]$, $x \tan x + \log \cos x - \frac{1}{2} x^2$.

13. $\int x^2 \sin x \, dx$, $\qquad 2x \sin x + 2 \cos x - x^2 \cos x$.

14. $\int_0^1 x \sin^{-1} x \, dx$, $\qquad \frac{1}{2} x^2 \sin^{-1} x \Big]_0^1 - \frac{1}{2} \int_0^{\frac{\pi}{2}} \sin^2 \theta \, d\theta = \frac{\pi}{8}$.

15. $\int x^2 \tan^{-1} x \, dx$, $\qquad \dfrac{x^3 \tan^{-1} x}{3} - \dfrac{x^2}{6} + \dfrac{\log(1+x^2)}{6}$.

16. $\int_0^1 x^2 \sin^{-1} x \, dx$, $\dfrac{1}{3} x^3 \sin^{-1} x + \dfrac{2+x^2}{9} \sqrt{(1-x^2)} \Big]_0^1 = \dfrac{\pi}{6} - \dfrac{2}{9}$.

17. $\int_0^\infty \varepsilon^{-x} \cos x \, dx$, $\qquad \dfrac{\varepsilon^{-x}(\sin x - \cos x)}{2} \Big]_0^\infty = \dfrac{1}{2}$.

18. $\int \varepsilon^{x \tan \beta} \cos x \, dx$, $\qquad \cos \beta \, \varepsilon^{x \tan \beta} \sin(\beta + x)$.

19. $\int \varepsilon^{-x} \sin^2 x \, dx \left[= \dfrac{1}{2} \int \varepsilon^{-x} (1 - \cos 2x) \, dx \right]$,

$\qquad \dfrac{\varepsilon^{-x}}{10} (\cos 2x - 2 \sin 2x - 5)$.

20. $\int_0^{\frac{\pi}{4}} \varepsilon^\theta \sin \theta \, d\theta$, $\qquad \dfrac{\varepsilon^\theta}{2} (\sin \theta - \cos \theta) \Big]_0^{\frac{\pi}{4}} = \dfrac{1}{2}$.

§ VI.] EXAMPLES. 93

21. $\int \epsilon^x \sin x \cos x \, dx,$ $\qquad \dfrac{\epsilon^x}{10}(\sin 2x - 2 \cos 2x).$

22. $\int_0 \sin^4 m\theta \, d\theta,$ $\qquad -\dfrac{\sin^3 m\theta \cos m\theta}{4m} + \dfrac{3\theta}{8} - \dfrac{3\sin m\theta \cos m\theta}{8m}.$

23. Derive a formula of reduction for $\int (\log x)^n x^m \, dx$, and deduce from it the value of $\int (\log x)^3 x^2 \, dx$.

$$\int (\log x)^n x^m \, dx = (\log x)^n \frac{x^{m+1}}{m+1} - \frac{n}{m+1} \int (\log x)^{n-1} x^m \, dx.$$

$$\int (\log x)^3 x^2 \, dx = (\log x)^3 \frac{x^3}{3} - (\log x)^2 \frac{x^3}{3} + \frac{2x^3 \log x}{9} - \frac{2x^3}{27}.$$

24. $\int x \cos^2 x \, dx,$ $\qquad \tfrac{1}{2} x \sin x \cos x - \tfrac{1}{4} \sin^2 x + \tfrac{1}{4} x^2.$

25. $\int_1 x^2 \sec^{-1} x \, dx,$ $\dfrac{x^3 \sec^{-1} x}{3} - \dfrac{x\sqrt{(x^2-1)}}{6} - \dfrac{\log[x + \sqrt{(x^2-1)}]}{6}.$

26. Derive a formula of reduction for $\int x^n \sin(x+\alpha) \, dx$, and deduce from it the value of $\int x^5 \cos x$.

$$\int x^n \sin(x+\alpha) \, dx = -x^n \sin\left[x + \alpha + \frac{\pi}{2}\right]$$
$$+ n \int x^{n-1} \sin\left[x + \alpha + \frac{\pi}{2}\right] dx.$$

$$\int x^5 \cos x \, dx = (x^5 - 20x^3 + 120x) \sin x + (5x^4 - 60x^2 + 120) \cos x.$$

27. $\int \cos^2\theta \sin^4\theta\, d\theta$, $\quad \dfrac{\sin^5\theta\cos\theta}{6} - \dfrac{\sin^3\theta\cos\theta}{24} + \dfrac{1}{16}[\theta - \sin\theta\cos\theta]$.

28. $\int_0^{\frac{\pi}{4}} \cos^4\theta \sin^4\theta\, d\theta$, $\quad \dfrac{1}{32}\int_0^{\frac{\pi}{2}} \sin^4\theta'\, d\theta' = \dfrac{\pi}{512}$.

29. $\int_0^{\frac{\pi}{4}} \cos^4\theta\, d\theta$, $\quad \dfrac{\sin\theta\cos^3\theta}{4} + \dfrac{3\sin\theta\cos\theta}{8} + \dfrac{3\theta}{8}\Big]_0^{\frac{\pi}{4}} = \dfrac{8 + 3\pi}{32}$.

30. $\int_0^{\frac{\pi}{3}} \cos^6\theta\, d\theta$,

$\dfrac{\sin\theta\cos\theta(8\cos^4\theta + 10\cos^2\theta + 15) + 15\theta}{48}\Big]_0^{\frac{\pi}{3}} = \dfrac{9\sqrt{3} + 10\pi}{96}$.

31. $\int \dfrac{\cos^4\theta}{\sin^3\theta}\, d\theta$, $\quad -\dfrac{\cos^3\theta}{2\sin^2\theta} - \dfrac{3\cos\theta}{2} - \dfrac{3\log\tan\frac{1}{2}\theta}{2}$.

32. $\int \dfrac{\sin^2\theta}{\cos^5\theta}\, d\theta$, $\quad \dfrac{\sin\theta}{4\cos^4\theta} - \dfrac{\sin\theta}{8\cos^2\theta} - \dfrac{1}{8}\log\tan\left[\dfrac{\pi}{4} + \dfrac{\theta}{2}\right]$.

33. $\int \dfrac{\sin^6\theta}{\cos^4\theta}\, d\theta$, $\quad \dfrac{\sin^5\theta}{3\cos^3\theta} - \dfrac{5\sin^3\theta}{3\cos\theta} + \dfrac{5}{2}[\theta - \sin\theta\cos\theta]$.

34. $\int_{\frac{\pi}{4}}^{\frac{\pi}{2}} \dfrac{\cos^6\theta}{\sin^2\theta}\, d\theta$, $\quad -\dfrac{\cos^5\theta}{\sin\theta}\Big]_{\frac{\pi}{4}}^{\frac{\pi}{2}} - 5\int_{\frac{\pi}{4}}^{\frac{\pi}{2}} \cos^4\theta\, d\theta = \dfrac{48 - 15\pi}{32}$.

35. $\int_0^{\frac{\pi}{2}} \dfrac{d\theta}{(1 + \cos\theta)^2}$, $\quad \dfrac{1}{2}\int_0^{\frac{\pi}{4}} \dfrac{d\theta'}{\cos^4\theta'} = \dfrac{2}{3}$.

36. $\int \dfrac{d\theta}{\sin\theta\cos^4\theta}$, $\quad \dfrac{1}{3\cos^3\theta} + \dfrac{1}{\cos\theta} + \log\tan\dfrac{\theta}{2}$.

§ VI.] EXAMPLES. 95

37. $\int \dfrac{d\theta}{\sin\theta \sin^2 2\theta}$, $\dfrac{1}{4\sin^2\theta\cos\theta} - \dfrac{3\cos\theta}{8\sin^2\theta} + \dfrac{3}{8}\log\tan\dfrac{\theta}{2}$.

38. Prove that when n is odd

$$\int \dfrac{d\theta}{\sin\theta \cos^n\theta} = \dfrac{\sec^{n-1}\theta}{n-1} + \dfrac{\sec^{n-3}\theta}{n-3} + \cdots\cdots + \log\tan\theta\;;$$

and when n is even

$$\int \dfrac{d\theta}{\sin\theta \cos^n\theta} = \dfrac{\sec^{n-1}\theta}{n-1} + \dfrac{\sec^{n-3}\theta}{n-3} + \cdots\cdots + \log\tan\dfrac{\theta}{2}.$$

39. $\int \dfrac{d\theta}{\sin^2\theta \cos^4\theta}$, $-\dfrac{1}{2\sin^2\theta\cos^3\theta} + \dfrac{5}{2}\left[\dfrac{\sec^2\theta}{3} + \sec\theta + \log\tan\dfrac{\theta}{2}\right]$.

40. $\int \dfrac{x^2\,dx}{\sqrt{(x^2-1)}}$, *Put* $x = \sec\theta$.

$\tfrac{1}{2}x\sqrt{(x^2-1)} + \tfrac{1}{2}\log[x + \sqrt{(x^2-1)}]$.

41. $\int (a^2 - x^2)^{\frac{3}{2}}\,dx$, $\dfrac{(5a^2 - 2x^2)\,x\sqrt{(a^2-x^2)}}{8} + \dfrac{3a^4}{8}\sin^{-1}\dfrac{x}{a}$.

42. $\int_0^a \dfrac{dx}{(a^2+x^2)^3}$, $\dfrac{1}{a^5}\int_0^{\frac{\pi}{4}}\cos^4\theta\,d\theta = \dfrac{3\pi+8}{32a^5}$.

43. $\int (a^2+x^2)\sqrt{(a^2-x^2)}\,dx$, $\dfrac{5a^4}{8}\sin^{-1}\dfrac{x}{a} + \dfrac{x\sqrt{(a^2-x^2)}[3a^2+2x^2]}{8}$.

44. $\int \dfrac{x^3\,dx}{(x^2-a^2)^{\frac{3}{2}}}$, $-\dfrac{x}{\sqrt{(x^2-a^2)}} + \log[x + \sqrt{(x^2-a^2)}]$.

45. $\int \dfrac{x^2\,dx}{(x^2+1)^3}$, $\dfrac{x(x^2-1)}{8(x^2+1)^2} + \dfrac{\tan^{-1}x}{8}$.

46. $\int \dfrac{\cos^2\theta - \sin^2\theta}{(\sin\theta + \cos\theta)^2} d\theta \left[= \int (1 + \sin\theta \cos\theta) \dfrac{\cos\theta - \sin\theta}{(\sin\theta + \cos\theta)^2} d\theta \right]$,

$$\dfrac{\sin\theta \cos\theta}{\sin\theta + \cos\theta}.$$

47. Derive a formula for the reduction of $\int x \sec^n x \, dx$; and referring to Ex. 11, thence show that this is an integrable form when n is an even integer. Give the result when $n = 4$.

$$\int x \sec^n x \, dx = \dfrac{x \sec^{n-2} x \tan x}{n-1} - \dfrac{\sec^{n-2} x}{(n-1)(n-2)}$$

$$+ \dfrac{n-2}{n-1} \int x \sec^{n-2} x \, dx.$$

$$\int x \sec^4 x \, dx = \dfrac{x \sec^2 x \tan x}{3} - \dfrac{\sec^2 x}{6} + \dfrac{2}{3} [x \tan x + \log \cos x].$$

48. Derive a formula of reduction for $\int x \cos^n x \, dx$, and deduce from it the value of $\int x \cos^3 x \, dx$.

$$\int x \cos^n x \, dx = \dfrac{x \cos^{n-1} x \sin x}{n} + \dfrac{\cos^n x}{n^2} + \dfrac{n-1}{n} \int x \cos^{n-2} x \, dx.$$

$$\int x \cos^3 x \, dx = \dfrac{x \sin x}{3} (\cos^2 x + 2) + \dfrac{\cos x}{9} (\cos^2 x + 6).$$

49. Find the area between the curve

$$y = \sec^{-1} x,$$

the axis of x, and the ordinate corresponding to $x = 2$.

$$\frac{2\pi}{3} - \log[2 + \sqrt{3}] = 0.77744.$$

50. Find the area between the axis of x, the curve

$$y = \tan^{-1} x,$$

and the ordinate corresponding to $x = 1$. $\dfrac{\pi}{4} - \dfrac{\log 2}{2} = 0.43882.$

VII.

Definite Integrals.

76. Before proceeding to transformations of definite integrals involving the values of the limits, it is necessary to resume the consideration of the relations between a definite integral and its limits, as defined in the first section.

By definition, the symbol

$$\int_a^X f(x)\,dx$$

denotes the quantity generated at the rate

$$f(x)\frac{dx}{dt},$$

while x passes from the initial value a to the final value X. The rate of x is arbitrary, and may be assumed constant; but in that case its sign must be the same as that of the increment received by x; that is, the sign of dx is the same as that of $X-a$.

These considerations often serve to determine the sign of an integral. Thus

$$\int_0^\pi \frac{\sin x\, dx}{x}$$

denotes a positive quantity, because dx is positive, and $\dfrac{\sin x}{x}$ is positive for all values of x between 0 and π.

77. Now let $F(x)$ denote a value of the indefinite integral, so that

$$d\{F(x)\} = f(x)\, dx;$$

thus $f(x)$ is the derivative of $F(x)$. Then, *supposing* F (x) *to vary continuously as* x *passes from* a *to* X; that is, to have no infinite or imaginary values for values of x between a and X, the integral is the actual increment received by $F(x)$, while x passes from a to X. In this case, therefore

$$\int_a^X f(x)\, dx = F(X) - F(a) \quad \ldots \quad (1).$$

If, on the other hand, there is any value, α, between a and X, such that

$$F(\alpha) = \infty,$$

equation (1) does not hold true. For example,

$$\int \frac{dx}{x^2} = -\frac{1}{x},$$

and in the case of the definite integral

$$\int_{-1}^{1} \frac{dx}{x^2}$$

x passes through the value zero, for which $F(x)$ is infinite; *we cannot therefore write*

$$\int_{-1}^{1} \frac{dx}{x^2} = -\frac{1}{x}\Big]_{-1}^{1} = -2.$$

This result indeed is obviously false, since dx is here positive, and x^2 is never negative for real values of x. The value of the integral is in fact infinite, since the increments received by $-\frac{1}{x}$, while x passes from -1 to 0, and while x passes from 0 to 1, are both infinite and positive.

78. Since the derivative of a function becomes infinite when the function becomes infinite, [Diff. Calc., Art. 104; Abridged Ed., Art. 89], we can have $F(a) = \infty$ only when $f(a) = \infty$; but it is to be noticed that $F(x)$ does not necessarily become infinite when $f(x)$ becomes infinite. Thus, in

$$\int_{-1}^{8} \frac{dx}{x^{\frac{1}{3}}}$$

$f(x) = x^{-\frac{1}{3}}$, which becomes infinite for $x = 0$, a value of x between the limits; but since

$$\int x^{-\frac{1}{3}} dx = \tfrac{3}{2} x^{\frac{2}{3}}$$

the indefinite integral $F(x)$ does not become infinite. Therefore equation (1) holds true, and

$$\int_{-1}^{8} \frac{dx}{x^{\frac{1}{3}}} = \frac{3}{2} x^{\frac{2}{3}}\Big]_{-1}^{8} = \frac{9}{2}.$$

79. We have, in the preceding articles, assumed that the independent variable varies uniformly in passing from the lower to the upper limit; but when a change of independent variable is made, the new variable does not generally vary uniformly between its limits. It is, however, obvious, that, in equation (1), Art. 77, x may vary in any manner whatever in passing from a to X, provided that $F(x)$ *remains throughout a continuous one-valued function;* x may even pass through infinity, provided $F(x)$ is finite and one-valued when $x = \infty$.

Multiple-Valued Integrals.

80. When the indefinite integral is a multiple-valued function, a particular value of this function must of course be employed, and it is necessary to take care that this value varies continuously while x passes from the lower to the upper limit. In the fundamental formula (j) it is sufficient (provided the radical $\sqrt{(1 - x^2)}$ does not change sign), to limit the meaning of the symbols $\sin^{-1} x$ and $\cos^{-1} x$ to the primary values of these symbols (see Diff. Calc., Arts. 54 and 55), since these values are so taken as to vary continuously while x passes through all its possible values from -1 to $+1$.

81. In the case of formula (k) the primary value of $\tan^{-1} x$ is so defined that, as x passes from $-\infty$ to $+\infty$, the primary value varies continuously from $-\frac{1}{2}\pi$ to $+\frac{1}{2}\pi$. We may therefore employ the primary value at both limits, *unless* x *passes through infinity*, as in the following example. Given the integral

$$\int_0^{\frac{5\pi}{6}} \frac{d\theta}{\cos^2\theta + 9\sin^2\theta} = \int_0^{\frac{5\pi}{6}} \frac{\sec^2\theta\, d\theta}{1 + 9\tan^2\theta},$$

if we put $\tan\theta = x$, this becomes

$$\int_0^{-\frac{\sqrt{3}}{3}} \frac{dx}{1 + 9x^2} = \frac{1}{3} \tan^{-1} 3x \Big]_0^{-\frac{\sqrt{3}}{3}} = \frac{1}{3}[\tan^{-1}(-\sqrt{3}) - \tan^{-1} 0].$$

But here it is to be noticed, that, as θ passes from 0 to $\frac{5}{6}\pi$, x

§ VII.] MULTIPLE-VALUED INTEGRALS. 101

passes through infinity when $\theta = \frac{1}{2}\pi$. Hence, if the value of $\tan^{-1} 3x$ is taken as 0 at the lower limit, it is to be regarded as increasing and passing through $\frac{1}{2}\pi$, when $x = \infty$, so that its value at the upper limit is $\frac{2}{3}\pi$, and not $-\frac{1}{3}\pi$. Hence

$$\int_0^{\frac{5\pi}{6}} \frac{d\theta}{\cos^2\theta + 9\sin^2\theta} = \frac{2\pi}{9}.$$

82. When the symbol $\cot^{-1} x$ is employed, the primary value, defined in the same manner as in the case of $\tan^{-1} x$, cannot be taken at both limits *when x passes through zero*. Thus, using the second form of (k), Art. 10, we have

$$\int_1^{-1} \frac{dx}{1 + x^2} = \cot^{-1} 1 - \cot^{-1}(-1),$$

in which, if $\cot^{-1} 1$ is taken as $\frac{1}{4}\pi$, $\cot^{-1}(-1)$ must be taken as $\frac{3}{4}\pi$. Thus

$$\int_1^{-1} \frac{dx}{1 + x^2} = -\frac{1}{2}\pi.$$

Formulas of Reduction for Definite Integrals.

83. The limits of a definite integral are very often such as to simplify materially the formula of reduction appropriate to it. For example, to reduce

$$\int_0^\infty x^n \varepsilon^{-x} dx,$$

we have by the method of parts

$$\int x^n \varepsilon^{-x} dx = -x^n \varepsilon^{-x} + n \int \varepsilon^{-x} x^{n-1} dx.$$

Now, supposing n positive, the quantity $x^n \varepsilon^{-x}$ vanishes when $x = 0$, and also when $x = \infty$ [See Diff. Calc., Art. 107; Abridged Ed., Art. 91]. Hence, applying the limits 0 and ∞,

$$\int_0^\infty x^n \varepsilon^{-x} dx = n \int_0^\infty x^{n-1} \varepsilon^{-x} dx.$$

By successive application of this formula we have, when n is an integer,

$$\int_0^\infty x^n \varepsilon^{-x} dx = n(n-1)\cdots\cdots 2 \cdot 1.$$

84. From equation (1) Art. 66, supposing $m > 1$, we have

$$\int_0^{\frac{\pi}{2}} \sin^m \theta \, d\theta = \frac{m-1}{m} \int_0^{\frac{\pi}{2}} \sin^{m-2} \theta \, d\theta.$$

If m is an integer, we shall, by successive application of this formula, finally arrive at $\int_0^{\frac{\pi}{2}} d\theta = \frac{\pi}{2}$ or $\int_0^{\frac{\pi}{2}} \sin \theta \, d\theta = 1$, according as m is even or odd. Hence

if m is even, $\quad \int_0^{\frac{\pi}{2}} \sin^m \theta \, d\theta = \frac{(m-1)(m-3)\cdots 1}{m(m-2)\cdots\cdots 2} \cdot \frac{\pi}{2}, \ldots (P)$

and if m is odd, $\quad \int_0^{\frac{\pi}{2}} \sin^m \theta \, d\theta = \frac{(m-1)(m-3)\cdots 2}{m(m-2)\cdots\cdots 1}. \ldots (P')$

§ VII.] *FORMULAS OF REDUCTION.* 103

85. From equations (3) and (4) Art. 69, we derive

$$\int_0^{\frac{\pi}{2}} \sin^m \theta \cos^n \theta \, d\theta = \frac{n-1}{m+n} \int_0^{\frac{\pi}{2}} \sin^m \theta \cos^{n-2} \theta \, d\theta,$$

and $\quad \int_0^{\frac{\pi}{2}} \sin^m \theta \cos^n \theta \, d\theta = \frac{m-1}{m+n} \int_0^{\frac{\pi}{2}} \sin^{m-2} \theta \cos^n \theta \, d\theta.$

By successive application of these formulas, we shall have for the final integral one of the four forms

$$\int_0^{\frac{\pi}{2}} d\theta, \quad \int_0^{\frac{\pi}{2}} \sin \theta \, d\theta, \quad \int_0^{\frac{\pi}{2}} \cos \theta \, d\theta, \quad \text{or} \quad \int_0^{\frac{\pi}{2}} \sin \theta \cos \theta \, d\theta.$$

The numerator of the final fraction $\left(\frac{n-1}{m+n} \text{ or } \frac{m-1}{m+n}\right)$ is in each case either 2 or 1. In the first case, the value of the final integral is $\frac{1}{2}\pi$, and the final denominator is 2: in the second and third cases, the value of the final integral is 1, and the final denominator is 3: in the fourth case, the value of the final integral is $\frac{1}{2}$, and the final denominator is 4. Therefore (since the factors in the denominator proceed by intervals of 2), it is readily seen that we may write

$$\int_0^{\frac{\pi}{2}} \sin^m \theta \cos^n \theta \, d\theta = \frac{(m-1)(m-3)\cdots(n-1)(n-3)\cdots}{(m+n)(m+n-2)\cdots\cdots\cdots} \alpha, \quad . (Q)$$

provided that each series of factors is carried to 2 or 1, *and α is taken equal to unity, except when* m *and* n *are both even, in which case* $\alpha = \frac{1}{2}\pi$.

Elementary Theorems Relating to Definite Integrals.

86. The following propositions are obvious consequences of equation (1), Art. 77.

$$\int_a^b f(x)\,dx = -\int_b^a f(x)\,dx \quad \ldots \ldots \quad (1)$$

$$\int_a^b f(x)\,dx = \int_a^c f(x)\,dx + \int_c^b f(x)\,dx \quad \ldots \quad (2)$$

Again, if we put $x = a + b - z$, we have

$$\int_a^b f(x)\,dx = -\int_b^a f(a+b-z)\,dz = \int_a^b f(a+b-z)\,dz$$

by (1), or since it is indifferent whether we write z or x for the variable in a definite integral,

$$\int_a^b f(x)\,dx = \int_a^b f(a+b-x)\,dx \quad \ldots \ldots \quad (3)$$

If $a = 0$, we have the particular case

$$\int_0^b f(x)\,dx = \int_0^b f(b-x)\,dx \quad \ldots \ldots \quad (4)$$

87. As an application of formula (4), we have

$$\int_0^{\frac{\pi}{2}} \cos^m \theta \, d\theta = \int_0^{\frac{\pi}{2}} \cos^m\left(\frac{\pi}{2} - \theta\right) d\theta = \int_0^{\frac{\pi}{2}} \sin^m \theta \, d\theta \quad \ldots \quad (1)$$

Hence the value of $\int_0^{\frac{\pi}{2}} \cos^m \theta \, d\theta$ as well as that of $\int_0^{\frac{\pi}{2}} \sin^m \theta \, d\theta$ is given by formulas (P) and (P'). The values of these integrals are readily found when the limits are any multiples of $\frac{1}{2}\pi$. For, by equation (2) of the preceding article, we may sum the values in the several quadrants. But, putting $\theta = k\dfrac{\pi}{2} + \theta'$, and employing equation (1), we have

$$\int_{k\frac{\pi}{2}}^{(k+1)\frac{\pi}{2}} \sin^m \theta \, d\theta = \pm \int_{k\frac{\pi}{2}}^{(k+1)\frac{\pi}{2}} \cos^m \theta \, d\theta = \pm \int_0^{\frac{\pi}{2}} \sin^m \theta \, d\theta, \quad \ldots \quad (2)$$

in which the sign to be used is determined by that of $\sin^m \theta$ or $\cos^m \theta$ in the given quadrant.

In like manner the value of the integral in formula (Q) is numerically the same in every quadrant, and its sign is the same as that of $\sin^m \theta \cos^n \theta$ in the given quadrant.

Change of Independent Variable in a Definite Integral.

88. It is often useful to make such a change of independent variable as will leave unchanged, or simply interchange, the values of the limits. As an illustration, let us take the definite integral

$$u = \int_0^\infty \frac{\log x}{1 + x + x^2} dx.$$

If we put $x = \dfrac{1}{y}$, whence $\log x = -\log y$, and $dx = -\dfrac{dy}{y^2}$,

$$u = \int_{\infty}^{0} \frac{\log y}{y^2 + y + 1} dy = -u;$$

whence we infer that

$$u = \int_{0}^{\infty} \frac{\log x}{1 + x + x^2} dx = 0.$$

89. Again, let

$$u = \int_{0}^{\infty} \frac{\log x}{a^2 + x^2} dx.$$

Putting $x = \dfrac{a^2}{y}$, we have

$$u = \int_{0}^{\infty} \frac{2 \log a - \log y}{a^2 + y^2} dy = 2 \log a \int_{0}^{\infty} \frac{dy}{a^2 + y^2} - u;$$

hence

$$\int_{0}^{\infty} \frac{\log x}{a^2 + x^2} dx = \frac{\pi \log a}{2a}.$$

Differentiation of an Integral.

90. The integral $\displaystyle\int_{a} f(x)\, dx$ is by definition a function of x, whose derivative, with reference to x, is $f(x)$. Thus, putting

$$U = \int_{a}^{x} f(x)\, dx,$$

$$\frac{dU}{dx} = f(x).$$

This gives the derivative of an integral with reference to its upper limit. By reversing the limits we have, in like manner,

$$\frac{dU}{da} = -f(a),$$

when the lower limit is regarded as variable.

91. Now writing the integral in the form

$$U = \int_a u\, dx, \quad \ldots \ldots \quad (1)$$

if u is a function of some other quantity, α, independent of x and a, U is also a function of α, and therefore admits of a derivative with reference to α. From (1) we have

$$\frac{dU}{dx} = u,$$

whence

$$\frac{d}{d\alpha}\frac{dU}{dx} = \frac{du}{d\alpha}.$$

By the principle of differentiation with respect to independent variables [See Diff. Calc., Art. 401; Abridged Ed., Art. 200].

$$\frac{d}{dx}\frac{dU}{d\alpha} = \frac{d}{d\alpha}\frac{dU}{dx}.$$

Therefore

$$\frac{d}{dx}\frac{dU}{d\alpha} = \frac{du}{d\alpha};$$

and by integration

$$\frac{dU}{d\alpha} = \int \frac{du}{d\alpha} dx + C \quad \ldots \ldots \quad (2)$$

Now, in equation (1), U is a function of x and α which, when $x = a$, is equal to zero, independently of the value of α. In other words, it is a constant with reference to α, when $x = a$; therefore $\dfrac{dU}{d\alpha} = 0$ when $x = a$. If, then, we use a as a lower limit in equation (2), we shall have $C = 0$. Therefore

$$\frac{dU}{d\alpha} = \int_a \frac{du}{d\alpha} dx \ . \quad \ldots \ldots \quad (3)$$

Substituting for x any value b independent of α, we have

$$\frac{d}{d\alpha}\int_a^b u\, dx = \int_a^b \frac{d}{d\alpha} u\, dx, \quad \ldots \ldots \quad (4)$$

which expresses that *an integral may be differentiated with reference to a quantity of which the limits are independent, by differentiating the expression under the integral sign.*

92. By means of this theorem, we may derive from an integral whose value is known, the values of certain other integrals. Thus, from the first fundamental integral,

$$\int x^n\, dx = \frac{x^{n+1}}{n+1}, \quad \ldots \ldots \quad (1)$$

we derive, by differentiating with reference to n,

$$\int x^n \log x\, dx = \frac{(n+1) x^{n+1} \log x - x^{n+1}}{(n+1)^2},$$

the result being the same as that which is obtained by the method of parts.

93. The principal application of this method, however, is to definite integrals, when the limits are such as materially to simplify the value of the original integral. Thus, equation (1) of the preceding article gives

$$\int_0^1 x^n \, dx = \frac{1}{n+1},$$

whence, by successive differentiation,

$$\int_0^1 x^n \log x \, dx = -\frac{1}{(n+1)^2},$$

$$\int_0^1 x^n (\log x)^2 \, dx = \frac{1 \cdot 2}{(n+1)^3},$$

....

$$\int_0^1 x^n (\log x)^r \, dx = (-1)^r \frac{1 \cdot 2 \cdots r}{(n+1)^{r+1}}.$$

Integration under the Integral Sign.

94. Let u be a function of x and α, and let a and α_0 be constants; then the integral

$$U = \int_{\alpha_0} \left[\int_a u \, dx \right] d\alpha, \quad \ldots \ldots (1)$$

is a function of x and α, which vanishes when $\alpha = \alpha_0$, independently of the value of x, and when $x = a$, independently of the value of α. From (1)

$$\frac{dU}{d\alpha} = \int_a u \, dx, \qquad \text{whence} \qquad \frac{d}{dx}\frac{dU}{d\alpha} = u;$$

therefore $\quad \dfrac{d}{d\alpha}\dfrac{dU}{dx} = u, \qquad$ whence $\qquad \dfrac{dU}{dx} = \int u \, d\alpha + C.$

Now $\dfrac{dU}{dx}$ must vanish when $\alpha = \alpha_0$, since this supposition makes U independent of x; therefore, if we use α_0 for a lower limit in the last equation, we must have $C = 0$; therefore

$$\frac{dU}{dx} = \int_{\alpha_0} u\, d\alpha,$$

and since u vanishes when $x = a$,

$$U = \int_a \left[\int_{\alpha_0} u\, d\alpha \right] dx. \quad \ldots \ldots \quad (2)$$

Comparing the values of U in equations (1) and (2), we have

$$\int_{\alpha_0} \int_a u\, dx\, d\alpha = \int_a \int_{\alpha_0} u\, d\alpha\, dx.$$

It is evident that we may also write

$$\int_{\alpha_0}^{\alpha_1} \int_a^b u\, dx\, dx = \int_a^b \int_{\alpha_0}^{\alpha_1} u\, d\alpha\, dx, \quad \ldots \ldots \quad (3)$$

provided that the limits of each integration are independent of the other variable.

95. By means of this formula, a new integral may be derived from the value of a given integral, provided we can integrate, with reference to the other variable, both the expressions under the integral sign and also the value of the integral. Thus, from

$$\int_0^1 x^n\, dx = \frac{1}{n+1},$$

§ VII.] INTEGRATION UNDER THE INTEGRAL SIGN.

by multiplying by dn, and integrating between the limits r and s, we derive

$$\int_0^1 \int_r^s x^n\, dn\, dx = \int_r^s \frac{dn}{n+1},$$

whence

$$\int_0^1 \frac{x^s - x^r}{\log x}\, dx = \log \frac{s+1}{r+1}.$$

96. When the derivative of a proposed integral with reference to α is a known integral, we can sometimes derive its value by integrating the latter with reference to α. Thus, let

$$u = \int_0^\infty \frac{\varepsilon^{-\alpha x} - \varepsilon^{-\beta x}}{x}\, dx. \quad \ldots \quad \ldots \quad (1)$$

In this case

$$\frac{du}{d\alpha} = \int_0^\infty -\varepsilon^{-\alpha x}\, dx = \left[\frac{\varepsilon^{-\alpha x}}{\alpha}\right]_0^\infty = -\frac{1}{\alpha};$$

hence, integrating, $u = -\log \alpha + C = \log \dfrac{\beta}{\alpha}$ . . . . (2)

since in (1) u vanishes when $\alpha = \beta$.

The Definite Integral Regarded as the Limiting Value of a Sum.

97. Let A denote the greatest, and B the least value assumed by $f(x)$, while x varies from a to b. Then it is evident that

$$\int_a^b f(x)\, dx < \int_a^b A\, dx; \quad \ldots \quad \ldots \quad (1)$$

for, while x passes from a to b, the rate of the former integral is generally less, and never greater than the rate of the latter. In like manner

$$\int_a^b f(x)\,dx > \int_a^b B\,dx. \quad \ldots \ldots \quad (2)$$

The values of the integrals in the second members of equations (1) and (2) are $A(b-a)$ and $B(b-a)$ respectively. Therefore, if we assume

$$\int_a^b f(x)\,dx = M(b-a), \quad \ldots \ldots \quad (3)$$

we shall have $\quad\quad A > M > B.$

The quantity M in equation (3) is called the *mean value* of the function $f(x)$ for the interval between a and b.

98. Let
$$b - a = n\,\Delta x; \quad \ldots \ldots \quad (4)$$

then the $n+1$ values of x,

$$a, \quad\quad a + \Delta x, \quad\quad a + 2\,\Delta x, \cdots \quad b,$$

define n equal intervals into which the whole interval $b-a$ is separated. Let $x_1, x_2, \ldots\ldots x_n$ be n values of x, one comprised in each of these intervals; also let $\Sigma_a^b f(x_r)\,\Delta x$ denote the sum of the n terms formed by giving to r the n values $1 \cdot 2 \cdots\cdot\ n$ in the typical term $f(x_r)\,\Delta x$; that is, let

$$\Sigma_a^b f(x_r)\,\Delta x = f(x_1)\,\Delta x + f(x_2)\,\Delta x \cdots + f(x_n)\,\Delta x. \quad (5)$$

§ VII.]   *AN INTEGRAL THE LIMIT OF A SUM.*   113

We shall now show that when n is indefinitely increased the limiting value of $\Sigma_a^b f(x_r) \Delta x$ is $\int_a^b f(x)\, dx$.

99. If we separate the integral into parts corresponding to the terms above mentioned; thus,

$$\int_a^b f(x)\,dx = \int_a^{a+\Delta x} f(x)\,dx + \int_{a+\Delta x}^{a+2\Delta x} f(x)\,dx \cdots$$

$$+ \int_{b-\Delta x}^{b} f(x)\,dx,$$

and let $M_1, M_2, \cdots\; M_n$ denote the mean values of $f(x)$ in the several intervals, we have, in accordance with equation (3), Art. 97,

$$\int_a^b f(x)\,dx = M_1 \Delta x + M_2 \Delta x \cdots + M_n \Delta x. \quad \ldots \quad (6)$$

Now, since $f(x_r)$ and M_r are both intermediate in value between the greatest and the least values of $f(x)$ in the interval to which they belong, their difference is less than the difference between these values of $f(x)$. Therefore, if we put

$$f(x_r) = M_r + e_r, \quad \ldots \ldots \ldots (7)$$

e_r is a quantity whose limit is zero when n, the number of intervals, is indefinitely increased, and Δx in consequence diminished indefinitely.

Comparing the terms in equations (5) and (6) we have, by means of equation (7),

$$\Sigma_a^b f(x) \Delta x = \int_a^b f(x)\,dx + (e_1 + e_2 \cdots + e_n) \Delta x. \quad \ldots \quad (8)$$

Denote by e the arithmetical mean of the n quantities $e_1, e_2, \cdots e_n$; that is, let

$$ne = e_1 + e_2 + e_3 \cdots e_n; \quad \ldots \ldots (9)$$

then, since e is an intermediate value between the greatest and the least value of e_r, it is also a quantity whose limit is zero when n is indefinitely increased. By equations (9) and (4), equation (8) becomes

$$\Sigma_a^b f(x_r) \Delta x = \int_a^b f(x)\, dx + e(b-a),$$

whence it follows that $\int_a^b f(x)\, dx$ is the limit of $\Sigma_a^b f(x_r)\, dx$ when n is indefinitely increased, since the limit of e is zero.

100. It was shown in the Differential Calculus, Art. 390 [Abridged Ed., Art. 193], that, in an expression for the ratio of finite differences, we may pass to the limit which the expression approaches, when the differences are diminished without limit, by substituting the symbol d for the symbol Δ. The theorem proved in the preceding articles shows that, in like manner, in the summation of an expression involving finite differences, we may pass to the limit approached when the differences are indefinitely diminished, by changing the symbols Σ and Δ into $\int$ and d.

The term *integral*, and the use of the long s, the initial of the word *sum*, as the sign of integration, have their origin in this connection between the processes of integration and summation.

Additional Formulas of Integration.

101. The formulas recapitulated below are useful in evaluating other integrals. (A) and (A') are demonstrated in Art. 17; (B) and (C) in Art. 29; (D) and (E) in Art. 30; (F) in Art. 31; (G) and (G') in Art. 35; (H) and (I) in Art. 50; (J) in Art. 51; (K) in Art. 52; (L) in Art. 53; (M) in Art. 55; (N) and (O) in Art. 58; (P) and (P') in Art. 84; and (Q) in Art. 85.

$$\int \frac{dx}{(x-a)(x-b)} = \frac{1}{a-b} \log \frac{x-a}{x-b}. \quad \ldots \ldots \quad (A)$$

$$\int \frac{dx}{x^2 - a^2} = \frac{1}{2a} \log \frac{x-a}{x+a}. \quad \ldots \ldots \quad (A')$$

$$\int \sin^2 \theta \, d\theta = \tfrac{1}{2}(\theta - \sin \theta \cos \theta). \quad \ldots \ldots \quad (B)$$

$$\int \cos^2 \theta \, d\theta = \tfrac{1}{2}(\theta + \sin \theta \cos \theta). \quad \ldots \ldots \quad (C)$$

$$\int \frac{d\theta}{\sin \theta \cos \theta} = \log \tan \theta. \quad \ldots \ldots \quad (D)$$

$$\int \frac{d\theta}{\sin \theta} = \log \tan \tfrac{1}{2}\theta = \log \frac{1 - \cos \theta}{\sin \theta}. \quad \ldots \ldots \quad (E)$$

$$\int \frac{d\theta}{\cos \theta} = \log \tan \left[\frac{\pi}{4} + \frac{\theta}{2}\right] = \log \frac{1 + \sin \theta}{\cos \theta}. \quad \ldots \quad (F)$$

$$\int \frac{d\theta}{a + b \cos \theta} = \frac{2}{\sqrt{(a^2 - b^2)}} \tan^{-1}\left[\sqrt{\frac{a-b}{a+b}} \tan \tfrac{1}{2}\theta\right]. \quad \ldots \quad (G)$$

$$\int \frac{d\theta}{a+b\cos\theta} = \frac{1}{\sqrt{(b^2-a^2)}} \log \frac{\sqrt{(b+a)} + \sqrt{(b-a)}\tan\frac{1}{2}\theta}{\sqrt{(b+a)} - \sqrt{(b-a)}\tan\frac{1}{2}\theta} \quad \ldots (G')$$

$$\int \frac{dx}{x\sqrt{(x^2+a^2)}} = \frac{1}{a} \log \frac{\sqrt{(x^2+a^2)} - a}{x} \quad \ldots \ldots \ldots (H)$$

$$\int \frac{dx}{x\sqrt{(a^2-x^2)}} = \frac{1}{a} \log \frac{a - \sqrt{(a^2-x^2)}}{x} \quad \ldots \ldots \ldots (I)$$

$$\int \frac{dx}{(ax^2+b)^{\frac{3}{2}}} = \frac{x}{b\sqrt{(ax^2+b)}} \quad \ldots \ldots \ldots \ldots (J)$$

$$\int \frac{dx}{\sqrt{(x^2 \pm a^2)}} = \log[x + \sqrt{(x^2 \pm a^2)}]. \quad \ldots \ldots \ldots (K)$$

$$\int \sqrt{(x^2 \pm a^2)}\, dx = \frac{x\sqrt{(x^2 \pm a^2)}}{2} \pm \frac{a^2}{2}\log[x + \sqrt{(x^2 \pm a^2)}] \quad \ldots (L)$$

$$\int \sqrt{(a^2-x^2)}\, dx = \frac{a^2}{2}\sin^{-1}\frac{x}{a} + \frac{x\sqrt{(a^2-x^2)}}{2} \quad \ldots \ldots (M)$$

$$\int \frac{dx}{\sqrt{[(x-\alpha)(x-\beta)]}} = 2\log[\sqrt{(x-\alpha)} + \sqrt{(x-\beta)}] \quad \ldots (N)$$

$$\int \frac{dx}{\sqrt{[(x-\alpha)(\beta-x)]}} = 2\sin^{-1}\sqrt{\frac{x-\alpha}{\beta-\alpha}} \quad \ldots \ldots \ldots (O)$$

$$\int_0^{\frac{\pi}{2}} \sin^m\theta\, d\theta = \int_0^{\frac{\pi}{2}} \cos^m\theta\, d\theta = \frac{(m-1)(m-3)\cdots 1}{m(m-2)\cdots 2}\cdot\frac{\pi}{2} \quad \ldots (P)$$

§ VII.]   ADDITIONAL FORMULAS OF INTEGRATION.   117

$$\int_0^{\frac{\pi}{2}} \sin^m \theta \, d\theta = \int_0^{\frac{\pi}{2}} \cos^m \theta \, d\theta = \frac{(m-1)(m-3)\cdots 2}{m(m-2)\cdots\cdots 1} \cdot \cdot \cdot (P')$$

$$\int_0^{\frac{\pi}{2}} \sin^m \theta \cos^n \theta \, d\theta = \frac{(m-1)(m-3)\cdots \times (n-1)(n-3)\cdots}{(m+n)(m+n-2)\cdots\cdots\cdots} \alpha, \quad (Q)$$

in which $\alpha = 1$, unless m and n are both even, when $\alpha = \dfrac{\pi}{2}$.

Examples VII.

1. $\displaystyle\int_0^{n\pi} \frac{d\theta}{a + b\cos\theta},$ $[a > b,$ and n an integer$]$ $\dfrac{n\pi}{\sqrt{(a^2 - b^2)}}.$

2. $\displaystyle\int_0^{2n\pi \pm \frac{\pi}{2}} \frac{d\theta}{2 + \cos\theta},$ $\dfrac{2n\pi \pm \frac{1}{3}\pi}{\sqrt{3}}.$

3. $\displaystyle\int_0^{\frac{\pi}{2}} \sin^6 \theta \, d\theta,$ $\dfrac{5\pi}{32}.$

4. $\displaystyle\int_0^{\pi} \sin^5 \theta \, d\theta,$ $\dfrac{16}{15}.$

5. $\displaystyle\int_{-\frac{\pi}{2}}^{\pi} \cos^7 \theta \, d\theta,$ $\dfrac{16}{35}.$

6. $\displaystyle\int_0^{\frac{\pi}{2}} \sin^4 \theta \cos^6 \theta \, d\theta,$ $\dfrac{3\pi}{512}.$

7. $\displaystyle\int_0^{\pi} \sin^2\theta \cos^4\theta\, d\theta,$ $\qquad\qquad\dfrac{4}{35}.$

8. $\displaystyle\int_0^{\frac{\pi}{2}} \sin^m\theta \cos^m\theta\, d\theta,$ $\qquad\dfrac{1}{2^m}\displaystyle\int_0^{\frac{\pi}{2}} \sin^m\theta\, d\theta.$

9. $\displaystyle\int_0^1 \dfrac{x^{2n}\,dx}{\sqrt{(1-x^2)}},$ $\qquad\dfrac{1\cdot 3\cdot 5\,\cdots\,(2n-1)}{2\cdot 4\cdot 6\,\cdots\,2n}\cdot\dfrac{\pi}{2}.$

10. $\displaystyle\int_0^1 \dfrac{x^{2n+1}\,dx}{\sqrt{(1-x^2)}},$ $\qquad\dfrac{2\cdot 4\cdot 6\,\cdots\,2n}{3\cdot 5\cdot 7\,\cdots\,(2n+1)}.$

11. $\displaystyle\int_0^a x^3(a^2-x^2)^{\frac{3}{2}}\,dx,$ $\qquad\dfrac{2a^7}{63}.$

12. $\displaystyle\int_a^{\infty} \dfrac{(x^2-a^2)^{\frac{3}{2}}\,dx}{x^6},$ $\qquad\dfrac{3\pi}{16a}.$

13. $\displaystyle\int_0^{\infty} \dfrac{x^5\,dx}{(a^2+x^2)^{\frac{7}{2}}},$ $\qquad\dfrac{8}{15a}.$

14. $\displaystyle\int_0^{\infty} \dfrac{x^4\,dx}{(a^2+x^2)^4},$ $\qquad\dfrac{\pi}{32a^3}.$

15. Prove that

$$\int_0^a x^{n-1}(a-x)^{m-1}\,dx = \int_0^a x^{m-1}(a-x)^{n-1}\,dx,$$

and derive a formula of reduction for this integral, supposing $n > 0$ and $m > 1$.

$$\int_0^a x^{n-1}(a-x)^{m-1}\,dx = \dfrac{m-1}{n}\int_0^a x^n(a-x)^{m-2}\,dx.$$

16. Deduce from the result of Ex. 15 the value of the integral when m is an integer.

$$\int_0^a x^{n-1}(a-x)^{m-1}dx = \frac{1\cdot 2\cdot 3 \cdots (m-1)}{n(n+1)\cdots(n+m-1)}a^{m+n-1}$$

17. $\int_{-a}^{a}(a+x)^5(a-x)^{\frac{3}{2}}dx$. See Ex. 16. $\qquad \dfrac{2^{16}\sqrt{2}}{45045}a^{\frac{15}{2}}$.

18. $\int_0^{\frac{\pi}{2}}\sin^7\theta(\cos\theta)^{\frac{5}{2}}d\theta$. Put $\sin^2\theta = x$, and see Ex. 16.

$$\frac{2^5}{5\cdot 7\cdot 11\cdot 19}.$$

19. Show by a change of independent variable that

$$\int_0^\infty \frac{x^2\, dx}{(a^2+x^2)^2} = \int_0^\infty \frac{a^2\, dx}{(a^2+x^2)^2},$$

and therefore $\qquad \displaystyle\int_0^\infty \frac{x^2\, dx}{(a^2+x^2)^2} = \frac{1}{2}\int_0^\infty \frac{dx}{a^2+x^2} = \frac{\pi}{4a}$.

20. $\displaystyle\int_0^\infty \frac{x\log x\, dx}{(x^2+a^2)^2}$, $\qquad\qquad\qquad\qquad \dfrac{\log a}{2a^2}$.

21. $\displaystyle\int_0^\infty \frac{\tan^{-1}x\, . \, dx}{x^2+x+1}$, $\qquad\qquad\qquad\qquad \dfrac{\pi^2}{6\sqrt{3}}$.

22. $\displaystyle\int_0^\infty \tan^{-1}\frac{x}{a}\frac{x\, dx}{x^4+a^4}$, $\qquad\qquad\qquad\qquad \dfrac{\pi^2}{16a^2}$.

23. Derive a series of integrals by successive differentiation of the definite integral $\displaystyle\int_0^\infty \varepsilon^{-\alpha x}dx$.

$$\int_0^\infty x^r\varepsilon^{-\alpha x}dx = \frac{1\cdot 2 \cdots n}{\alpha^{n+1}}.$$

24. Derive from the result of Art. 63 the definite integrals

$$\int_0^\infty \varepsilon^{-mx} \sin nx \, dx = \frac{n}{m^2 + n^2}, \quad \text{and} \quad \int_0^\infty \varepsilon^{-mx} \cos nx \, dx = \frac{m}{m^2 + n^2};$$

and thence derive by differentiation the integrals

$$\int_0^\infty x\varepsilon^{-mx} \sin nx \, dx = \frac{2mn}{(m^2 + n^2)^2}, \quad \text{and} \quad \int_0^\infty x\varepsilon^{-mx} \cos nx \, dx = \frac{m^2 - n^2}{(m^2 + n^2)^2}.$$

25. From the results of Ex. 24 derive

$$\int_0^\infty x^2 \varepsilon^{-mx} \sin nx \, dx = \frac{2n(3m^2 - n^2)}{(m^2 + n^2)^3};$$

$$\int_0^\infty x^2 \varepsilon^{-mx} \cos nx \, dx = \frac{2m(m^2 - 3n^2)}{(m^2 + n^2)^3}.$$

26. From the fundamental formula (k') derive

$$\int_0^\infty \frac{dx}{\alpha + \beta x^2} = \frac{\pi}{2\alpha^{\frac{1}{2}}\beta^{\frac{1}{2}}};$$

and thence derive a series of formulas by differentiation with reference to α.

$$\int_0^\infty \frac{dx}{(\alpha + \beta x^2)^n} = \frac{\pi}{2^n \beta^{\frac{1}{2}}} \cdot \frac{1 \cdot 3 \cdots (2n-3)}{1 \cdot 2 \cdots (n-1)} \cdot \frac{1}{\alpha^{n-\frac{1}{2}}}.$$

27. Derive a series of integrals by differentiating with reference to β, the integral used in Ex. 26.

$$\int_0^\infty \frac{x^{2n-2} \, dx}{(\alpha + \beta x^2)^n} = \frac{\pi}{2^n \alpha^{\frac{1}{2}}} \cdot \frac{1 \cdot 3 \cdot 5 \cdots (2n-3)}{1 \cdot 2 \cdot 3 \cdots (n-1)} \cdot \frac{1}{\beta^{n-\frac{1}{2}}}.$$

§ VII.] EXAMPLES. 121

28. From the integral employed in examples 26 and 27, derive the value of
$$\int_0^\infty \frac{x^4 \, dx}{(\alpha + \beta x^2)^6}.$$

Differentiate twice with reference to β, and once with reference to α.

$$\int_0^\infty \frac{x^4 \, dx}{(\alpha + \beta x^2)^6} = \frac{1 \cdot 3 \cdot 1}{1 \cdot 2 \cdot 3} \cdot \frac{\pi}{16 \alpha^{\frac{3}{2}} \beta^{\frac{5}{2}}}.$$

29. Derive an integral by differentiation, from the result of Ex. II., 67

$$\int_0^\infty \frac{dx}{(x^2 + b^2)(x^2 + a^2)^2} = \frac{\pi(2a + b)}{4a^3 b(a + b)^2}.$$

30. Derive an integral by integrating $\int_0^\infty \frac{dx}{a^2 + x^2} = \frac{\pi}{2a}$.

$$\int_0^\infty \left[\tan^{-1} \frac{p}{x} - \tan^{-1} \frac{q}{x} \right] \frac{dx}{x} = \frac{\pi}{2} \log \frac{p}{q}.$$

31. Derive a definite integral by integrating

$$\int_0^\infty \varepsilon^{-mx} \sin nx \, dx = \frac{n}{m^2 + n^2}$$

with reference to n.

$$\int_0^\infty \frac{\varepsilon^{-mx}}{x} (\cos ax - \cos bx) \, dx = \frac{1}{2} \log \frac{m^2 + b^2}{m^2 + a^2}.$$

32. Derive a definite integral from the integral employed in Ex. 31 by integration with reference to m.

$$\int_0^\infty \frac{\sin nx}{x} \left[\varepsilon^{-ax} - \varepsilon^{-bx} \right] dx = \left[\tan^{-1} \frac{b}{n} - \tan^{-1} \frac{a}{n} \right].$$

33. Derive an integral by integrating with respect to m

$$\int_0^\infty \varepsilon^{-mx} \cos nx \, dx = \frac{m}{m^2 + n^2}.$$

$$\int_0^\infty \frac{\varepsilon^{-ax} - \varepsilon^{-bx}}{x} \cos nx \, dx = \frac{1}{2} \log \frac{b^2 + n^2}{a^2 + n^2}.$$

34. Derive an integral by integrating with respect to n the integral used in the preceding example.

$$\int_0^\infty \frac{\varepsilon^{-mx}}{x} (\sin ax - \sin bx) \, dx = \tan^{-1} \frac{m(a-b)}{m^2 + ab}.$$

35. Show by means of the result of Ex. 32 that

$$\int_0^\infty \frac{\sin nx}{x} \, dx = \frac{\pi}{2}.$$

36. Derive an integral by integration from the result of Ex. II., 67.

$$\int_0^\infty \frac{1}{x} \left[\tan^{-1} \frac{p}{x} - \tan^{-1} \frac{q}{x} \right] \frac{dx}{x^2 + b^2} = \frac{\pi}{2b^2} \log \frac{p(q+b)}{q(p+b)}.$$

37. Evaluate $\displaystyle\int_0^\infty \log \frac{x^2 + a^2}{x^2 + b^2} dx$ by the method of Art. 96.

$$\pi(a-b).$$

38. Evaluate $\displaystyle\int_0^\infty \log\left[1 + \frac{a^2}{x^2}\right] \log x \, dx.$ $\pi a (\log a - 1).$

CHAPTER III.

GEOMETRICAL APPLICATIONS.

VIII.

Plane Areas.

102. THE first step in making an application of the Integral Calculus is to express the required magnitude in the form of an integral. In the geometrical applications, the magnitude is regarded as generated while some selected independent variable undergoes a given change of value. The independent variable is usually a straight line or an angle, varying between known limits; the required magnitude is either a line regarded as generated by the motion of a point, an area generated by the motion of a line, or a solid generated by the motion of an area. A plane area may be generated by the motion of a straight line, generally of variable length, the method selected depending upon the mode in which the boundaries of the area are defined.

An Area Generated by a Variable Line having a Fixed Direction.

103. The differential of the area generated by the ordinate of a curve, whose equation is given in rectangular coordinates, has been derived in Art. 3. The same method may be employed in the case of any area generated by a straight line whose direction is invariable.

Let AB be the generating line, and let R be its intersection with a fixed line CD, to which it is always perpendicular. Suppose R to move uniformly along CD, and let RS be the space described by R in the interval of time, dt. Then the value of the differential of the area, at the instant when the generating line passes the position AB, is the area which would be generated in the time dt, if the rate of the area were constant. This rate would evidently become constant if the generating line were made constant in length; and therefore the differential is the rectangle, represented in the figure, whose base and altitude are AB and RS; that is, it is *the product of the generating line, and the differential of its motion in a direction perpendicular to its length.*

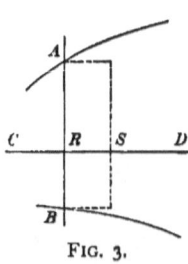

FIG. 3.

104. In the algebraic expression of this principle, the independent variable is the distance of R from some fixed origin upon CD, and the length of AB is to be expressed in terms of this independent variable.

When the curve or curves defining the length of AB are given in rectangular coordinates, CD is generally one of the axes; thus, if the generating line is the ordinate of a curve, the differential is $y\,dx$, as shown in Art. 3. It is often, however, convenient to regard the area as generated by some other line.

For example, given the curve known as the witch, whose equation is

$$y^2 x - 2ay^2 + 4a^2 x = 0 \quad . \quad . \quad . \quad . \quad . \quad . \quad (1)$$

This curve passes through the origin, is symmetrical to the axis of x, and has the line $x = 2a$ for an asymptote, since $x = 2a$ makes $y = \pm \infty$.

Let the area between the curve and its asymptote be re-

§ VIII.] AREAS GENERATED BY VARIABLE LINES.

quired. We may regard this area as generated by the line PQ parallel to the axis of x, y being taken as the independent variable. Now

$$PQ = 2a - x,$$

hence the required area is

$$A = \int_{-\infty}^{\infty} (2a - x)\, dy. \quad \ldots \quad (2)$$

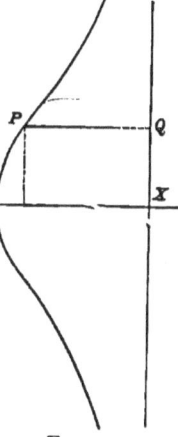

From the equation (1) of the curve, we have

$$x = \frac{2ay^2}{y^2 + 4a^2};$$

whence $\quad 2a - x = \dfrac{8a^3}{y^2 + 4a^2},$

Fig. 4.

and equation (2) becomes

$$A = 8a^3 \int_{-\infty}^{\infty} \frac{dy}{y^2 + 4a^2} = 4a^2 \tan^{-1}\frac{y}{2a}\bigg]_{-\infty}^{\infty} = 4\pi a^2.$$

Oblique Coordinates.

105. When the coordinate axes are oblique, if α denotes the angle between them, and the ordinate is the generating line, the differential of its motion in a direction perpendicular to its length is evidently $\sin \alpha \cdot dx$; therefore, the expression for the area is

$$A = \sin \alpha \int y\, dx.$$

As an illustration let the area between a parabola and a chord passing through the focus be required. It is shown in treatises on conic sections, the expression for a focal chord is

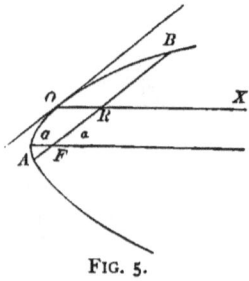

FIG. 5.

$$AB = 4a\,\text{cosec}^2\alpha, \quad \ldots \quad (1)$$

where α is the inclination of the chord to the axis of the curve, and a is the distance from the focus to the vertex. It is also shown that the equation of the curve referred to the diameter which bisects the chord, and the tangent at its extremity which is parallel to the chord is

$$y^2 = 4a\,\text{cosec}^2\alpha \cdot x. \quad \ldots \ldots \quad (2)$$

The required area may be generated by the double ordinate in this equation; and since from (1) the final value of y is $\pm 2a\,\text{cosec}^2\alpha$, equation (2) gives for the final value of x

$$OR = a\,\text{cosec}^2\alpha.$$

Hence we have

$$A = 2\sin\alpha \int_0^{a\,\text{cosec}^2\alpha} y\,dx,$$

or by equation (2)

$$A = 4\sqrt{a}\int_0^{a\,\text{cosec}^2\alpha} \sqrt{x}\,dx = \frac{8a^2\,\text{cosec}^3\alpha}{3}.$$

Employment of an Auxiliary Variable.

106. We have hitherto assumed that, in the expression

$$A = \int y\,dx,$$

x is taken as the independent variable, so that dx may be assumed constant; and it is usual to take the limits in such a manner that dx is positive. The resulting value of A will then have the sign of y, and will change sign if y changes sign.

It is frequently desirable, however, as in the illustration given below, to express both y and dx in terms of some other variable. When this is done, it is to be noticed that it is not necessary that dx should retain the same sign throughout the entire integral. The limits may often be so taken that the extremity of the generating ordinate must pass completely around a closed curve, and in that case it is easily seen that the complete integral, which represents the algebraic sum of the areas generated positively and negatively, will be the whole area of the closed curve.

107. As an illustration, let the whole area of the closed curve

$$\left(\frac{x}{a}\right)^{\frac{2}{3}} + \left(\frac{y}{b}\right)^{\frac{2}{3}} = 1,$$

represented in Fig. 6, be required. If in this equation we put

$$\left(\frac{x}{a}\right)^{\frac{1}{3}} = \sin \psi,$$

we shall have

$$\left(\frac{y}{b}\right)^{\frac{1}{3}} = \cos \psi\ ;$$

whence $\quad x = a \sin^3 \psi, \quad$ and $\quad y = b \cos^3 \psi. \quad . \quad .$ (1)

Therefore $\quad \int y\, dx = 3ab \int \cos^4 \psi \sin^2 \psi\, d\psi.$

Now if in this integral we use the limits 0 and 2π, the point determined by equation (1) describes the whole curve in the direction $ABCDA$. Hence we have for the whole area

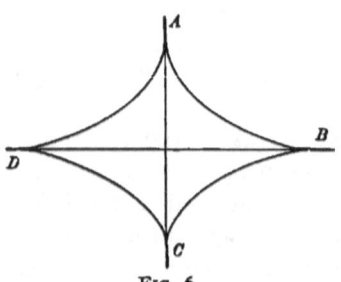

FIG. 6.

$$A = 3ab \int_0^{2\pi} \cos^4 \psi \sin^2 \psi \, d\psi,$$

and by formula (Q)

$$A = 3ab \frac{3 \cdot 1 \cdot 1}{6 \cdot 4 \cdot 2} 2\pi = \frac{3\pi ab}{8}.$$

The areas in this case are all generated with the positive sign, since when y is negative dx is also negative. Had the generating point moved about the curve in the opposite direction, the result would have been negative.

Area generated by a Rotating Line or Radius Vector.

108. The radius vector of a curve given in polar coordinates is a variable line rotating about a fixed extremity. The angular rate is denoted by $\dfrac{d\theta}{dt}$ and may be regarded as constant, although the rate at which area is generated by the radius vector OP, Fig. 7, is not constant, because the length of OP is not constant. The differential of this area is the area which would be generated in the time dt, if the rate of the area were constant; that is to say, if the radius vector were of constant

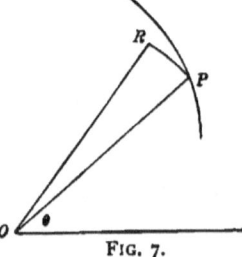

FIG. 7.

§ VIII.] *AREAS GENERATED BY ROTATING LINES.* 129

length. It is therefore the circular sector OPR of which the radius is r and the angle at the centre is $d\theta$.

Since $\quad\quad\quad$ arc $PR = r\, d\theta$,

$$\text{sector } OPR = \frac{1}{2} r^2\, d\theta;$$

therefore the expression for the generated area is

$$A = \frac{1}{2}\int r^2\, d\theta \quad \cdots \quad \cdots \quad (1)$$

109. As an illustration, let us find the area of the right-hand loop of the lemniscata

$$r^2 = a^2 \cos 2\theta.$$

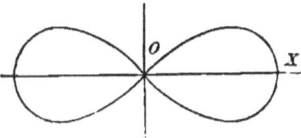

Fig. 8.

The limits to be employed are those values of θ which make $r = 0$; that is $-\frac{\pi}{4}$ and $\frac{\pi}{4}$.
Hence the area of the loop is

$$A = \frac{a^2}{2}\int_{-\frac{\pi}{4}}^{\frac{\pi}{4}} \cos 2\theta\, d\theta = \frac{a^2}{4} \sin 2\theta \Big]_{-\frac{\pi}{4}}^{\frac{\pi}{4}} = \frac{a^2}{2}.$$

110. When the radii vectores, r_2 and r_1 corresponding to the same value of θ in two curves, have the same sign, the area generated by their difference is the difference of the polar areas generated by r_1 and r_2. Hence the expression for this area is

$$A = \frac{1}{2}\int (r_2^2 - r_1^2)\, d\theta. \quad \cdots \quad \cdots \quad (2)$$

III. Let us apply this formula to find the whole area between the cissoid

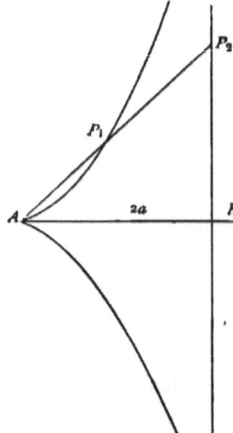

$$r_1 = 2a (\sec \theta - \cos \theta),$$

Fig. 9, and its asymptote BP_2, whose polar equation is

$$r_2 = 2a \sec \theta.$$

One half of the required area is generated by the line P_1P_2, while θ varies from 0 to $\frac{1}{2}\pi$. Hence by the formula

$$A = 2a^2 \int_0^{\frac{\pi}{2}} (2 - \cos^2\theta) \, d\theta = \frac{3}{2}\pi a^2.$$

Fig. 9.

Therefore the whole area required is $3\pi a^2$.

Transformation of the Polar Formulas.

112. In the case of curves given in rectangular coordinates, it is sometimes convenient to regard an area as generated by a radius vector, and to use the transformations deduced below in place of the polar formulas.

Put
$$y = mx; \quad \ldots \ldots \quad (1)$$

now taking the origin as pole and the initial line as the axis of x, we have

$$x = r \cos \theta, \qquad y = r \sin \theta; \quad \ldots \quad (2)$$

therefore
$$m = \frac{y}{x} = \tan \theta,$$

and
$$dm = \sec^2 \theta \, d\theta. \quad \ldots \ldots \quad (3)$$

§ VIII.] *TRANSFORMATION OF THE POLAR FORMULAS.* 131

From equations (2) and (3),

$$x^2 \, dm = r^2 \, d\theta \,;$$

therefore equation (1) of Art. 108 gives

$$A = \frac{1}{2} \int x^2 \, dm. \quad \cdots \cdots \quad (4)$$

In like manner, equation (2) Art. 110 becomes

$$A = \frac{1}{2} \int (x_2^2 - x_1^2) \, dm. \quad \cdots \cdots \quad (5)$$

113. As an illustration, let us take the folium

$$x^3 + y^3 - 3axy = 0. \quad \cdots \cdots \quad (1)$$

Putting $y = mx$, we have

$$x^3 (1 + m^3) - 3amx^2 = 0. \quad \cdots \cdots \quad (2)$$

This equation gives three roots or values of x, of which two are always equal zero, and the third is

$$x = \frac{3am}{1 + m^3}; \quad \cdots \cdots \quad (4)$$

whence

$$y = \frac{3am^2}{1 + m^3}. \quad \cdots \cdots \quad (5)$$

These are therefore the coordinates of the point P in Fig. 10. Since the values of x and y vanish when $m = 0$, and when $m = \infty$, the curve has a loop in the first quadrant. To find the area of this loop we therefore have, by equation (4) of the preceding article,

$$A = \frac{9a^2}{2} \int_0^\infty \frac{m^2\,dm}{(1+m^3)^2} = -\frac{3a^2}{2} \frac{1}{1+m^3}\Big]_0^\infty = \frac{3a^2}{2}.$$

114. The area included between this curve and its asymptote may be found by means of equation (5), Art. 112. The equation of a straight line is of the form

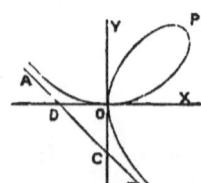

FIG. 10.

$$y = mx + b,$$

and since this line is parallel to $y = mx$, the value of m for the asymptote must be that which makes x and y in equations (4) and (5) infinite: that is, $m = -1$; hence the equation of the asymptote is

$$y + x = b, \quad \ldots \ldots \quad (6)$$

in which b is to be determined. Since when $m = -1$, the point P of the curve approaches indefinitely near to the asymptote, equation (6) must be satisfied by P when $m = -1$. From (4) and (5) we derive

$$y + x = 3a\frac{m^2 + m}{1 + m^3} = \frac{3am}{1 - m + m^2};$$

whence, putting $m = -1$, and substituting in equation (6)

$$-a = b,$$

the equation of the asymptote AB, Fig. 10, is

$$y + x = -a. \quad \ldots \ldots \quad (7)$$

§ VIII.] *TRANSFORMATION OF THE POLAR FORMULAS* 133

Now, as m varies from $-\infty$ to 0, the difference between the radii vectores of the asymptote and curve will generate the areas OBC and ODA, hence the sum of these areas is represented by

$$A = \frac{1}{2}\int_{-\infty}^{0} (x_2^2 - x_1^2)\, dm,$$

in which x_2 is taken from the equation of the asymptote (7), and x_1 from that of the curve.

Putting $y = mx$, in (7), we have

$$x_2 = -\frac{a}{1+m},$$

and the value of x_1 is given in equation (4). Hence

$$A = \frac{a^2}{2}\int_{-\infty}^{0}\left[\frac{1}{(1+m)^2} - \frac{9m^2}{(1+m^3)^2}\right] dm$$

$$= \frac{a^2}{2}\left[\frac{3}{1+m^3} - \frac{1}{1+m}\right]_{-\infty}^{0}$$

$$= \frac{a^2}{2}\frac{2+m-m^2}{1+m^3}\bigg]_{-\infty}^{0} = \frac{a^2}{2}\frac{2-m}{1-m+m^2}\bigg]_{-\infty}^{0}{}^{*} = a^2.$$

Adding the triangle OCD, whose area is $\frac{1}{2}a^2$, we have for the whole area required $\frac{3}{2}a^2$.

---

* This reduction is given to show that the integral is not infinite for the value $m = -1$, which is between the limits. See Art. 77.

Examples VIII.

1. Find the area included between the curve
$$a^2y = x^3 + ax^2,$$
and the axis of x. $\dfrac{a^2}{12}$.

2. Find the whole area of the curve
$$a^4y^2 = x^4(a^2 - x^2).$$ $\dfrac{\pi a^3}{4}$.

3. Find the area of a loop of the curve
$$x^2(a^2 + y^2) = y^2(a^2 - y^2).$$ $\dfrac{a^2}{2}(\pi - 2)$.

4. Find the area between the axes and the curve
$$y(x^2 + a^2) = b^2(a - x).$$ $b^2\left[\dfrac{\pi}{4} - \dfrac{\log 2}{2}\right]$.

5. Find the area between the curve
$$x^2y^2 + a^2y^2 - a^2x^2 = 0,$$
and one of its asymptotes. $2a^2$.

6. Find the area between the parabola $y^2 = 4ax$ and the straight line $y = x$. $\dfrac{8a^2}{3}$.

7. Find the area of the ellipse whose equation is
$$ax^2 + 2bxy + cy^2 = 1.$$ $\dfrac{\pi}{\sqrt{(ac - b^2)}}$.

8. Find the area of the loop of the curve

$$cy^2 = (x-a)(x-b)^2,$$

in which $c > 0$ and $b > a$.
$$\frac{8(b-a)^{\frac{5}{2}}}{15\sqrt{c}}.$$

9. Find the area of the loop of the curve

$$a^2 y^2 = x^4(b+x).$$
$$\frac{32b^{\frac{7}{2}}}{105 a^2}$$

10. Find the area included between the axes and the curve

$$\left(\frac{x}{a}\right)^{\frac{1}{3}} + \left(\frac{y}{b}\right)^{\frac{1}{3}} = 1.$$
$$\frac{ab}{20}.$$

11. If n is an integer, prove that the area included between the axes and the curve

$$\left(\frac{x}{a}\right)^{\frac{1}{n}} + \left(\frac{y}{b}\right)^{\frac{1}{n}} = 1$$

is
$$A = \frac{n(n-1)\cdots 1}{2n(2n-1)\cdots(n+1)} ab.$$

12. If n is an odd integer, prove that the area included between the axes and the curve

$$\left(\frac{x}{a}\right)^{\frac{2}{n}} + \left(\frac{y}{b}\right)^{\frac{2}{n}} = 1$$

is
$$A = \frac{[n(n-2)\cdots 1]^2}{2n(2n-2)\cdots 2} \frac{\pi ab}{2}.$$

13. In the case of the curtate cycloid

$$x = a\psi - b \sin \psi, \qquad y = a - b \cos \psi,$$

find the area between the axis of x and the arc below this axis.

$$(2a^2 + b^2) \cos^{-1}\frac{a}{b} - 3a\sqrt{(b^2 - a^2)}.$$

14. If $b = \tfrac{1}{2}a\pi$, show that the area of the loop of the curtate cycloid is

$$\pi a^2 \left[\frac{\pi^2}{8} - 1\right].$$

15. Find the area of the segment of the hyperbola

$$x = a \sec \psi, \qquad y = b \tan \psi,$$

cut off by the double ordinate whose length is $2b$.

$$ab\left[\sqrt{2} - \log \tan \frac{3\pi}{8}\right].$$

16. Find the whole area of the curve

$$r^2 = a^2 \cos^2 \theta + b^2 \sin^2 \theta. \qquad \frac{\pi}{2}(a^2 + b^2).$$

17. Find the area of a loop of the curve

$$r^2 = a^2 \cos^2 \theta - b^2 \sin^2 \theta. \qquad \frac{ab}{2} + \frac{(a^2 - b^2)}{2} \tan^{-1}\frac{a}{b}.$$

18. Find the areas of the large and of each of the small loops of the curve

$$r = a \cos \theta \cos 2\theta;$$

and show that the sum of the loops may be expressed by a single integral.

$$\frac{\pi a^2}{16} + \frac{a^2}{4}, \quad \text{and} \quad \frac{\pi a^2}{32} - \frac{a^2}{8}.$$

19. In the case of the spiral of Archimedes,

$$r = a\theta,$$

find the area generated by the radius vector of the first whorl and that generated by the difference between the radii vectores of the nth and $(n + 1)$th whorl.

$$\frac{8a^2\pi^3}{6}, \quad \text{and} \quad 8na^2\pi^3.$$

20. Find the area of a loop of the curve

$$r = a \sin 3\theta.$$

$$\frac{\pi a^2}{12}.$$

21. Find the area of the cardioid

$$r = 4a \sin^2 \tfrac{1}{2}\theta.$$

$$6\pi a^2.$$

22. Find the area of the loop of the curve

$$r = a\frac{\cos 2\theta}{\cos \theta}.$$

$$\frac{a^2(4 - \pi)}{2}.$$

23. In the case of the hyperbolic spiral,

$$r\theta = a,$$

show that the area generated by the radius vector is proportional to the difference between its initial and its final value.

24. Find the area of a loop of the curve

$$r = a \cos n\theta.$$

$$\frac{\pi a^2}{4n}.$$

25. Find the area of a loop of the curve

$$r^2 = a^2 \frac{\sin 3\theta}{\cos^5 \theta}.$$

$$\frac{a^2}{2}.$$

26. Find the area of a loop of the curve

$$r^2 \sin\theta = a^2 \cos 2\theta.$$

Notice that r *is real and finite from* $\theta = \frac{5\pi}{4}$ *to* $\theta = \frac{7\pi}{4}$, *and that* $\int \frac{d\theta}{\sin\theta}$ *is negative in this interval.*

$$a^2 \left[\sqrt{2} - \log(1 + \sqrt{2}) \right].$$

27. Find the area of a loop of the curve

$$(x^2 + y^2)^2 = a^2 xy.$$

Transform to polar coordinates.

$$\frac{a^2}{4}.$$

28. In the case of the limaçon

$$r = 2a \cos\theta + b,$$

find the whole area of the curve when $b > 2a$ and show that the same expression gives the sum of the loops when $b < 2a$.

$$(2a^2 + b^2)\pi.$$

29. Find separately the areas of the large and small loops of the limaçon when $b < 2a$.

If $\alpha = \cos^{-1}\left(-\dfrac{b}{2a}\right)$,

$$\text{large loop} = (2a^2 + b^2)\,\alpha + \frac{3b}{2}\sqrt{(4a^2 - b^2)};$$

$$\text{small loop} = (2a^2 + b^2)(\pi - \alpha) - \frac{3b}{2}\sqrt{(4a^2 - b^2)}.$$

30. Find the area of a loop of the curve

$$r^2 = a^2 \cos n\theta + b^2 \sin n\theta. \qquad \frac{\sqrt{(a^4 + b^4)}}{n}.$$

31. Find the area of the loop of the curve

$$r = a\,\frac{2\cos 2\theta - 1}{\cos\theta}, \qquad \left[5\sqrt{3} - \frac{8}{3}\pi\right]a^2.$$

32. Show that the sectorial area between the axis of x, the equilateral hyperbola

$$x^2 - y^2 = 1,$$

and the radius vector making the angle θ at the centre is represented by the formula

$$A = \frac{1}{4}\log\frac{1 + \tan\theta}{1 - \tan\theta};$$

and hence show that

$$x = \frac{\varepsilon^{2A} + \varepsilon^{-2A}}{2}, \qquad \text{and} \qquad y = \frac{\varepsilon^{2A} - \varepsilon^{-2A}}{2}.$$

If A denotes the corresponding area in the case of the circle

$$x^2 + y^2 = 1,$$

we have

$$x = \cos 2A, \qquad \text{and} \qquad y = \sin 2A.$$

In accordance with the analogy thus presented, the values of x and y given above are called the hyperbolic cosine and the hyperbolic sine of $2A$. Thus

$$\frac{\varepsilon^{2A} + \varepsilon^{-2A}}{2} = \cosh(2A), \qquad \frac{\varepsilon^{-2A} - \varepsilon^{2A}}{2} = \sinh(2A).$$

33. Find the area of the loop of the curve

$$x^4 - 3axy^2 + 2ay^3 = 0. \qquad \frac{3^5 a^2}{35 \cdot 2^{11}}.$$

34. Find the area of the loop of the curve

$$x^{2n+1} + y^{2n+1} = (2n+1)\,ax^n y^n. \qquad \frac{2n+1}{2}a^2.$$

35. Find the area between the curve

$$x^{2n+1} + y^{2n+1} = (2n+1)\,ax^n y^n$$

and its asymptote. $\qquad \dfrac{2n+1}{2}a^2.$

36. Find the area of the loop of the curve

$$y^3 + ax^2 - axy = 0. \qquad \frac{a^2}{60}.$$

37. Find the area of a loop of the curve

$$x^4 + y^4 = a^2 xy. \qquad \frac{\pi a^2}{8}.$$

38. Trace the curve

$$x = 2a \sin\frac{y}{x},$$

and find the area of one loop. $\qquad \pi a^2.$

IX.

Volumes of Geometric Solids.

115. A geometric solid whose volume is required is frequently defined in such a way that the area of the plane section parallel to a fixed plane may be expressed in terms of the perpendicular distance of the section from the fixed plane. When this is the case, the solid is to be regarded as generated by the motion of the plane section, and its differential, when thus considered, is readily expressed.

116. For example, let us consider the solid whose surface is formed by the revolution of the curve APB, Fig. 11, about the axis OX. The plane section perpendicular to the axis OX is a circle; and if APB be referred to rectangular coordinates, the distance of the section from a parallel plane passing through the origin is x, while the radius of the circle is y. Supposing the centre of the section to move uniformly along the axis, the rate at which the volume is generated is not uniform, but its differential is the volume which would be generated while the centre is describing the distance dx, if the rate were made constant. This differential volume is therefore the cylinder whose altitude is dx, and the radius of whose base is y. Hence, if V denote the volume,

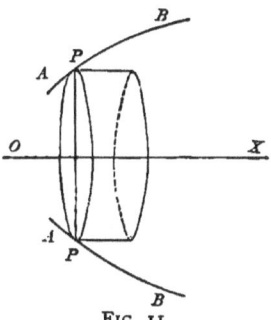

Fig. 11.

$$dV = \pi y^2 \, dx.$$

117. As an illustration, let it be required to find the volume of the paraboloid, whose height is h, and the radius of whose base is b.

The revolving curve is in this case a parabola, whose equation is of the form

$$y^2 = 4ax;$$

and since $y = b$ when $x = h$,

$$b^2 = 4ah, \qquad \text{whence} \qquad 4a = \frac{b^2}{h};$$

the equation of the parabola is therefore

$$y^2 = \frac{b^2}{h} x.$$

Hence the volume required is

$$V = \pi \int_0^h y^2\, dx = \pi \frac{b^2}{h} \int_0^h x\, dx = \frac{\pi b^2 h}{2}.$$

118. It can obviously be shown, by the method used in Art. 116, that whatever be the shape of the section parallel to a fixed plane, *the differential of the volume is the product of the area of the generating section and the differential of its motion perpendicular to its plane.*

If the volume is completely enclosed by a surface whose equation is given in the rectangular coordinates x, y, z, and if we denote the areas of the sections perpendicular to the axes by A_x, A_y, and A_z, we may employ either of the formulas

$$V = \int A_x\, dx, \qquad V = \int A_y\, dy, \qquad V = \int A_z\, dz.$$

The equation of the section perpendicular to the axis of x is determined by regarding x as constant in the equation of the surface, and its area A_x is of course a function of x.

§ IX.] *VOLUMES OF GEOMETRIC SOLIDS.* 143

For example, the equation of the surface of an ellipsoid is

$$\frac{x^2}{a^2} + \frac{y^2}{b^2} + \frac{z^2}{c^2} = 1.$$

The section perpendicular to the axis of x is the ellipse

$$\frac{y^2}{b^2} + \frac{z^2}{c^2} = \frac{a^2 - x^2}{a^2},$$

whose semi-axes are $\frac{b}{a}\sqrt{(a^2 - x^2)}$ and $\frac{c}{a}\sqrt{(a^2 - x^2)}$.

Since the area of an ellipse is the product of π and its semi-axes,

$$A_x = \frac{\pi bc}{a^2}(a^2 - x^2).$$

The limits for x are $\pm a$, the values between which x must lie to make the ellipse possible. Hence

$$V = \frac{\pi bc}{a^2}\int_{-a}^{a}(a^2 - x^2)\,dx = \frac{4\pi abc}{3}.$$

119. The area A_x can frequently be determined by the conditions of the problem without finding the equation of the surface. For example, let it be required to find the volume of the solid generated by so moving an ellipse with constant major axis, that its center shall describe the major axis of a fixed ellipse, to whose plane it is perpendicular, while the extremities of its minor axis describe the fixed ellipse. Let the equation of the fixed ellipse be

$$\frac{x^2}{a^2} + \frac{y^2}{b^2} = 1,$$

and let c be the major semi-axis of the moving ellipse. The minor semi-axis of this ellipse is y. Since the area of an ellipse is equal to π multiplied by the product of its semi-axes, we have

$$A_x = \pi c y = \frac{\pi c b}{a} \sqrt{(a^2 - x^2)}.$$

Therefore $\quad V = \dfrac{\pi b c}{a} \displaystyle\int_{-a}^{a} \sqrt{(a^2 - x^2)}\, dx;$

hence, see formula (M),

$$V = \frac{\pi^2 a b c}{2}.$$

The Solid of Revolution regarded as Generated by a Cylindrical Surface.

120. A solid of revolution may be generated in another manner, which is sometimes more convenient than the employment of a circular section, as in Art. 116. For example, let the cissoid POR, Fig. 12, whose equation is

$$y^2 (2a - x) = x^3,$$

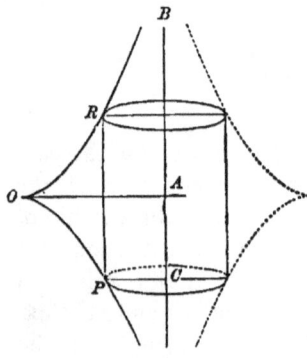

FIG. 12.

revolve about its asymptote AB. The line PR, parallel to AB and terminated by the curve, describes a cylindrical surface. If we conceive the radius of this cylinder to pass from the value $OA = 2a$ to zero, the cylindrical surface will evidently generate the solid of revolution. Now every point of this cylindrical surface moves with a rate equal to that of the radius; therefore the differential of the solid is the product of the cylindrical surface, and the differential of the radius. The radius and altitude in this case are

$$PC = 2a - x, \qquad \text{and} \qquad PR = 2y,$$

therefore
$$V = 4\pi \int_0^{2a} (2ax - x^2)^{\frac{1}{2}} x \, dx.$$

Putting
$$x - a = a \sin \theta,$$

$$V = 4\pi a^3 \int_{-\frac{\pi}{2}}^{\frac{\pi}{2}} (\cos^2 \theta + \cos^2 \theta \sin \theta) \, d\theta = 2\pi^2 a^3.$$

Double Integration.

121. When rectangular coordinates are used, the expression for the area generated by a line parallel to the axis of y and terminated by two curves is

$$A = \int_a^b (y_2 - y_1) \, dx. \quad \ldots \quad (1)$$

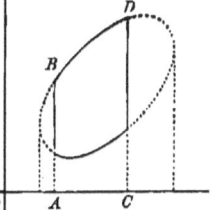

Fig. 13.

Let AB, in Fig. 13, be the initial, and CD the final position of the generating line, then the area is $ABDC$, which is enclosed by the curves

$$y = y_1, \qquad y = y_2,$$

and by the straight lines

$$x = a, \qquad x = b.$$

If in equation (1) we substitute for $y_2 - y_1$ the equivalent expression $\int_{y_1}^{y_2} dy$, we have

$$A = \int_a^b \int_{y_1}^{y_2} dy\, dx, \quad \ldots \ldots \quad (2)$$

which expresses the area in the form of a double integral. In this double integral the limits y_1 and y_2 for y, are functions of x, while a and b, the limits for x, are constants.

122. If the area is that of a closed curve y_1 and y_2 are two values of y corresponding to the same value of x in the equation of the curve, and a and b are the values of x for which y_1 and y_2 become equal, as represented by the dotted lines in Fig. 13. It is evident that the entire area may also be expressed in the form

$$A = \int_p^q \int_{x_1}^{x_2} dx\, dy; \quad \ldots \ldots \quad (3)$$

and that when either of the forms (2) or (3) is applied to the area of a closed curve the limits are completely determined by the equation of the curve.

123. The limits in either of the expressions (1) or (2) define a certain closed boundary, and since either of these integrals represents the included area, it is evident that we may write

$$\iint dy\, dx = \iint dx\, dy;$$

provided it is understood that the limits in the two expressions are such as to represent the same boundary. It should however be noticed that if the boundary is like that represented by the full lines in Fig. 13, or if the arcs $y = y_1$ and $y = y_2$ *do not belong to the same curve*, we cannot make a practical application of the form (3) without breaking up the integral into several parts.

§ IX.] DOUBLE INTEGRATION. 147

124. Let $\phi(x,y)$ be any function of x and y. In the double integral

$$\int_a^b \int_{y_2}^{y_1} \phi(x, y)\, dy\, dx, \quad \ldots \ldots (1)$$

x is considered as a constant or independent of y in the first integration, but the limits of this integration are functions of x. The double integration is then said to *extend over the area* which is represented by the expression

$$\int_a^b \int_{y_1}^{y_2} dy\, dx, \quad \text{or} \quad \int_a^b (y_2 - y_1)\, dx. \quad \ldots \ldots (2)$$

125. Now let the surface, of which

$$z = \phi(x, y) \quad \ldots \ldots \ldots (3)$$

is the equation in rectangular coordinates, be constructed; and let a cylindrical surface be formed by moving a line perpendicular to the plane of xy about the boundary of the area (2) over which the integration extends. Let us suppose the value of z to be positive for all values of x and y which represent points within this boundary. Then the cylindrical surface, together with the plane of xy and the surface (3), encloses a solid, of which the base is the area (2) in the plane xy, or $ASBR$ in Fig. 14, and the upper surface is $CQDP$ a portion of the surface (3).

Let $SRPQ$ be a section of this solid perpendicular to the axis of x. In this section x has a constant value, and the ordinates of R and S are the corresponding values of y_1 and y_2. The area of this section, which denote

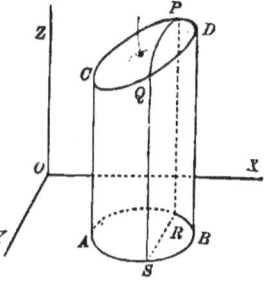

FIG. 14.

by A_x, as in Art. 117, may be regarded as generated by the line z, hence

$$A_x = \int_{y_1}^{y_2} z\, dy;$$

and therefore

$$V = \int_a^b \int_{y_1}^{y_2} z\, dy\, dx, \quad \ldots \ldots \quad (1)$$

which is identical with expression (1) Art. 124.

126. Now it is evident that the same volume may be expressed by

$$V = \iint z\, dx\, dy,$$

provided that the double integration extends over the same area. Hence, with this understanding, we may write

$$\iint \phi(x, y)\, dy\, dx = \iint \phi(x, y)\, dx\, dy.$$

In this formula x and y may be regarded as taking the places of any two variables, the limits of integration being determined by a given relation between the variables. Thus we may write

$$\iint \phi(u, v)\, dv\, du = \iint \phi(u, v)\, du\, dv,$$

provided the limits of integration are determined in each case by the same relation between u and v.

127. For example, if this relation is

$$u^2 + v^2 - c^2 = 0,$$

the range of values in the first integration is between

$$v = \pm \sqrt{(c^2 - u^2)};$$

that is, we must have

$$v^2 < c^2 - u^2,$$

or
$$u^2 + v^2 - c^2 < 0 \quad \ldots \ldots \quad (1)$$

But this condition also expresses the limits for u, since v is only possible when $u^2 < c^2$. Now, putting rectangular coordinates, x and y, in place of u and v, it is convenient to express the restriction (1), by saying that the range of values of x and y is such as to represent every point *within* the circle

$$x^2 + y^2 - c^2 = 0.$$

Volumes by Double and Triple Integration.

128. As an application of formula (1), Art. 125, let us suppose the curve $ASBR$ to be the circle

$$(x - h)^2 + (y - k)^2 = c^2, \quad \ldots \ldots \quad (1)$$

and the equation of the surface $CQDP$ to be

$$xy = pz. \quad \ldots \ldots \ldots \quad (2)$$

Then
$$V = \frac{1}{p}\int_a^b \int_{y_1}^{y_2} xy \, dy \, dx = \frac{1}{2p}\int_a^b (y_2^2 - y_1^2) x \, dx,$$

in which the limits y_1 and y_2 are derived from equation (1). Hence

$$y_2 = k + \sqrt{[c^2 - (x-h)^2]}, \qquad y_1 = k - \sqrt{[c^2 - (x-h)^2]},$$

and

$$V = \frac{2k}{p}\int_a^b \sqrt{[c^2 - (x-h)^2]}\, x\, dx.$$

The limits for x are the extreme values of x which make y possible; that is,

$$a = h - c \qquad \text{and} \qquad b = h + c$$

To evaluate the integral, put

$$x - h = c \sin\theta;$$

then

$$V = \frac{2kc^2}{p}\int_{-\frac{\pi}{2}}^{\frac{\pi}{2}} \cos^2\theta\,(h + c\sin\theta)\, d\theta.$$

Since, by Art. 87,

$$\int_{-\frac{\pi}{2}}^{\frac{\pi}{2}} \cos^2\theta \sin\theta\, d\theta = 0,$$

we have finally

$$V = \frac{\pi k h c^2}{p}.$$

129. A volume in general may be represented by the triple integral

$$V = \iiint dz\, dy\, dx, \quad \ldots \ldots \quad (1)$$

which is equivalent to

$$V = \iint (z_2 - z_1)\, dy\, dx\,; \quad \ldots \ldots (2)$$

for $\int (z_2 - z_1)\, dy = A_x$, the area of a section perpendicular to the axis of x. We may regard this formula as expressing the difference between two cylindrical solids of the form represented in Fig. 14.

130. When the volume is that of a closed surface, z_2 and z_1 are two values of z in terms of x and y found from the equation of the surface. The area over which the integration extends is in this case the projection of the solid upon the plane of xy; in other words, the base of a circumscribing cylinder. Thus, if the volume is that of the sphere

$$x^2 + y^2 + (z - c)^2 = a^2, \quad \ldots \ldots (1)$$

z_1 and z_2 are the two values of z derived from this equation; that is

$$c \pm \sqrt{(a^2 - x^2 - y^2)}.$$

Hence

$$z_2 - z_1 = 2\sqrt{(a^2 - x^2 - y^2)},$$

and

$$V = 2 \iint \sqrt{(a^2 - x^2 - y^2)}\, dy\, dx. \quad \ldots (2)$$

The integration here extends over the circle

$$x^2 + y^2 - a^2 = 0. \quad \ldots \ldots (3)$$

since $z_2 - z_1$ is real only when

$$a^2 - x^2 - y^2 > 0.$$

From equation (3) we find the limits for y to be

$$\pm \sqrt{(a^2 - x^2)},$$

hence, by formula (M), equation (2) becomes

$$V = \pi \int (a^2 - x^2) \, dx.$$

Finally the limits for x are $\pm a$, since y is real only when x is between these limits;

therefore
$$V = \pi \left[a^2 x - \frac{1}{3} x^3 \right]_{-a}^{a} = \frac{4}{3} \pi a^3.$$

Elements of Area and Volume.

131. In accordance with Art. 100, the expression for an area,

$$\int_a^b \int_{y_1}^{y_2} dy \, dx, \quad \ldots \ldots \ldots (1)$$

is the limit of the sum

$$\sum_a^b \left[\sum_{y_1}^{y_2} \Delta y \right] \Delta x.$$

Since each of the terms included in $\sum_{y_1}^{y_2} \Delta y$ is multiplied by the common factor Δx, this sum may be written in the form

$$\sum_a^b \sum_{y_1}^{y_2} \Delta y \, \Delta x. \quad \ldots \ldots \ldots (2)$$

§ IX.] ELEMENTS OF AREA AND VOLUME. 153

The sum (2) consists of terms of the form

$$\Delta y \, \Delta x ;$$

and this product is called *the element of the sum;* in like manner, the product

$$dy \, dx,$$

which takes the place of $\Delta y \, \Delta x$ when we pass to the limit by substituting integration for summation, is called *the element of the integral* (1), or of the area represented by it.

132. We may now regard the process of double integration as a process of double summation, as indicated by expression (2), followed by the act of passing to the limiting value. In the first summation indicated, the elemental rectangles corresponding to the same value of x are combined into the term $(y_2 - y_1) \Delta x$, which may be called a *linear element of area*, since its length is independent of the symbol Δ.

133. It is easy to see that, in a similar manner, when rectangular coordinates are used, a volume may be regarded as the limiting value of the sum of terms of the form

$$\Delta x \, \Delta y \, \Delta z ;$$

and hence $\qquad dx \, dy \, dz,$

which takes its place when we pass to the limiting value by substituting integration for summation, is called *the element of volume.*

If the summation is effected in the order z, y, x, the first operation combines the elements which have common values of y and x into the *linear element of volume*,

$$(z_2 - z_1) \, \Delta x \, \Delta y.$$

The second operation combines the linear elements corresponding to a common value of x, over a certain range of values of y, into a term whose limiting value takes the form

$$A_x \triangle x.$$

This last expression represents a *lamina* perpendicular to the axis of x, whose area is A_x a section of the solid, and whose thickness is $\triangle x$.

Polar Elements.

134. If in the formula for a polar area,

$$A = \frac{1}{2}\int (r_2^2 - r_1^2)\, d\theta, \quad \ldots \ldots \quad (1)$$

[equation (2), Art. 110], we substitute for $\frac{1}{2}(r_2^2 - r_1^2)$ the equivalent expression $\int_{r_1}^{r_2} r\, dr$, we obtain

$$A = \int_\alpha^\beta \int_{r_1}^{r_2} r\, dr\, d\theta, \quad \ldots \ldots \quad (2)$$

in which α and β are fixed limits for θ.

Now it follows, from Art. 126, that the limits being determined by a certain relation between r and θ, this integral may also be put in the form

$$A = \int_a^b r \int_{\theta_1}^{\theta_2} d\theta \cdot dr = \int_a^b r\, (\theta_2 - \theta_1)\, dr, \quad \ldots \quad (3)$$

in which a and b are the limiting values of r, between which θ is possible.

The expression $\qquad r\,dr\,d\theta,$

in equation (2), is called the *polar element of area.**

135. The formula

$$A = \int r\,(\theta_2 - \theta_1)\,dr$$

may also be derived geometrically; for $r\,(\theta_2 - \theta_1)$ is the length of an arc whose radius is r. As r increases, this arc generates the surface, and it is plain that every point has a motion, whose differential is dr, in a direction perpendicular to the arc.

136. In determining the volume of a solid, it is sometimes convenient to express z as a function of the polar coordinates of its projection in the plane of xy. In this case we employ the linear element of volume,

$$(z_2 - z_1)\,r\,dr\,d\theta,$$

corresponding to the polar element of area.

---

* It is easily shown that the area included between the circles whose radii are r and $r + \triangle r$, and the radii whose inclinations to the initial line are θ and $\theta + \triangle \theta$ is

$$(r + \tfrac{1}{2}\triangle r)\,\triangle r\,\triangle \theta.$$

Since $r + \tfrac{1}{2}\triangle r$ is intermediate between r and $r + \triangle r$, the limiting value of the sum, of which this is the element, is, by Art. 99, the integral of the element

$$r\,dr\,d\theta.$$

In the summation corresponding to equation (1), the elements are first combined into the *sectorial element*

$$\tfrac{1}{2}(r_2^2 - r_1^2)\,\triangle\theta\,;$$

while in the summation corresponding to equation (3), they are first combined into the *arc-shaped element*

$$(r + \tfrac{1}{2}\triangle r)(\theta_2 - \theta_1)\,\triangle r.$$

As an illustration, let us determine the volume cut from a sphere by a right cylinder, having a radius of the sphere for one of its diameters. Taking the centre of the sphere as the origin, the diameter of the cylinder as initial line, and the axis of z parallel to the axis of the cylinder, we have for every point on the surface of the sphere

$$z^2 + r^2 = a^2, \quad \ldots \ldots \ldots \quad (1)$$

where a is the radius of the sphere. Hence

$$z_2 - z_1 = 2\sqrt{(a^2 - r^2)},$$

and $\quad V = 2 \iint_{r_1}^{r_2} (a^2 - r^2)^{\frac{1}{2}} r\, dr\, d\theta = \int \left[-\frac{2}{3}(a^2 - r^2)^{\frac{3}{2}} \right]_{r_1}^{r_2} d\theta.$

The circular base passes through the pole, and its equation is

$$r = a \cos \theta, \quad \ldots \ldots \ldots \quad (2)$$

hence the limits for r are 0 and $a \cos \theta$, and by substitution we obtain

$$V = \frac{2a^3}{3} \int (1 - \sin^3 \theta)\, d\theta.$$

The limits for θ are $\pm \dfrac{\pi}{2}$, the values which make r vanish in equation (2); but it is to be noticed that the expression $(a^2 - r^2)^{\frac{3}{2}}$, for which we have substituted $a^3 \sin^3 \theta$, is *always positive*, whereas $\sin^3 \theta$ is negative in the fourth quadrant. Hence the value of V is double the value of the integral in the first quadrant; that is,

$$V = \frac{4a^3}{3} \int_0^{\frac{\pi}{2}} (1 - \sin^3 \theta)\, d\theta = \frac{2\pi a^3}{3} - \frac{8a^3}{9}.$$

If a second cylinder whose diameter is the opposite radius of the sphere be constructed, the whole volume removed from the sphere is $\frac{4\pi a^3}{3} - \frac{16a^3}{9}$, and the portion of the sphere which remains is $\frac{16a^3}{9}$, a quantity commensurable with the cube of the diameter.

Polar Coordinates in Space.

137. A point in space may be determined by the polar coordinates ρ, ϕ, and θ, of which ρ denotes the radius vector OP, Fig. 15, ϕ the inclination POR of ρ to a fixed plane passing through the pole, and θ the angle ROA, which the projection of ρ upon this plane makes with a fixed line in the plane. The angles ϕ and θ thus correspond to the latitude and longitude of the point P considered as situated upon the surface of a sphere whose radius is ρ. The radius of the circle of latitude BP is

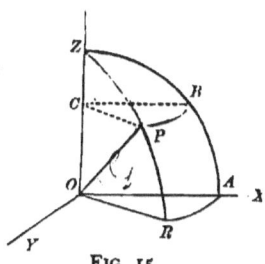

Fig. 15.

$$PC = \rho \cos \phi.$$

The motions of P, when ρ, ϕ, and θ independently vary, are in the directions of the radius vector OP and of the tangents at P to the arcs PR and PB. The differentials of these motions are respectively

$$d\rho, \qquad \rho\, d\phi, \qquad \text{and} \qquad \rho \cos\phi\, d\theta\,;$$

and since these motions are mutually rectangular, the element of volume is their product,

$$\rho^2 \cos \phi \, d\rho \, d\phi \, d\theta,$$

and $$V = \iiint \rho^2 \cos \phi \, d\rho \, d\phi \, d\theta. \quad \ldots \quad (1)$$

138. Performing the integration with respect to ρ, the formula becomes

$$V = \frac{1}{3} \iint (\rho_2^3 - \rho_1^3) \cos \phi \, d\phi \, d\theta. \quad \ldots \quad (2)$$

When the radius vector lies entirely within the solid, the lower limit ρ_1 must be taken equal zero, and we may write

$$V = \frac{1}{3} \iint \rho^3 \cos \phi \, d\phi \, d\theta. \quad \ldots \quad (3)$$

The element of this double integral has the form of a pyramid with vertex at the pole.

If, on the other hand, in formula (1) we perform first the integration with respect to ϕ, we have

$$V = \iint (\sin \phi_2 - \sin \phi_1) \rho^2 \, d\rho \, d\theta. \quad \ldots \quad (4)$$

Taking the lower limit $\phi_1 = 0$, so that the solid is bounded by the plane ORA, we have the simpler formula

$$V = \iint \sin \phi \, \rho^2 \, d\rho \, d\theta. \quad \ldots \quad (5)$$

139. The formulas of the preceding article take simpler

§ IX.] POLAR COORDINATES IN SPACE.

forms when applied to solids of revolution. Let OZ, Fig. 15, be the axis of revolution, then ρ and ϕ are polar coordinates of the revolving curve, OR being the initial line. Now θ is in this case independent of ρ and ϕ, and its limits are 0 and 2π. The integration with reference to θ may therefore be performed at once. Thus from (3) we obtain

$$V = \frac{2\pi}{3}\int \rho^3 \cos\phi \, d\phi; \quad \ldots \ldots \quad (6)$$

and in each of the formulas the factor 2π may take the place of the integration with reference to θ.

140. As an example of the use of equation (6), let us find the volume generated by a circle revolving about one of its tangents. The initial line, being perpendicular to the axis of revolution, is a diameter; hence if a is the radius of the circle its equation is

$$\rho = 2a \cos\phi,$$

and the limits for ϕ are $-\frac{\pi}{2}$ and $\frac{\pi}{2}$. Substituting in (6)

$$V = \frac{16\pi a^3}{3}\int_{-\frac{\pi}{2}}^{\frac{\pi}{2}} \cos^4\phi \, d\phi = 2\pi^2 a^3.$$

141. The following example of the use of equation (4), Art. 138, is added to illustrate the necessity of drawing a figure in each case to determine the limits to be employed.

Let it be required to find the volume generated by the revolution of the cardioid about its axis, the equation of the curve being

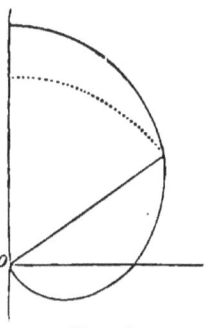

FIG. 16.

$$\rho = a(1 + \sin\phi), \quad \ldots \ldots \quad (1)$$

when the initial line is perpendicular to the axis of the curve, as in Fig. 16. The figure shows that the upper limit for ϕ is $\frac{1}{2}\pi$, while the lower limit is the value of ϕ given by equation (1); therefore

$$\sin \phi_2 = 1, \quad \text{and} \quad \sin \phi_1 = \frac{\rho}{a} - 1.$$

The limits for ρ are evidently 0 and $2a$. Substituting in equation (4) Art. 138,

$$V = 2\pi \int_0^{2a} \left(2 - \frac{\rho}{a}\right) \rho^2 \, d\rho$$

$$= 2\pi \left[\frac{2\rho^3}{3} - \frac{\rho^4}{4a}\right]_0^{2a} = \frac{8\pi a^3}{3}.$$

Examples IX.

1. Find the volume of the spheroid produced by the revolution of the ellipse,

$$\frac{x^2}{a^2} + \frac{y^2}{b^2} = 1,$$

about the axis of x. $\qquad \frac{4\pi ab^2}{3}$.

2. Find the volume of a right cone whose altitude is a, and the radius of whose base is b. $\qquad \frac{\pi ab^2}{3}$.

3. Find the volume of the solid produced by the revolution about the axis of x of the area between this axis, the cissoid

$$y^2(2a - x) = x^3,$$

and the ordinate of the point (a, a). $\qquad 8a^3\pi (\log 2 - \tfrac{2}{3})$.

4. Find the volume generated by the revolution of the witch,
$$y^2x - 2ay^2 + 4a^2x = 0,$$
about its asymptote.

See *Art.* 104. $\qquad 4\pi^2a^2.$

5. The equilateral hyperbola
$$x^2 - y^2 = a^2$$
revolves about the axis of x: show that the volume cut off by a plane cutting the axis of x perpendicularly at a distance a from the vertex is equal to a sphere whose radius is a.

6. An anchor ring is formed by the revolution of a circle whose radius is b about a straight line in its plane at a distance a from its centre: find its volume. $\qquad 2\pi^2ab^2.$

7. Express the volume of a segment of a sphere in terms of the altitude h and the radii a_1 and a_2 of the bases.
$$\frac{\pi h}{6}(h^2 + 3a_1^2 + 3a_2^2).$$

8. Find the volume generated by the revolution of the cycloid,
$$x = a(\psi - \sin\psi), \qquad y = a(1 - \cos\psi),$$
about its base. $\qquad 5\pi^2a^3.$

9. The area included between the cycloid and tangents at the cusp and at the vertex revolves about the latter; find the volume generated.
$\qquad \pi^2a^3.$

10. Find the volume generated by the revolution of the part of the curve
$$y = \varepsilon^x,$$
which is on the left of the origin, about the axis of x.
$\qquad \dfrac{\pi}{2}.$

11. The axes of two equal right circular cylinders, whose common radius is a, intersect at the angle α; find the volume common to the cylinders.

The section parallel to the axes is a rhombus. $\quad\dfrac{16a^3}{3\sin\alpha}$.

12. Find the volume generated by the revolution of one branch of the sinusoid,
$$y = b \sin \frac{x}{a},$$
about the axis of x. $\quad\dfrac{\pi^2 b^2}{2a}$.

13. Find the volume enclosed by the surface generated by the revolution of an arc of a parabola about a chord, whose length is $2c$, perpendicular to the axis, and at a distance b from the vertex.

$\dfrac{16\pi b^2 c}{15}$.

14. Find the volume generated by the revolution of the tractrix, whose differential equation is
$$\frac{dy}{dx} = \pm \frac{y}{\sqrt{(a^2 - y^2)}},$$
about the axis of x.

Express $\pi y^2\, dx$ in terms of y. $\quad\dfrac{2\pi a^3}{3}$.

15. Find the volume cut from a right circular cylinder whose radius is a, by a plane passing through the centre of the base, and making the angle α with the plane of the base.

$\dfrac{2a^3 \tan \alpha}{3}$.

16. Find the volume generated by the curve
$$xy^2 = 4a^2 (2a - x)$$
revolving about its asymptote. $\quad 4\pi^2 a^3$.

17. Express the volume of a frustum of a cone in terms of its height h, and the radii a_1 and a_2 of its bases.
$$\frac{\pi h}{3}(a_1^2 + a_1 a_2 + a_2^2).$$

18. Find the volume generated by the revolution of the cardioid,
$$r = a(1 - \cos\theta),$$
about the initial line.
Express y and dx in terms of θ.
$$\frac{8\pi a^3}{3}.$$

19. Find the volume of a barrel whose height is $2h$, and diameter $2b$, the longitudinal section through the centre being a segment of an ellipse whose foci are in the ends of the barrel.
$$2\pi b^2 h \frac{2h^2 + 3b^2}{3(b^2 + h^2)}.$$

20. Find the volume generated by the superior and by the inferior branch of the conchoid each revolving about the directrix; the equation, when the axis of y is the directrix, being
$$x^2 y^2 = (a + x)^2 (b^2 - x^2).$$
$$\pi^2 a b^2 \pm \frac{4\pi b^3}{3}.$$

21. On two opposite lateral faces of a rectangular parallelopiped whose base is ab, oblique lines are drawn, cutting off the distances c_1, c_2, c_3, c_4 on the lateral edges. A straight line intersecting each of these lines moves across the parallelopiped, remaining always parallel to the other lateral faces: find the volume cut off.
$$\frac{ab(c_1 + c_2 + c_3 + c_4)}{4}.$$

22. Find the volume enclosed by the surface generated by an arc of a circle whose radius is a, about a chord whose length is $2c$.
$$\frac{2\pi c(3a^2 - c^2)}{3} - 2\pi a^2 \sqrt{a^2 - b^2} \sin^{-1}\frac{c}{a}.$$

23. The area included between a quadrant of the ellipse

$$x = a \cos \phi, \qquad y = b \sin \phi,$$

and the tangents at its extremities revolves about the tangent at the extremity of the minor axis; find the volume generated.
$$\frac{\pi a b^2 (10 - 3\pi)}{6}.$$

24. An ellipse revolves about the tangent at the extremity of its major axis; express the entire volume in the form of an integral, whose limits are 0 and 2π, and find its value. $\qquad 2\pi^2 a^2 b.$

25. Show that the volume between the surface,

$$z^n = a^2 x^2 + b^2 y^2,$$

and any plane parallel to the plane of xy is equal to the circumscribing cylinder divided by $n + 1$.

26. A straight line of fixed length $2c$ moves with its extremities in two fixed perpendicular straight lines not in the same plane, and at a distance $2b$. Prove that every point in the moving line describes an ellipse in a plane parallel to both the fixed lines, and find the volume enclosed by the generated surface.
$$\frac{4\pi (c^2 - b^2) b}{3}.$$

27. Find the volume enclosed by the surface whose equation is

$$\frac{x^2}{a^2} + \frac{y^2}{b^2} + \frac{z^4}{c^4} = 1. \qquad \frac{8\pi a b c}{5}.$$

28. A moving straight line, which is always perpendicular to a fixed straight line through which it passes, passes also through the circumference of a circle whose radius is a, in a plane parallel to the fixed straight line and at a distance b from it; find the volume enclosed by the surface generated and the circle.
$$\frac{\pi a^2 b}{2}.$$

§ IX.] EXAMPLES. 165

29. Find the volume enclosed by the surface
$$\frac{y^2}{b^2} + \frac{z^2}{c^2} = \frac{x}{a}$$
and the plane $x = a$. $\dfrac{\pi abc}{2}$.

30. Find the volume enclosed by the surface
$$x^{\frac{2}{3}} + y^{\frac{2}{3}} + z^{\frac{2}{3}} = a^{\frac{2}{3}}.$$

Find A_z as in Art. 107, and then evaluate V by a similar method.
$$\frac{4\pi a^3}{35}.$$

31. Find the volume between the coordinate planes and the surface
$$\left(\frac{x}{a}\right)^{\frac{1}{2}} + \left(\frac{y}{b}\right)^{\frac{1}{2}} + \left(\frac{z}{c}\right)^{\frac{1}{2}} = 1.$$ $\dfrac{abc}{90}$.

32. Find the volume cut from the paraboloid of revolution
$$y^2 + z^2 = 4ax$$
by the right circular cylinder
$$x^2 + y^2 = 2ax,$$
whose axis intersects the axis of the paraboloid perpendicularly at the focus, and whose surface passes through the vertex. $2\pi a^3 + \dfrac{16a^3}{3}$.

33. The paraboloid of revolution
$$x^2 + y^2 = cz$$
is pierced by the right circular cylinder
$$x^2 + y^2 = ax,$$

whose diameter is a, and whose surface contains the axis of the paraboloid; find the volume between the plane of xy and the surfaces of the paraboloid and of the cylinder. $\quad\dfrac{3\pi a^4}{32c}$.

34. Find the volume cut from a sphere whose radius is a by a right circular cylinder whose radius is b, and whose axis passes through the centre of the sphere. $\quad\dfrac{4\pi}{3}\left[a^3-(a^2-b^2)^{\frac{3}{2}}\right]$.

35. Find the volume cut from a sphere whose radius is a by the cylinder whose base is the curve

$$r = a \cos 3\theta. \qquad \dfrac{2a^3\pi}{3} - \dfrac{8a^3}{9}.$$

36. Find the volume cut from a sphere whose radius is a by the cylinder whose base is the curve

$$r^2 = a^2 \cos^2\theta + b^2 \sin^2\theta,$$

supposing $b < a$. $\qquad \dfrac{4\pi a^3}{3} - \dfrac{16}{9}(a^2-b^2)^{\frac{3}{2}}$.

37. A right cone, the radius of whose base is a and whose altitude is b, is pierced by a cylinder whose base is a circle having for diameter a radius of the base of the cone; find the volume common to the cone and the cylinder. $\qquad \dfrac{ba^2}{36}(9\pi - 16)$.

38. The axis of a right cone whose semi-vertical angle is α coincides with a diameter of the sphere whose radius is a, the vertex being on the surface of the sphere; find the volume of the portion of the sphere which is outside of the cone. $\qquad \dfrac{4\pi a^3 \cos^4\alpha}{3}$.

39. Find the volume produced by the revolution of the lemniscata

$$r^2 = a^2 \cos 2\theta,$$

about a perpendicular to the initial line. $\qquad \dfrac{\pi^2 a^3 \sqrt{2}}{8}$.

40. Find the volumes generated by the revolution of the large loop and by one of the small loops of the curve

$$r = a \cos \theta \cos 2\theta$$

about a perpendicular to the initial line.

$$\frac{\pi^2 a^3}{16} + \frac{\pi a^3}{5}, \text{ and } \frac{\pi^2 a^3}{32} - \frac{\pi a^3}{10}.$$

41. From the element

$$r\, dr\, d\theta\, dz$$

derive the formulas for determining the volume of a solid of revolution whose axis is the axis of z.

$$V = 2\pi \iint r\, dr\, dz,$$

$$V = \pi \int (r_2^2 - r_1^2)\, dz, \quad \text{and} \quad V = 2\pi \int (z_2 - z_1) r\, dr.$$

Interpret the elements in these integrals.

42. Find the volume generated by the revolution of the curve

$$(x^2 + y^2)^2 = a^2 x^2 + b^2 y^2,$$

in which $a > b$, about the axis of y.

Transform to polar coordinates, and use the method of Art. 139.

$$\frac{\pi b(2b^2 + 3a^2)}{6} + \frac{\pi a^4}{2\sqrt{a^2 - b^2}} \cos^{-1}\frac{b}{a}.$$

43. Find the volume generated by the curve given in the preceding example, when revolving about the axis of x.

$$\frac{\pi a(2a^2 + 3b^2)}{6} + \frac{\pi b^4}{2\sqrt{a^2 - b^2}} \cdot \log \frac{a + \sqrt{a^2 - b^2}}{b}.$$

44. Find the volume common to the sphere whose radius is $\rho = a_1$ and to the solid formed by the revolution of the cardioid,

$$r = a(1 + \cos\theta),$$

about the initial line.

See Art. 141. $\qquad \dfrac{5\pi a^3}{6}.$

45. Find the whole volume enclosed by the surface

$$(x^2 + y^2 + z^2)^3 = a^6 xyz.$$

Transform to the coordinates ρ, ϕ, θ, *and show that the solid consists of four equal detached parts.* $\qquad \dfrac{a^3}{6}.$

X.

Rectification of Plane Curves.

142. A curve is said to be *rectified* when its length is determined, the unit of measure to which it is referred being a right line.

It is shown in Diff. Calc., Art. 314 [Abridged Ed., Art. 164], that, if s denotes the length of the arc of a curve given in rectangular coordinates, we shall have

$$ds = \sqrt{(dx^2 + dy^2)}.$$

If the abscissas of the extremities of the arc are known, s is found by substituting for dy in this expression its value in terms of x and dx, and integrating the result between the given values of x as limits. Thus, to express the arc measured from the vertex of the semi-cubical parabola

$$ay^2 = x^3$$

§ X.] RECTIFICATION OF PLANE CURVES. 169

in terms of the abscissa of its other extremity, we derive, from the equation of the curve,

$$dy = \frac{3\sqrt{x}\,dx}{2\sqrt{a}},$$

whence
$$ds = \frac{\sqrt{(9x+4a)}}{2\sqrt{a}}\,dx.$$

Integrating,

$$s = \frac{1}{2\sqrt{a}} \int_0 \sqrt{(9x+4a)}\,dx$$

$$= \frac{1}{27\sqrt{a}}(9x+4a)^{\frac{3}{2}} - \frac{8a}{27}.$$

143. When x and y are given in terms of a third variable, ds is generally expressed in terms of this variable. For example, from the equations of the four-cusped hypocycloid,

$$x = a\cos^3\psi, \qquad\qquad y = a\sin^3\psi, \quad \ldots \quad (1)$$

we derive

$$dx = -3a\cos^2\psi \sin\psi\,d\psi, \quad\text{and}\quad dy = 3a\sin^2\psi \cos\psi\,d\psi;$$

whence
$$ds = 3a\sin\psi\cos\psi\,d\psi. \quad \ldots \ldots \quad (2)$$

The length of the arc between the point $(a, 0)$, corresponding to $\psi = 0$, and $(0, a)$ corresponding to $\psi = \frac{1}{2}\pi$, is therefore

$$\left.\frac{3a}{2}\sin^2\psi\right]_0^{\frac{\pi}{2}} = \frac{3a}{2}.$$

Change of the Sign of ds.

144. We have hitherto assumed ds to be positive, but it is to be remarked that an expression substituted for ds, as in the illustration given in the preceding article, may change sign. Thus, in equation (2), ds, which is so written as to be positive while ψ passes from 0 to $\frac{1}{2}\pi$, becomes negative while ψ passes from $\frac{1}{2}\pi$ to π. Thus the integral gives a negative result for the arc between the points $(0, a)$ and $(-a, 0)$, corresponding to $\frac{1}{2}\pi$ and π. This change of sign in ds indicates a *cusp* or *stationary point* of the curve; and the existence of such points must be considered before we can properly interpret the resulting values of s. For instance, if in this example we integrate between the limits 0 and $\frac{3\pi}{4}$, we get the result $s = \frac{3a}{4}$, which is *the algebraic sum*, but *the numerical difference* of the arcs between the points corresponding to the limits.

Polar Coordinates.

145. It is proved in Diff. Calc., Art. 317 [Abridged Ed., Art. 167], that when the curve is given in polar coordinates

$$ds = \sqrt{(dr^2 + r^2 d\theta^2)}.$$

This is usually expressed in terms of θ. For example, the equation of the cardioid is

$$r = a(1 - \cos\theta) = 2a \sin^2 \tfrac{1}{2}\theta;$$

whence
$$dr = 2a \sin \tfrac{1}{2}\theta \cos \tfrac{1}{2}\theta \, d\theta,$$

and by substitution
$$ds = 2a \sin \tfrac{1}{2}\theta \, d\theta.$$

The limits for the whole perimeter of the curve are o and 2π, and ds remains positive for the whole interval. Therefore

$$s = 2a \int_0^{2\pi} \sin\frac{\theta}{2} d\theta = -4a \cos\frac{\theta}{2} \Big]_0^{2\pi} = 8a.$$

Rectification of Curves of Double Curvature.

146. Let σ denote the length of the arc of *a curve of double curvature*; that is, one which does not lie in a plane, and suppose the curve to be referred to rectangular coordinates x, y and z. If at any point of the curve the differentials of the coordinates be drawn in the directions of their respective axes, a rectangular parallelopiped will be formed, whose sides are dx, dy and dz, and whose diagonal is $d\sigma$. Hence

$$d\sigma = \sqrt{(dx^2 + dy^2 + dz^2)}.$$

The curve is determined by means of two equations connecting x, y and z, one of which usually expresses the value of y in terms of x, and the other that of z in terms of x. We can then express $d\sigma$ in terms of x and dx.

If the given equations contain all the variables, equations of the required form may be obtained by elimination.

147. An equation containing the two variables x and y only is evidently the equation of *the projection upon the plane of xy* of a curve traced upon the surface determined by the other equation. Let s denote the length of this projection; then, since $ds^2 = dx^2 + dy^2$,

$$d\sigma = \sqrt{(ds^2 + dz^2)},$$

in which ds may, if convenient, be expressed in polar coordinates; thus,

$$d\sigma = \sqrt{(dr^2 + r^2 d\theta^2 + dz^2)}.$$

148. As an illustration, let us use this formula to determine the length of the loxodromic curve from the equation of the sphere,
$$x^2 + y^2 + z^2 = a^2, \quad \ldots \ldots \quad (1)$$
upon which it is traced, and its projection upon the plane of the equator, of which the equation is
$$2a = \sqrt{(x^2 + y^2)} \left(\varepsilon^{n \tan^{-1} \frac{y}{x}} + \varepsilon^{-n \tan^{-1} \frac{y}{x}} \right),$$
or in polar coordinates
$$2a = r \left(\varepsilon^{n\theta} + \varepsilon^{-n\theta} \right). \quad \ldots \ldots \quad (2)$$
Equation (1) is equivalent to
$$r^2 + z^2 = a^2;$$
and, denoting the latitude of the projected point by ϕ, this gives
$$z = a \sin \phi, \qquad r = a \cos \phi. \quad \ldots \quad (3)$$
In order to express $d\theta$ in terms of ϕ, we substitute the value of r in (2); whence
$$\varepsilon^{n\theta} + \varepsilon^{-n\theta} = 2 \sec \phi, \quad \ldots \ldots \quad (4)$$
and by differentiation
$$\varepsilon^{n\theta} - \varepsilon^{-n\theta} = \frac{2}{n} \sec \phi \tan \phi \frac{d\phi}{d\theta}. \quad \ldots \quad (5)$$
Squaring and subtracting equation (5) from equation (4),
$$4 = \frac{4 \sec^2 \phi}{n^2} \left[n^2 - \tan^2 \phi \frac{d\phi^2}{d\theta^2} \right],$$
which reduces to
$$d\theta^2 = \frac{\sec^2 \phi \, d\phi^2}{n^2}. \quad \ldots \ldots \quad (6)$$

§ X.] LENGTH OF THE LOXODROMIC CURVE.

From equations (3) and (6)

$$r^2 d\theta^2 = \frac{a^2}{n^2} d\phi^2,$$

$$dr^2 = a^2 \sin^2 \phi \, d\phi^2,$$

$$dz^2 = a^2 \cos^2 \phi \, d\phi^2;$$

whence substituting in the value of $d\sigma$ (p. 171)

$$d\sigma = a \sqrt{\left(1 + \frac{1}{n^2}\right)} d\phi.$$

Integrating,

$$\sigma = a \frac{\sqrt{(n^2 + 1)}}{n} \int_\alpha^\beta d\phi = a \frac{\sqrt{(n^2 + 1)}}{n} (\beta - \alpha),$$

where α and β denote the latitudes of the extremities of the arc.

Examples X.

1. Find the length of an arc measured from the vertex of the catenary

$$y = \frac{c}{2}\left(\varepsilon^{\frac{x}{c}} + \varepsilon^{-\frac{x}{c}}\right),$$

and show that the area between the coordinate axes and any arc is proportional to the arc.

$$s = \frac{c}{2}\left(\varepsilon^{\frac{x}{c}} - \varepsilon^{-\frac{x}{c}}\right).$$

$$A = cs.$$

2. Find the length of an arc measured from the vertex of the parabola

$$y^2 = 4ax.$$

$$\sqrt{(ax + x^2)} + a \log \frac{\sqrt{x} + \sqrt{(x + a)}}{\sqrt{a}}.$$

3. Find the length of the curve

$$y = \frac{\varepsilon^x + 1}{\varepsilon^x - 1},$$

between the points whose abscissas are a and b.

$$\log \frac{\varepsilon^{2b} - 1}{\varepsilon^{2a} - 1} + a - b.$$

4. Find the length, measured from the origin, of the curve

$$y = a \log \frac{a^2 - x^2}{a^2}.$$

$$a \log \frac{a + x}{a - x} - x.$$

5. Given the differential equation of the tractrix,

$$\frac{dy}{dx} = -\frac{y}{\sqrt{(a^2 - y^2)}},$$

and, assuming $(0, a)$ to be a point of the curve, find the value of s as measured from this point, and also the value of x in terms of y; that is, find the rectangular equation of the curve.

$$s = a \log \frac{y}{a}.$$

$$x = a \log \frac{a + \sqrt{(a^2 - y^2)}}{y} - \sqrt{(a^2 - y^2)}.$$

6. Find the length of one branch of the cycloid

$$x = a(\psi - \sin \psi), \qquad y = a(1 - \cos \psi).$$
$$8a.$$

7. When the cycloid is referred to its vertex, the equations being

$$x = a(1 - \cos \psi), \qquad y = a(\psi + \sin \psi),$$

prove that $\quad s = \sqrt{(8ax)}.$

8. Find the length from the point $(a, 0)$ of the curve

$$x = 2a\cos\psi - a\cos 2\psi,$$

$$y = 2a\sin\psi - a\sin 2\psi.$$

$$4a\,(\psi - \sin\psi).$$

9. Show that the curve,

$$x = 3a\cos\psi - 2a\cos^3\psi, \qquad y = 2a\sin^3\psi,$$

has cusps at the points given by $\psi = 0$ and $\psi = \pi$; and find the whole length of the curve. $12a$.

10. Find the length of a quadrant of the curve

$$\left(\frac{x}{a}\right)^{\frac{2}{3}} + \left(\frac{y}{b}\right)^{\frac{2}{3}} = 1.$$

See Fig. 6, Art. 107.
$$\frac{a^2 + ab + b^2}{a + b}.$$

11. Show that the curve

$$x = 2a\cos^2\theta\,(3 - 2\cos^2\theta), \qquad y = 4a\sin\theta\cos^3\theta$$

has three cusps, and that the length of each branch is $\dfrac{8a}{3}$.

12. Find the length of the arc between the points at which the curve

$$x = a\cos^2\theta\cos 2\theta, \qquad y = a\sin^2\theta\sin 2\theta$$

cuts the axes.
$$\frac{2 - \sqrt{2}}{3}a.$$

$$\frac{2 + \sqrt{2}}{3}a$$

13. Show that the curve

$$x = a \cos \psi (1 + \sin^2 \psi),$$
$$y = a \sin \psi \cos^2 \psi$$

is symmetrical to the axes, and find the length of the arcs between the cusps.

$$a \left(\sqrt{2} - \sin^{-1} \frac{1}{\sqrt{3}} \right);$$

$$a \left(\sqrt{2} + \cos^{-1} \frac{1}{\sqrt{3}} \right).$$

14. Find the length of one branch of the epicycloid

$$x = (a + b) \cos \psi - b \cos \frac{a+b}{b} \psi,$$

$$y = (a + b) \sin \psi - b \sin \frac{a+b}{b} \psi.$$

$$\frac{8b(a+b)}{a}.$$

15. Show that the curve

$$x = 9a \sin \psi - 4a \sin^3 \psi,$$
$$y = -3a \cos \psi + 4a \cos^3 \psi$$

is symmetrical to the axes, and has double points and cusps: find the lengths of the arcs, (α) between the double points, (β) between a double point and a cusp, and (γ) the arc connecting two cusps, and not passing through the double points.

(α), $a(\pi + 3\sqrt{3})$;

(β), $\dfrac{\pi a}{2}$;

(γ), $a(3\sqrt{3} - \pi)$.

16. Find the whole length of the curve

$$x = 3a \sin \psi - a \sin^3 \psi,$$
$$y = a \cos^3 \psi.$$

$3\pi a.$

EXAMPLES.

17. Find the length, measured from the pole, of any arc of the equiangular spiral
$$r = a\varepsilon^{n\theta},$$
in which $n = \cot \alpha$.

$r \sec \alpha$.

18. Prove by integration that the arc subtending the angle θ at the circumference in a circle whose radius is a, is $2a\theta$.

19. Find the length, measured from the origin, of the curve defined by the equations
$$y = \frac{x^2}{2a}, \qquad z = \frac{x^3}{6a^2}.$$

$$x + \frac{x^3}{6a^2}.$$

20. Find the length, measured from the origin, of the intersection of the surfaces
$$y = 4n \sin x, \qquad z = 2n^2 (2x + \sin 2x).$$

$$(4n^2 + 1)x + 2n^2 \sin 2x.$$

21. Find the length, measured from the origin, of the intersection of the cylindrical surfaces
$$(y - x)^2 = 4ax, \qquad 9a(z - x)^2 = 4x^3.$$

$$\frac{2x^{\frac{3}{2}}}{3\sqrt{a}} + 2\sqrt{(ax)} + x.$$

22. If upon the hyperbolic cylinder
$$\frac{y^2}{c^2} - \frac{z^2}{b^2} = 1,$$

a curve whose projection upon the plane of xy is the catenary
$$y = \frac{c}{2}(\varepsilon^{\frac{x}{c}} + \varepsilon^{-\frac{x}{c}})$$

be traced, prove that any arc of the curve bears to the corresponding arc of its projection the constant ratio $\sqrt{(b^2 + c^2)} : c$.

XI.

Surfaces of Solids of Revolution.

149. The surface of a solid of revolution may be generated by the circumference of the circular section made by a plane perpendicular to the axis of revolution. Thus in Fig. 17, the surface produced by the revolution of the curve AB about the axis of x is regarded as generated by the circumference PQ. The radius of this circumference is y, and its *plane* has a motion whose differential is dx, but every point in the circumference itself has a motion whose differential is ds, s denoting an arc of the curve AB.

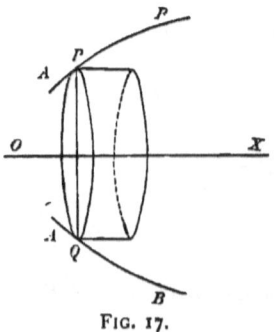

Fig. 17.

Hence, denoting the required surface by S, we have

$$dS = 2\pi y\, ds = 2\pi y\, \sqrt{(dx^2 + dy^2)}.$$

The value of dS must of course be expressed in terms of a single variable before integration.

150. For example, let us determine the area of the zone of spherical surface included between any two parallel planes. The radius of the sphere being a, the equation of the revolving curve is

$$x^2 + y^2 = a^2;$$

whence
$$y = \sqrt{(a^2 - x^2)},$$
$$dy = -\frac{x\, dx}{\sqrt{(a^2 - x^2)}},$$
$$ds = \frac{a\, dx}{\sqrt{(a^2 - x^2)}},$$

and
$$dS = 2\pi a\, dx;$$

§ XI.] SURFACES OF SOLIDS OF REVOLUTION. 179

therefore

$$S = 2\pi a \int dx = 2\pi a (x_2 - x_1).$$

Since $x_2 - x_1$ is the distance between the parallel planes, the area of a zone is the product of its altitude by $2\pi a$, the circumference of a great circle, and the area of the whole surface of the sphere is $4\pi a^2$.

151. When the curve is given in polar coordinates, it is convenient to transform the expression for S to polar coordinates. Thus, if the curve revolves about the initial line,

$$S = 2\pi \int y\, ds = 2\pi \int r \sin\theta \sqrt{(dr^2 + r^2 d\theta^2)}.$$

For example, if the curve is the cardioid

$$r = 2a \sin^2 \tfrac{1}{2}\theta,$$

we find, as in Art. 145,

$$ds = 2a \sin \tfrac{1}{2}\theta\, d\theta.$$

Hence

$$S = 16\pi a^2 \int_0^\pi \sin^4 \tfrac{1}{2}\theta \cos \tfrac{1}{2}\theta\, d\theta$$

$$= \frac{32\pi a^2}{5} \sin^5 \tfrac{1}{2}\theta \Big]_0^\pi = \frac{32\pi a^2}{5}.$$

Areas of Surfaces in General.

152. Let a surface be referred to rectangular coordinates x, y and z; the projection of a given portion of the surface upon the plane of xy is a plane area determined by a given relation between x and y. We may take as the elements of the surface the portions which are projected upon the corresponding elements of area in the plane of xy. If at a point within the element of surface, which is projected upon a given element $\triangle x \triangle y$, a tangent plane be passed, and if γ denote the inclination of this plane to the plane of xy, the area of the corresponding element *in the tangent plane* is

$$\sec \gamma \; \triangle x \triangle y.$$

The surface is evidently the limit of the sum of the elements in the tangent planes when $\triangle x$ and $\triangle y$ are indefinitely diminished. Now $\sec \gamma$ is a function of the coordinates of the point of contact of the tangent plane; and since these coordinates are values of x and y which lie respectively between x and $x + \triangle x$ and between y and $y + \triangle y$, the theorem proved in Art. 99 shows that this limit is

$$S = \iint \sec \gamma \, dx \, dy.$$

153. The value of $\sec \gamma$ may be derived by the following method. Through the point P of the surface let planes be passed parallel to the coordinate planes, and let PD, and PE, Fig. 17, be the intersections of the tangent plane with the planes parallel to the planes of xz and yz. Then PD and PE are tangents at P to the sections of the surface made by these planes. The equations of these sections are found by regarding y and x in turn as constants in the equation of the surface; therefore denoting the inclinations of these tangent lines to the plane of xy by ϕ and ψ, we have

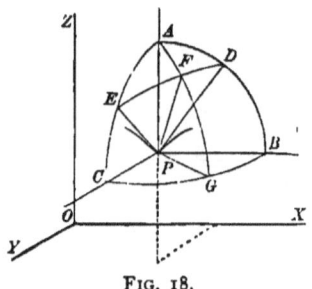

Fig. 18.

$$\tan \phi = \frac{dz}{dx}, \quad \text{and} \quad \tan \psi = \frac{dz}{dy},$$

in which $\dfrac{dz}{dx}$ and $\dfrac{dz}{dy}$ are partial derivatives derived from the equation of the surface.

If the planes be intersected by a spherical surface whose centre is P, ADE is a spherical triangle right angled at A, whose sides are the complements of ϕ and ψ. Moreover, if a plane perpendicular to the tangent plane PED be passed through AP, the angle FPG will be γ, and the perpendicular from the right angle to the base of the triangle the complement of γ.

Denoting the angle EAF by θ, the formulas for solving spherical right triangles give

$$\cos \theta = \frac{\tan \psi}{\tan \gamma}, \quad \text{and} \quad \sin \theta = \frac{\tan \phi}{\tan \gamma}.$$

Squaring and adding,

$$1 = \frac{\tan^2 \psi + \tan^2 \phi}{\tan^2 \gamma},$$

or $\tan^2 \gamma = \tan^2 \psi + \tan^2 \phi$;

whence
$$\sec^2 \gamma = 1 + \left(\frac{dz}{dx}\right)^2 + \left(\frac{dz}{dy}\right)^2.$$

Substituting in the formula derived in Art. (152), we have

$$S = \iint \sqrt{\left[1 + \left(\frac{dz}{dx}\right)^2 + \left(\frac{dz}{dy}\right)^2\right]}\, dx\, dy.$$

154. It is sometimes more convenient to employ the polar element of the projected area. Thus the formula becomes

$$S = \iint \sec \gamma \, r \, dr \, d\theta,$$

where $\sec \gamma$ has the same meaning as before.

For example, let it be required to find the area of the surface of a hemisphere intercepted by a right cylinder having a radius of the hemisphere for one of its diameters. From the equation of the sphere,

$$x^2 + y^2 + z^2 = a^2, \quad \ldots \ldots \quad (1)$$

we derive

$$\frac{dz}{dx} = -\frac{x}{z}, \qquad \frac{dz}{dy} = -\frac{y}{z};$$

whence

$$\sec \gamma = \sqrt{\left[1 + \sqrt{\left(\frac{dz}{dx}\right)^2 + \left(\frac{dz}{dy}\right)^2}\right]} = \frac{a}{z};$$

therefore

$$S = a \iint \frac{r \, dr \, d\theta}{z},$$

the integration extending over the area of the circle

$$r = a \cos \theta. \quad \ldots \ldots \quad (2)$$

Since equation (1) is equivalent to

$$z^2 + r^2 = a^2,$$

$$S = a \iint_{r_1}^{r_2} \frac{r \, dr}{\sqrt{(a^2 - r^2)}} \, d\theta = a \int [\sqrt{(a^2 - r_1^2)} - \sqrt{(a^2 - r_2^2)}] \, d\theta.$$

§ XI.] AREAS OF SURFACES IN GENERAL. 183

From (2) the limits for r are $r_1 = 0$, and $r_2 = a \cos \theta$, hence

$$S = a^2 \int (1 - \sin \theta) \, d\theta,$$

in which $a \sin \theta$ is put for the *positive* quantity $\sqrt{(a^2 - r_2^2)}$. The limits for θ are $-\frac{1}{2}\pi$ and $\frac{1}{2}\pi$, but since $\sin \theta$ is in this case to be regarded as invariable in sign, we must write

$$S = 2a^2 \int_0^{\frac{\pi}{2}} (1 - \sin \theta) \, d\theta = \pi a^2 - 2a^2.$$

If another cylinder be constructed, having the opposite radius of the hemisphere for diameter, the surface removed is $2\pi a^2 - 4a^2$, and the surface which remains is $4a^2$, a quantity commensurable with the square of the radius. This problem was proposed in 1692, in the form of an enigma, by Viviani, a Florentine mathematician.

Examples XI.

1. Find the surface of the paraboloid whose altitude is a, and the radius of whose base is b.

$$\frac{\pi b}{6a^2} \left[(4a^2 + b^2)^{\frac{3}{2}} - b^3 \right]$$

2. Prove that the surface generated by the arc of the catenary given in Ex. X., 1, revolving about the axis of x, is equal to

$$\pi(cx + sy).$$

3. Find the whole surface of the oblate spheroid produced by the

revolution of an ellipse about its minor axis, a denoting the major, b the minor semi-axis, and e the excentricity, $\dfrac{\sqrt{(a^2-b^2)}}{a}$.

$$2\pi a^2 + \pi \dfrac{b^2}{e}\log\dfrac{1+e}{1-e}.$$

4. Find the whole surface of the prolate spheroid produced by the revolution of the ellipse about its major axis, using the same notation as in Ex. 3.

$$2\pi b^2 + 2\pi ab\dfrac{\sin^{-1}e}{e}.$$

5. Find the surface generated by the cycloid

$$x = a(\psi - \sin\psi), \qquad y = a(1 - \cos\psi)$$

revolving about its base.

$$\dfrac{64}{3}\pi a^2.$$

6. Find the surface generated when the cycloid revolves about the tangent at its vertex.

$$\dfrac{32}{3}\pi a^2.$$

7. Find the surface generated when the cycloid revolves about its axis.

$$8\pi a^2\left(\pi - \dfrac{4}{3}\right).$$

8. Find the surface generated by the revolution of one branch of the tractrix (see Ex. X., 5) about its asymptote.

$$2\pi a^2.$$

9. Find the surface generated by the revolution about the axis of x of the portion of the curve

$$y = \epsilon^x,$$

which is on the left of the axis of y.

$$\pi[\sqrt{2} + \log(1 + \sqrt{2})].$$

10. Find the surface generated by the revolution about the axis of x of the arc between the points for which $x = a$ and $x = b$ in the hyperbola

$$xy = k^2.$$

$$\pi k^2 \left[\log \frac{b^2 + \sqrt{(k^4 + b^4)}}{a^2 + \sqrt{(k^4 + a^4)}} + \frac{\sqrt{(k^4 + a^4)}}{a^2} - \frac{\sqrt{(k^4 + b^4)}}{b^2} \right].$$

11. Show that the surface of a cylinder whose generating lines are parallel to the axis of z is represented by the integral

$$S = \int z\, ds,$$

where s denotes the arc of the base in the plane of xy. Hence, deduce the surface cut from a right circular cylinder whose radius is a, by a plane passing through the centre and making the angle α with the plane of the base. $2a^2 \tan \alpha$.

12. Find the surface of that portion of the cylinder in the problem solved in Art. 154, which is within the hemisphere. $2a^2$.

13. Find the surface of a circular spindle, a being the radius and $2c$ the chord.

$$4\pi a \left[c - (a^2 - c^2)\sin^{-1}\frac{c}{a} \right].$$

XII.

The Area generated by a Straight Line moving in any Manner in a Plane.

155. If a straight line of indefinite length moves in any manner whatever in a plane, there is at each instant a point of the line about which it may be regarded as rotating. This point we shall call *the centre of rotation* for the instant. The rate of motion of every point of the line in a direction perpendicular to the line itself is at the instant the same as it would be if the line were rotating at the same angular rate about this point as a fixed centre.* Hence it follows that the area generated by a definite portion of the line has at the instant the same rate as if the line were rotating about a fixed instead of a variable centre.

156. Suppose at first that the centre of rotation is on the generating line produced, ρ_1 and ρ_2 denoting the distances from the centre of the extremities of the generating line, and let ϕ denote its inclination to a fixed line. By substitution in the general formula derived in Art. 110, we have

$$dA = \frac{1}{2}(\rho_2^2 - \rho_1^2)\,d\phi.$$

---

* Compare Diff. Calc., Art. 332 [Abridged Ed., Art. 176], where the moving line is the normal to a given curve, and the centre of rotation is the centre of curvature of the given curve. If the line is moving without change of direction, the centre is of course at an infinite distance.

When the line is regarded as forming a part of a rigidly connected system in motion, its centre of rotation is the foot of a perpendicular dropped upon it from the *instantaneous centre* of the motion of the system. Thus, if the tangent and normal in the illustration cited are rigidly connected, the centre of curvature, C, is the *instantaneous centre* of the motion of the system, and the point of contact, P, is the centre of rotation for the tangent line.

Applications.

157. The area between a curve and its evolute may be generated by the radius of curvature ρ, whose inclination to the axis of x is $\phi + \frac{1}{2}\pi$, in which ϕ denotes the inclination of the tangent line. Since the centre of rotation is one extremity of the generating line ρ, the differential of this area is found by substituting in the general expression $\rho_1 = 0$ and $\rho_2 = \rho$. Hence when ρ is expressed in terms of ϕ,

$$A = \frac{1}{2}\int \rho^2 \, d\phi$$

expresses the area between an arc of a given curve, its evolute, and the radii of curvature of its extremities, the limits being the values of ϕ at the ends of the given arc.

158. For example, in the case of the cardioid

$$r = a(1 - \cos\theta),$$

it is readily shown, from the results obtained in Art. 145, that the angle between the tangent and the radius vector is $\frac{1}{2}\theta$; and therefore $\phi = \frac{3}{2}\theta$, and

$$\rho = \frac{ds}{d\phi} = \frac{4a}{3}\sin\frac{\phi}{3}.$$

To obtain the whole area between the curve and its evolute, the limits for θ are 0 and 2π; hence the limits for ϕ are 0 and 3π. Therefore

$$A = \frac{1}{2}\int_0^{3\pi} \rho^2 \, d\phi = \frac{8a^2}{9}\int_0^{3\pi}\sin^2\frac{\phi}{3}\, d\phi = \frac{4\pi a^2}{3}.$$

159. As another application of the general formula of Art. 156, let one end of a line of fixed length a be moved along a given line in a horizontal plane, while a weight attached to the other extremity is drawn over the plane by the line, and is therefore always moving in the direction of the line itself. The line of fixed length in this case turns about the weight as a moving centre of rotation. Hence the area generated while the line turns through a given angle is the same as that of the corresponding sector of a circle whose radius is a.

The curve described by the weight is called *a tractrix*, and the line along which the other extremity is moved is *the directrix*. When the axis of x is the directrix, and the weight starts from the point $(0, a)$, the common tractrix is described; hence the area between this curve and the axis is $\frac{1}{4}\pi a^2$.

160. Again, in the generation of the cycloid, Diff. Calc., Art. 288 [Abridged Ed., Art. 156], the variable chord RP may be regarded as generating the area. The point R has a motion in the direction of the tangent RX; the point P partakes of this motion, which is the motion of the centre C, and also has an equal motion, due to the rotation of the circle in the direction of the tangent to the circle at P. Since the tangents at P and R are equally inclined to PR, the motion of P in a direction perpendicular to PR is double the component, in this direction, of the motion of R. Therefore the centre of rotation of PR is beyond R at a distance from it equal to PR. Hence, denoting PRO by ϕ,

$$\rho_1 = PR = 2a \sin \phi, \qquad \rho_2 = 2PR = 4a \sin \phi.$$

Substituting in the formula of Art. 156, we have for the area of the cycloid, since PRO varies from 0 to π,

$$A = 6a^2 \int_0^\pi \sin^2 \phi \, d\phi = 3\pi a^2.$$

Sign of the Generated Area.

161. Let AB be the generating line, and C the centre of rotation. The expression,

$$dA = \tfrac{1}{2}(\rho_2^2 - \rho_1^2)\,d\phi, \quad \ldots \ldots \quad (1)$$

for the differential of the area, was obtained upon the supposition that A and B were on the same side of C. Then supposing $\rho_2 > \rho_1$, and that the line rotates in the positive direction, as in figure 19, *the differential of the area is positive;* and we notice that every point in the area generated is swept over by the line AB, *the left hand side as we face in the direction AB preceding.*

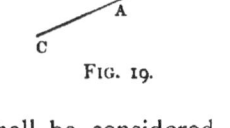

FIG. 19.

162. We shall now show that in every case, the formula requires that an area swept over with the left side preceding, shall be considered as positively generated, and one swept over in the opposite direction as negatively generated.

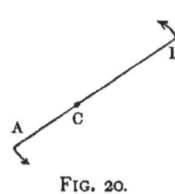

FIG. 20.

In the first place, if C is between A and B so that ρ_1 is negative, as in figure 20, ρ_1^2 is still positive, and formula (1) still gives the difference between the areas generated by AB and AC. Hence the latter area, which is now generated by a part of the line AB, must be regarded as generated negatively, but the *right hand side* as we face in the direction AB of this portion of the line is now preceding, which agrees with the rule given in Art. 161.

Again, if C is beyond B, the formula gives the difference of the generated areas; but since ρ_1^2 is numerically greater than ρ_2^2, in this case, dA is negative, and the area generated by AB is the difference of the areas, and is negative by the rule.

Finally, if the direction of rotation be reversed, $d\phi$ and therefore dA change sign, but the opposite side of each portion of the line becomes in this case the preceding side.

163. We may now put the expression for the area in another form. For

$$dA = \frac{1}{2}(\rho_2^2 - \rho_1^2) \, d\phi = (\rho_2 - \rho_1) \frac{\rho_2 + \rho_1}{2} d\phi;$$

whatever be the signs of ρ_2 and ρ_1, the first factor is the length of AB, which we shall denote by l, and the second factor is the distance of the middle point of AB from the centre of rotation, which we shall denote by ρ_m. Hence, putting

$$\rho_2 - \rho_1 = l, \quad \text{and} \quad \frac{\rho_2 + \rho_1}{2} = \rho_m,$$

we have
$$A = \int l \rho_m \, d\phi. \quad \ldots \ldots \ldots (2)$$

Since $\rho_m \, d\phi$ is the differential of the motion of the middle point in a direction perpendicular to AB, this expression shows that the differential of the area is the product of this differential by the length of the generating line.

Areas generated by Lines whose Extremities describe Closed Circuits.

164. Let us now suppose the generating line AB to move from a given position, and to return to the same position, each of the extremities A and B describing a closed curve in the positive direction, as indicated by the arrows in figure 21. It is readily seen that every point which is in the area described by B, and not in that described by A, will be swept over at least once by the line AB, the left side preceding, and if passed over more than once, there will be

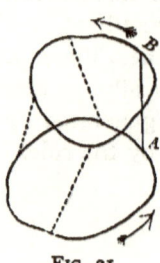

FIG. 21.

§ XII.] AREAS GENERATED BY MOVING LINES. 191

an excess of one passage, the left side preceding. Therefore the area within the curve described by B, and not within that described by A, will be generated positively. In like manner the area within the curve described by A, and not within that described by B, will be generated negatively. Furthermore, all points within both or neither of these curves are passed over, if at all, an equal number of times in each direction, so that the area common to the two curves and exterior to both disappears from the expression for the area generated by AB.

Hence it follows that, *regarding a closed area whose perimeter is described in the positive direction as positive, the area generated by a line returning to its original position is the difference of the areas described by its extremities.* This theorem is evidently true generally, if areas described in the opposite direction are regarded as negative.

Amsler's Planimeter.

165. The theorem established in the preceding article may be used to demonstrate the correctness of the method by which an area is measured by means of the *Polar Planimeter*, invented by Professor Amsler, of Schaffhausen.

This instrument consists of two bars, OA and AB, Fig. 22, jointed together at A. The rod OA turns on a fixed pivot at O, while a tracer at B is carried in the positive direction completely around the perimeter of the area to be measured. At some point C of the bar AB a small wheel is fixed, having its axis parallel to AB, and its circumference resting upon the paper. When B is moved, this wheel has a sliding and a rolling motion; the latter motion is recorded by an attachment by means of which the number of turns and parts of a turn of the wheel are registered.

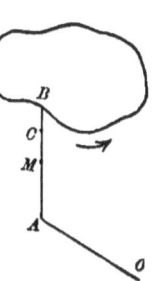

FIG. 22.

166. Let M be the middle point of AB, and let

$$OA = a, \qquad AB = b, \qquad MC = c.$$

Since b is constant, the area described by AB is by equation (2), Art. 163,

$$\text{Area } AB = b \int \rho_m \, d\phi. \qquad \ldots \quad (1)$$

Denoting the linear distance registered on the circumference of the wheel by s, ds is the differential of the motion of the point C, in a direction perpendicular to AB, and since the distance of this point from the centre of rotation is $\rho_m + c$,

$$ds = (\rho_m + c) \, d\phi :$$

substituting in (1) the value of $\rho_m \, d\phi$,

$$\text{Area } AB = b \int ds - bc \int d\phi. \quad \ldots \quad (2)$$

167. Two cases arise in the use of the instrument. When, as represented in Fig. 22, O is outside the area to be measured, the point A describes no area, and by the theorem of Art. 164, equation (2) represents simply the area described by B. In this case ϕ returns to its original value, hence $\int d\phi$ vanishes, and denoting the area to be measured by A, equation (2) becomes

$$A = bs. \quad \ldots \quad \ldots \quad (3)$$

In the second case, when O is within the curve traced by B, the point A describes a circle whose area is πa^2, and the limiting values of ϕ differ by a complete revolution. Hence in this case equation (2) becomes

$$A - \pi a^2 = bs - 2\pi bc,$$

or $\quad A = bs + \pi(a^2 - 2bc).*$ . . . . . (4)

In another form of the planimeter the point A moves in a straight line, and the same demonstration shows that the area is always equal to bs.

Examples XII.

1. The involute of a circle whose radius is a is drawn, and a tangent is drawn at the opposite end of the diameter which passes through the cusp; find the area between the tangent and the involute.

$$\frac{a^2\pi(3 + \pi^2)}{3}.$$

2. Two radii vectores of a closed oval are drawn from a fixed point within, one of which is parallel to the tangent at the extremity of the other; if the parallelogram be completed, the area of the locus of its vertex is double the area of the given oval.

3. Show that the area of the locus of the middle point of the chord joining the extremities of the radii vectores in Ex. 2, is one half the area of the given oval.

---

* The planimeter is usually so constructed that the positive direction of rotation is with the hands of a watch. The bar b is adjustable, but the distance AC is fixed so that c varies with b. Denoting AC by q, we have $c = q - \frac{1}{2}b$, and the constant to be added becomes $C = \pi(a^2 - 2bq + b^2)$ in which a and q are fixed and b adjustable. In some instruments q is negative.

It is to be noticed that in the second case s may be negative; the area is then the numerical difference between the constant and bs.

4. Prove that the difference of the perimeters of two parallel ovals, whose distance is b, is $2\pi b$, and that the difference of their areas is the product of b and the half sum of their perimeters.

5. A limaçon is formed by taking a fixed distance be on the radius vector from a point on the circumference of a circle whose radius is a; show that the area generated by b when $b > 2a$ is the area of the limaçon diminished by twice the area of the circle, and thence determine the area of the limaçon.
$$\pi(2a^2 + b^2).$$

6. Verify equation (4), Art. 167, when the tracer describes the circle whose radius is $a + b$.

7. Verify the value of the constant in equation (4), Art. 167, by determining the circle which may be described by the tracer without motion of the wheel.

8. If, in the motion of a crank and connecting rod (the line of motion of the piston passing through the centre of the crank), Amsler's recording wheel be attached to the connecting rod at the piston end, determine s geometrically, and verify by means of the area described by the other end of the rod.

9. The length of the crank in Ex. 8 being a, and that of the connecting rod b, find the area of the locus of a point on the connecting rod at a distance c from the piston end.
$$\frac{\pi a^2 c}{b}.$$

10. If a line AB of fixed length move in a plane, returning to its original position *without making a complete revolution*, denoting the areas of the curves described by its extremities by (A) and (B), determine the area of the curve described by a point cutting AB in the ratio $m : n$.
$$\frac{n(A) + m(B)}{m + n}.$$

§ XII.] EXAMPLES.

11. If the line in Ex. 10 return to its original position after *making a complete revolution*, prove *Holditch's Theorem*; namely, that the area of the curve described by a point at the distance c and c' from A and B is

$$\frac{c'(A) + c(B)}{c + c'} - \pi c c'.$$

12. Show by means of Ex. 11 that, if a chord of fixed length move around an oval, and a curve be described by a point at the distances c and c' from its ends, the area between the curves will be $\pi c c'$.

XIII.

Approximate Expressions for Areas and Volumes.

168. When the equation of a curve is unknown, the area between the curve, the axis of x, and two ordinates may be approximately expressed in terms of the base and a limited number of ordinates, which are supposed to have been measured.

Let $ABCDE$ be the area to be determined; denote the length of the base by $2h$; and let the ordinates at the extremities and middle point of the base be measured and denoted by $y_1, y_2,$ and y_3.

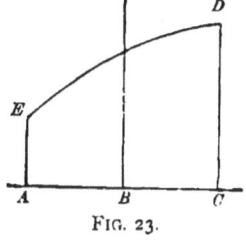

FIG. 23.

Taking the base for the axis of x, and the middle point as origin, let it be assumed that the curve has an equation of the form

$$y = A + Bx + Cx^2 + Dx^3; \quad \ldots \ldots (1)$$

then the area required is

$$A = \int_{-h}^{h} y\,dx = Ax + \frac{Bx^2}{2} + \frac{Cx^3}{3} + \frac{Dx^4}{4} \Big]_{-h}^{h} = \frac{h}{3}(6A + 2Ch^2), \quad . \ (2)$$

in which which A and C are unknown.

In order to express the area in terms of the measured ordinates, we have from equation (1),

$$y_1 = A + Bh + Ch^2 + Dh^3,$$
$$y_2 = A,$$
$$y_3 = A - Bh + Ch^2 - Dh^3;$$

whence we derive

$$y_1 + y_3 = 2A + 2Ch^2,$$
$$y_1 + 4y_2 + y_3 = 6A + 2Ch^2;$$

and substituting in (2),

$$A = \frac{h}{3}(y_1 + 4y_2 + y_3).$$

It will be noticed that this formula gives a perfectly accurate result when the curve is really a parabolic curve of the third or a lower degree.

169. If the base be divided into three equal intervals, each denoted by h, and the ordinates at the extremities and at the points of division measured, we have, by assuming the same equation,

$$A = \int_{-\frac{3h}{2}}^{\frac{3h}{2}} y\, dx = \frac{3h}{4}(4A + 3Ch^2) \quad \cdots \quad (1)$$

From the equation of the curve,

$$y_1 = A - \frac{3Bh}{2} + \frac{9Ch^2}{4} - \frac{27Dh^3}{8},$$
$$y_2 = A - \frac{Bh}{2} + \frac{Ch^2}{4} - \frac{Dh^3}{8},$$
$$y_3 = A + \frac{Bh}{2} + \frac{Ch^2}{4} + \frac{Dh^3}{8},$$
$$y_4 = A + \frac{3Bh}{2} + \frac{9Ch^2}{4} + \frac{27Dh^3}{8};$$

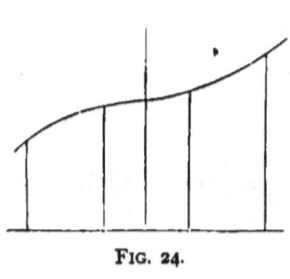

FIG. 24.

whence
$$y_1 + y_4 = 2A + \frac{9Ch^2}{2},$$

$$y_2 + y_3 = 2A + \frac{Ch^2}{2}.$$

From these equations we obtain

$$A = \frac{-y_1 + 9y_2 + 9y_3 - y_4}{16},$$

and
$$Ch^2 = \frac{y_1 - y_2 - y_3 + y_4}{4}.$$

Substituting in equation (1),

$$A = \frac{3h}{8}(y_1 + 3y_2 + 3y_3 + y_4).$$

Simpson's Rules.

170. The formulas derived in Articles 168 and 169, although they were first given by Cotes and Newton, are usually known as *Simpson's Rules*, the following extensions of the formulas having been published in 1743, in his *Mathematical Dissertations*.

If the whole base be divided into an even number n of parts, each equal to h, and the ordinates at the points of division be numbered in order from end to end, then by applying the first formula to the areas between the alternate ordinates, we have

$$A = \frac{h}{3}(y_1 + 4y_2 + 2y_3 + 4y_4 \cdots + 4y_n + y_{n+1}).$$

That is to say, the area is equal to the product of the sum of the extreme ordinates, four times the sum of the even-numbered ordinates, and twice the sum of the remaining odd-numbered ordinates, multiplied by one third of the common interval.

Again, if the base be divided into a number of parts divisible by three, we have, by applying the formula derived in Art. 169, to the areas between the ordinates $y_1 y_4$, $y_4 y_7$, and so on,

$$A = \frac{3h}{8}(y_1 + 3y_2 + 3y_3 + 2y_4 + 3y_5 \cdots + 3y_n + y_{n+1}).$$

Cotes' Method of Approximation.

171. The method employed in Articles 168 and 169 is known as *Cotes' Method*. It consists in assuming the given curve to be a parabolic curve of the highest order which can be made to pass through the extremities of a series of equidistant measured ordinates.

The equation of the parabolic curve of the nth order contains $n + 1$ unknown constants; hence, in order to eliminate these constants from the expression for an area defined by the curve, it is in general necessary to have $n + 1$ equations connecting them with the measured ordinates. Hence, if n denote the number of intervals between measured ordinates over which the curve extends, the curve will in general be of the nth degree.*

---

* If H denotes the whole base, the first factor is always equivalent to H divided by the sum of the coefficients of the ordinates; for if all the ordinates are made equal, the expression must reduce to Hy_1. Thus, each of the rules for an approximate area, including those derived by repeated applications, as in Art. 170, may be regarded as giving an expression for the *mean ordinate*. The coefficients of the ordinates, according to Cotes' method, for all values of n up to $n = 10$, may be found in Bertrand's *Calcul Intégral*, pages 333 and 334. For example (using detached coefficients for brevity), we have, when $n = 4$,

$$A = \frac{H}{90}[7, 32, 12, 32, 7];$$

and when $n = 6$,

$$A = \frac{H}{840}[41, 216, 27, 272, 27, 216, 41].$$

§ XIII.] *THE FIVE-EIGHT RULE.* 199

172. For example, let it be required to determine the area between the ordinates y_1 and y_2, in terms of the three equidistant ordinates y_1, y_2 and y_3, the common interval being h.

We must assume
$$y = A + Bx + Cx^2;$$

then taking the origin at the foot of y_1,
$$A = \int_0^h y\,dx = h\left[A + \frac{Bh}{2} + \frac{Ch^2}{3} \right],$$

from which A, B and C must be eliminated by means of the equations
$$y_1 = A,$$
$$y_2 = A + Bh + Ch^2,$$
$$y_3 = A + 2Bh + 4Ch^2.$$

Solving these equations, we obtain
$$A = y_1,$$
$$Bh = \frac{-3y_1 + 4y_2 - y_3}{2},$$
$$Ch^2 = \frac{y_1 - 2y_2 + y_3}{2};$$

If we make a slight modification in the ratios of these last coefficients by substituting for each the nearest multiple of 42, we have
$$A = \frac{H}{840}[42,\ 210,\ 42,\ 252,\ 42,\ 210,\ 42],$$

(the denominator remaining unchanged, since the sum of the coefficients is still 840), which reduces to
$$A = \frac{H}{20}[1,\ 5,\ 1,\ 6,\ 1,\ 5,\ 1].$$

This result is known as *Weddles' Rule* for six intervals. The value thus given to the mean ordinate is evidently a very close approximation to that resulting from Cotes' method, the difference being
$$\frac{1}{840}[v_1 + y_7 + 15(y_3 + y_5) - 6(y_2 + y_6) - 20y_4].$$

and substituting

$$A = \frac{h}{12}(5y_1 + 8y_2 - y_3).$$

173. It is, however, to be noticed, that when the ordinates are symmetrically situated with respect to the area, if n is *even*, the parabolic curve may be assumed of the $(n + 1)$th degree. For example, in Art. 168, $n = 2$, but the curve was assumed of the third degree. Inasmuch as A, B, C and D cannot all be expressed in terms of y_1, y_2, and y_3, we see that a variety of parabolic curves of the third degree can be passed through the extremities of the measured ordinates, but all of these curves have the same area.*

Application to Solids.

174. If y denotes the area of the section of a solid perpendicular to the axis of x, the volume of the solid is $\int y\, dx$, and

---

* This circumstance indicates a probable advantage in making n an even number when repeated applications of the rules are made. Thus, in the case of six intervals, we can make three applications of Simpson's first rule, giving

$$A = \frac{H}{18}[1, 4, 2, 4, 2, 4, 1], \quad \ldots \ldots \quad (1)$$

or two of Simpson's second rule, giving

$$A = \frac{H}{16}[1, 3, 3, 2, 3, 3, 1]. \quad \ldots \ldots \quad (2)$$

In the first case, we assume the curve to consist of three arcs of the third degree, meeting at the extremities of the ordinates y_3 and y_5; but, since each of these arcs contains an undetermined constant, we can assume them to have common tangents at the points of meeting. We have therefore a *smooth*, though not a continuous curve. In the second case, we have two arcs of the third degree containing no arbitrary constants, and therefore making an angle at the extremity of y_4. It is probable, therefore, that the smooth curve of the first case will in most cases form a better approximation than the broken curve of the second case.

In confirmation of this conclusion, it will be noticed that the ratios of the coefficients in equation (1) are nearer to those of Cotes' coefficients for $n = 6$, given in the preceding foot-note, than are those in equation (2).

therefore the approximate rules deduced in the preceding articles apply to solids as well as to areas. Indeed, they may be applied to the approximate computation of any integral, by putting y equal to the coefficient of x under the integral sign.

The areas of the sections may of course be computed by the approximate rules.

Woolley's Rule.

175. When the base of the solid is rectangular, and the ordinates of the sections necessary to the application of Simpson's first rule are measured, we may, instead of applying that rule, introduce the ordinates directly into the expression for the area in the following manner.

Taking the plane of the base for the plane of xy, and its centre for the origin, let the equation of the upper surface be assumed of the form

$$z = A + Bx + Cy + Dx^2 + Exy + Fy^2 + Gx^3 + Hx^2y + Ixy^2 + Jy^3.$$

Let $2h$ and $2k$ be the dimensions of the base, and denote the measured values of z as indicated in Fig. 25. The required volume is

$$V = \int_{-h}^{h} \int_{-k}^{k} z \, dy \, dx.$$

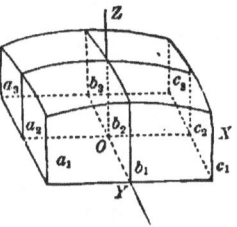

FIG. 25.

This double integral vanishes for every term containing an odd power of x or an odd power of y: hence

$$V = 4Ahk + \frac{4Dh^3k}{3} + \frac{4Fhk^3}{3},$$
$$= \frac{hk}{3}[12A + 4Dh^2 + 4Fk^2]. \quad \ldots \ldots (1)$$

By substituting the values of x and y in the equation of the surface, we readily obtain

$$b_2 = A, \quad \dots \dots \dots \quad (2)$$

$$a_1 + a_3 + c_1 + c_3 = 4A + 4Dh^2 + 4Fk^2, \quad \dots \quad (3)$$

$$a_2 + c_2 + b_1 + b_3 = 4A + 2Dh^2 + 2Fk^2. \quad \dots \quad (4)$$

From these equations two very simple expressions for the volume may be derived; for, employing (2) and (4), equation (1) becomes

$$V = \frac{2hk}{3}(a_2 + b_1 + 2b_2 + b_3 + c_2); \quad \dots \quad (4)$$

and employing (2) and (3),

$$V = \frac{hk}{3}(a_1 + a_3 + 8b_2 + c_1 + c_3). \quad \dots \quad (5)$$

Equation (4) is known as *Woolley's Rule*; the ordinates employed are those at the middles of the sides and at the centre; in (5), they are at the corners and at the centre.

Examples XIII.

1. Apply Simpson's Rule to the sphere, the hemisphere, and the cone, and explain why the results are perfectly accurate.

2. Apply Simpson's Second Rule to the larger segment of a sphere made by a plane bisecting at right angles a radius of the sphere.

$$\frac{9\pi a^3}{8}.$$

3. Find by Simpson's Rule the volume of a segment of a sphere, b and c being the radii of the bases, and h the altitude.

$$\frac{\pi h}{6}(3b^2 + 3c^2 + h^2).$$

4. Find by Simpson's Rule the volume of the frustum of a cone, b and c being the radii of the bases, and h the altitude.

$$\frac{\pi h}{3}(b^2 + bc + c^2).$$

5. Compute by Simpson's First and Second Rules, the value of $\int_0^1 \frac{dx}{1+x}$, the common interval being $\frac{1}{12}$ in each case.

The first rule gives 0.6931487, and the second rule gives 0.6931505. The correct value is obviously $\log_e 2 = 0.6931472$.

6. Find the volume considered in Art. 175, directly by Simpson's Rule, and show that the result is consistent with equations (4) and (5).

$$V = \frac{hk}{9}[a_1 + a_3 + c_1 + c_3 + 4(a_2 + b_1 + b_3 + c_2) + 16b_2].$$

7. Find, by elimination, from equations (4) and (5), Art. 175, a formula which can be used when the centre ordinate is unknown.

$$V = \frac{hk}{3}[4(a_2 + b_1 + b_3 + c_2) - (a_1 + a_3 + c_1 + c_3)].$$

CHAPTER IV.

Mechanical Applications.

XIV.

Definitions.

176. We shall give in this chapter a few of the applications of the Integral Calculus to mechanical questions.

The *mass* or quantity of matter contained in a body is proportional to its weight. When the masses of all parts of equal volume are equal, the body is said to be *homogeneous*. The factor by which it is necessary to multiply the unit of volume to produce the unit of mass is called the *density*, and usually denoted by γ.

In the following articles it will be assumed, when not otherwise stated, that the body is homogeneous, and that the density is equal to unity, so that the unit of mass is identical with the unit of volume. When the mass of an area is spoken of, it is regarded as a lamina of uniform thickness and density, and the unit of mass is taken to correspond with the unit of surface. In like manner the unit of mass for a line is taken as identical with the unit of length.

Statical Moments.

177. The *moment* of a force, with reference to a point, is the measure of the effectiveness of the force in producing motion about the point. It is shown in treatises on Mechanics, that this is *the product of the force and the perpendicular from the point upon the line of application of the force.*

§ XIV.] STATICAL MOMENTS. 205

The moment of the sum of a number of forces about a given point is the sum of the moments of the forces.

The *statical moment* of a body about a given point is the moment of its gravity; the force of gravity being supposed to act upon every part of the body, and in parallel lines.

178. In order to find the statical moment of a continuous body, we regard the body as generated geometrically in some convenient manner, and determine the corresponding differential of the moment.

In the case of a plane area, let the body be referred to rectangular axes, and let gravity be supposed to act in the direction of the axis of y. Then the abscissa of the point of application is the *arm* of the force when we consider the moment about the origin. Let us first suppose the area to be generated by the motion of the ordinate y. The differential of the area is then $y\,dx$. The corresponding element of the sum, of which the integral $\int_a^b y\,dx$ is the limiting value, see Art. 99, is

$$y_r \triangle x, \quad \ldots \ldots \ldots (1)$$

in which y_r is the ordinate corresponding to *any* value of x intermediate between $a + (r-1)\triangle x$, and $a + r\triangle x$. It is evident that the arm of the weight of the element (1) is such an intermediate value of x; hence the moment of the element is

$$x_r y_r \triangle x. \quad \ldots \ldots \ldots (2)$$

The whole moment is therefore the limiting value of a sum of the form

$$\sum_a^b x_r y_r \triangle x.$$

In other words, it is the integral

$$\int_a^b xy\,dx, \quad \ldots \ldots \ldots (3)$$

in which the differential of the moment is the product of the differential of the area and the arm of the force, which in this case is the same for every point of the element. In other words, *the moment of the differential is the differential of the moment*.

179. As an illustration, we find the moment of a semicircle (Fig. 26) about its centre. The area may be generated by the line $2y$, moving from $x = 0$ to $x = a$. The equation of the circle being

$$x^2 + y^2 = a^2,$$

the differential of the area is

$$2 \sqrt{(a^2 - x^2)}\, dx.$$

The moment of this differential is

$$2 \sqrt{(a^2 - x^2)}.x\, dx\,;$$

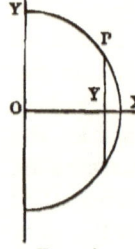

FIG. 26.

hence the whole moment is

$$2 \int_0^a \sqrt{(a^2 - x^2)} x\, dx = -\frac{2}{3}(a^2 - x^2)^{\frac{3}{2}} \Big]_0^a = \frac{2a^3}{3}.$$

Centres of Gravity.

180. If a force equal to the whole weight of a body be applied with an arm properly determined, its moment may be made equivalent to the whole statical moment of the body. If the force is in the direction of the axis of y, as in Fig. 26, we have, denoting this arm by $\bar{x}$,

$$\bar{x} \cdot \text{Area} = \text{Moment},$$

$$\bar{x} = \frac{\text{Moment}}{\text{Area}}.$$

In like manner, supposing the force to act in the direction of the axis of x, we may determine y for the same body.

It is shown in treatises on Mechanics that the point determined by the two coordinates $\bar{x}$ and $\bar{y}$, is independent of the position of the coordinate axis. This point is called the *centre of gravity* of the area. The centre of gravity of a volume is defined in like manner.

181. The symmetry of the form of a body may determine one or more of the coordinates of its centre of gravity. Thus the centre of gravity of a circle or a sphere coincides with the geometrical centre, and the centre of gravity of a solid of revolution is on the axis of revolution. The centre of gravity of the semicircle in Fig. 26, is on the axis of x; hence to determine its position we have only to find $\bar{x}$. Dividing the moment of the semicircle found in Art. 179 by the area $\frac{1}{2}\pi a^2$, we have

$$\bar{x} = \frac{4a}{3\pi}.$$

182. In finding the moment of the semicircle (Art. 179), we regarded the area as generated by the double ordinate $2y$, and the differential of the moment was found by multiplying the differential of the area by x, which is the arm of the force for every point of the generating line.

We may, however, derive the moment from the differential of area,

$$x\,dy, \quad \ldots \ldots \ldots (1)$$

since the area may be generated by the motion of the abscissa x from $y = -a$ to $y = a$. But in this case to find the moment of the differential we must multiply it by the distance of its centre of gravity from the given axis. The centre of gravity of the line x is evidently its middle point, hence the required arm is $\frac{1}{2}x$. Therefore the differential of the moment is

$$\frac{x^2\,dy}{2}; \quad \ldots \ldots \ldots (2)$$

and consequently the whole moment is

$$\frac{1}{2}\int_{-a}^{a} x^2\, dy = \frac{1}{2}\int_{-a}^{a} (a^2 - y^2)\, dy = \frac{2a^3}{3}.$$

This result is identical with that derived in Art. 179.

Polar Formulas.

183. When polar formulas are employed, r and θ being coordinates of the curved boundary of the area, the element is $\frac{1}{2}r^2\, d\theta$. Since this element is ultimately a triangle, we employ the well known property of triangles; that the centre of gravity is on a medial line at two-thirds the distance from the vertex to the base.

The coordinates of the centre of gravity of the element are, therefore,

$$\frac{2}{3}r \sin \theta \quad \text{and} \quad \frac{2}{3}r \cos \theta.$$

Hence we have the formula

$$\bar{x} = \frac{\int \frac{2}{3} r \cos \theta \cdot \frac{1}{2} r^2\, d\theta}{\frac{1}{2}\int r^2\, d\theta} = \frac{2}{3} \cdot \frac{\int r^3 \cos \theta\, d\theta}{\int r^2\, d\theta},$$

and similarly

$$\bar{y} = \frac{2}{3} \cdot \frac{\int r^3 \sin \theta\, d\theta}{\int r^2\, d\theta}.$$

184. To illustrate, let us find the centre of gravity of the area enclosed by the lemniscata

$$r^2 = a^2 \cos 2\theta.$$

Whence $\bar{x} = \dfrac{2a \displaystyle\int_0^{\frac{\pi}{4}} (\cos 2\theta)^{\frac{3}{2}} \cos\theta\, d\theta}{\displaystyle\int_0^{\frac{\pi}{4}} \cos 2\theta\, d\theta} = \dfrac{4a}{3}\displaystyle\int_0^{\frac{\pi}{4}} (\cos 2\theta) \cos\theta\, d\theta.$

Put $\quad \cos 2\theta = \cos^2\phi, \quad$ whence $\quad \sin\phi = \sqrt{2}\sin\theta,$

and $\quad\quad\quad \sqrt{2}\cos\theta\, d\theta = \cos\phi\, d\phi,$

$\therefore\quad \bar{x} = \dfrac{2\sqrt{2}}{3} a \displaystyle\int_0^{\frac{\pi}{2}} \cos^4\phi\, d\phi = \dfrac{2\sqrt{2}}{3} \cdot \dfrac{3\cdot 1}{4\cdot 2} \cdot \dfrac{\pi}{2} a = \dfrac{\sqrt{2}}{8}\pi a.$

Solids of Revolution.

185. To find the centre of gravity of a solid of revolution, we take the axis of revolution as the axis of x, and the circle whose area is πy^2 as the generating element. Replacing y in equation (3), Art. 178, by this expression, we have for the statical moment

$$\pi \int_a^b xy^2\, dx,$$

and for the abscissa of the centre of gravity

$$\bar{x} = \dfrac{\displaystyle\int_a^b xy^2\, dx}{\displaystyle\int_a^b y^2\, dx}$$

186. To illustrate, we find the centre of gravity of a spherical segment whose height is h. In this case, taking the origin at the vertex of the segment, and denoting the radius of the sphere by a, we have

$$y^2 = 2ax - x^2.$$

Hence $\bar{x} = \dfrac{\int_0^h (2ax - x^2)\,dx}{\int_0^h (2ax - x^2)\,dx} = \dfrac{\left[\dfrac{2}{3}ax^3 - \dfrac{1}{4}x^4\right]_0^h}{\left[ax^2 - \dfrac{1}{3}x^3\right]_0^h} = \dfrac{h}{4} \cdot \dfrac{8a - 3h}{3a - h}.$

If the centre of gravity of the surface of the segment be required, since the differential of the surface is $2\pi y\,ds$, we easily obtain the general formula

$$x = \dfrac{\int_0^h xy\,ds}{\int_0^h y\,ds},$$

and, in this case the curve being a circle, $y\,ds = a\,dx$; hence, substituting, we have

$$\bar{x} = \tfrac{1}{2}h.$$

The Properties of Pappus.

187. Let a solid be generated by the revolution of any plane figure about an exterior axis in its own plane. It is required to determine the volume and the surface thus generated.

It is evident that this solid may also be generated by a variable circular ring whose centre moves along the axis of revolution; denoting by y_1 and y_2 corresponding ordinates of

§ XIV.] THE PROPERTIES OF PAPPUS. 211

the outer and inner circles respectively, the area of this ring is $\pi(y_1^2 - y_2^2)$. Hence

$$V = \pi \int (y_1^2 - y_2^2)\, dx = 2\pi \int \frac{y_1 + y_2}{2} (y_1 - y_2)\, dx.$$

But this integral is the statical moment of the given figure, since $y_1 - y_2$ is the generating element of its area, and $\frac{y_1 + y_2}{2}$ is the corresponding arm. Denoting the area of the figure by A, we may therefore write

$$V = 2\pi \bar{y} A\,;$$

that is, *the volume is the product of the area of the figure and the path described by its centre of gravity.*

The surface (S) of this solid is, by Art. 149,

$$S = 2\pi \int y\, ds = 2\pi \int ds,$$

if $\bar{y}$ denotes the ordinate of the centre of gravity of the arc s.

Hence we have $\qquad S = 2\pi\bar{y}\cdot arc\,;$

that is, *the surface is the product of the length of the arc into the path described by the centre of gravity.*

These theorems are frequently called the properties of Guldinus; they are, however, due to Pappus, who published them 1588.

It is obvious that both theorems are true for any part of a revolution of the generating figure.

Examples XIV.

1. Find the centre of gravity of the area enclosed between the parabola $y^2 = 4mx$ and the double ordinate corresponding to the abscissa a.

$$\bar{x} = \frac{3a}{5}.$$

2. Find the centre of gravity of the area between the semi-cubical parabola $ay^2 = x^3$ and the double ordinate which corresponds to the abscissa a.

$$\bar{x} = \frac{5a}{7}.$$

3. Find the ordinate of the centre of gravity of the area between the axis of x and the sinusoid $y = \sin x$, the limits being $x = 0$ and $x = \pi$.
$$\bar{y} = \tfrac{1}{8}\pi.$$

4. Find the coordinates of the centre of gravity of the area between the axes and the parabola

$$\left(\frac{x}{a}\right)^{\frac{1}{2}} + \left(\frac{y}{b}\right)^{\frac{1}{2}} = 1.$$

$$\bar{x} = \frac{a}{5}, \text{ and } \bar{y} = \frac{b}{5}.$$

5. Find the centre of gravity of the area between the *cissoid* $y^2(a - x) = x^3$ and its asymptote.

Solution :—

Denoting the statical moment by M and the area by A,

$$M = \int_0^a \frac{x^{\frac{5}{2}}\, dx}{(a-x)^{\frac{1}{2}}} = -\left. 2x^{\frac{5}{2}}(a-x)^{\frac{1}{2}} \right]_0^a + 5\int_0^a x^{\frac{3}{2}}(a-x)^{\frac{1}{2}}\, dx$$

$$= 5a \cdot A - 5M ;$$

$$\therefore M = \frac{5a}{6} A, \qquad \text{hence} \qquad \bar{x} = \frac{5a}{6}.$$

§ XIV.]   EXAMPLES.   213

6. Find the centre of gravity of the area between the parabola $y^2 = 4ax$ and the straight line $y = mx$.

$$\bar{x} = \frac{8a}{5m^2}, \text{ and } \bar{y} = \frac{2a}{m}.$$

7. Find the centre of gravity of the segment of an ellipse cut off by a quadrantal chord.

$$\bar{x} = \frac{2}{3} \cdot \frac{a}{\pi - 2}, \text{ and } \bar{y} = \frac{2}{3} \cdot \frac{b}{\pi - 2}.$$

8. Given the cycloid,

$$y = a(1 - \cos\psi), \qquad x = a(\psi - \sin\psi),$$

find the distance of its centre of gravity from the base.

$$\bar{y} = \frac{5a}{6}.$$

9. Find the centre of gravity of the area enclosed between the positive directions of the coordinate axes and the four-cusped hypocycloid

$$x^{\frac{2}{3}} + y^{\frac{2}{3}} = a^{\frac{2}{3}}.$$

Put $x = a\cos^3\theta$, and $y = a\sin^3\theta$.

$$\bar{x} = \bar{y} = \frac{256a}{315\pi}.$$

10. Find the centre of gravity of the area enclosed by the *cardioid*

$$r = a(1 - \cos\theta).$$

$$\bar{x} = -\frac{5a}{6}.$$

11. Find the centre of gravity of the sector of a circle whose radius is a, the angle of the sector being 2α.

Use the method of Art. 183.

$$\bar{x} = \frac{2}{3} \frac{a\sin\alpha}{\alpha}.$$

12. Find the centre of gravity of the segment of a circle, the angle subtended being 2α and the radius of the circle a.

Solution :—

$$\bar{x} = \frac{2\int_{a\cos\alpha}^{a}(a^2-x^2)^{\frac{1}{2}}x\,dx}{\text{Area}} = \frac{2a^3\sin^3\alpha}{3\text{ Area}} = \frac{\text{Chord}^3}{12\text{ Area}}.$$

13. Find the centre of gravity of a circular ring, the radii being a and a_1, and the angle subtended 2α.

$$\bar{x} = \frac{2}{3} \cdot \frac{a^3 - a_1^3}{a^2 - a_1^2} \cdot \frac{\sin\alpha}{\alpha}.$$

14. Find the centre of gravity of a circular arc, whose length is $2s$.

Solution :—

We have in this case, taking the origin at the centre and the axis of x bisecting the arc,

$$\bar{x} = \frac{\int_{-s}^{s} x\,ds}{\int_{-s}^{s} ds}.$$

Put $x = a\cos\theta$, then $ds = a\,d\theta$, and denoting by α the angle subtended by s, we have

$$\bar{x} = \frac{a^2\int_{-\alpha}^{\alpha}\cos\theta\,d\theta}{2s} = \frac{a\sin\alpha}{\alpha} = \frac{c}{\alpha},$$

$2c$ being the chord.

15. Find the coordinates of the centre of gravity of arc of the semi-cycloid whose equations, referred to the vertex, are

$$x = a(1 - \cos \psi), \quad \text{and} \quad y = a(\psi + \sin \psi).$$

$$\bar{x} = \frac{2a}{3}, \text{ and } \bar{y} = \left(\pi - \frac{4}{3}\right)a.$$

16. Find the centre of gravity of the arc between two successive cusps of the four-cusped hypocycloid

$$x^{\frac{2}{3}} + y^{\frac{2}{3}} = a^{\frac{2}{3}}.$$

$$\bar{x} = \bar{y} = \frac{2a}{5}.$$

17. Find the position of the centre of gravity of the arc of the semi-cardioid

$$r = a(1 - \cos \theta).$$

$$\bar{x} = -\frac{4a}{5}, \text{ and } \bar{y} = \frac{4a}{5}.$$

18. A semi-ellipsoid is formed by the revolution of a semi-ellipse about its major axis; find the distance of the centre of gravity of the solid from the centre of the ellipse.

$$\bar{x} = \frac{3a}{8}.$$

19. Find the centre of gravity of a frustum of a paraboloid of revolution having a single base, h denoting the height of the frustum.

$$\bar{x} = \frac{2h}{3}.$$

20. A paraboloid and a cone have a common base and vertices at the same point; find the centre of gravity of the solid enclosed between them.

The centre of gravity is the middle point of the axis.

21. Find the centre of gravity of a hyperboloid whose height is h, the generating curve being
$$y^2 = m(2ax + x^2).$$
$$\bar{x} = \frac{h}{4} \cdot \frac{8a + 3h}{3a + h}.$$

22. Find the centre of gravity of the solid formed by the revolution of the sector of a circle about one of its extreme radii.

The height of the cone being denoted by h, and the radius of the circle by a, we have
$$\bar{x} = \frac{3}{8}(a + h).$$

23. Find the centre of gravity of the solid formed by the revolution about the axis of x of the curve
$$a^2 y = ax^2 - x^3,$$
between the limits o and a.
$$\bar{x} = \frac{5a}{8}.$$

24. A solid is formed by revolving about its axis the cardioid
$$r = a(1 - \cos \theta);$$
find the distance of the cusp from the centre of gravity.
$$\bar{x} = \frac{16a}{15}.$$

25. Determine the position of the centre of gravity of the volume included between the surfaces generated by revolving about the axis of x the two parabolas
$$y^2 = mx, \quad \text{and} \quad y^2 = m'(a - x).$$
$$\bar{x} = \frac{a}{3} \cdot \frac{m + 2m'}{m + m'}.$$

§ XIV.] *EXAMPLES.* 217

26. Find the centre of gravity of a rifle bullet consisting of a cylinder two calibers in length, and a paraboloid one and a half calibers in length having a common base, the opposite end of the cylinder containing a conical cavity one caliber in depth with a base equal in size to that of the cylinder.

The distance of the centre of gravity from the base of the bullet is $1\frac{33}{38}$ calibers.

27. A solid formed by the revolution of a circular segment about its chord is cut in halves by a plane perpendicular to the chord; determine the centre of gravity of one of the halves. This solid is called an *ogival*.

Denoting by 2α the angle subtended by the chord, and by a the radius of the circle, the distance of the centre of gravity from the base is

$$\bar{x} = \frac{a}{16} \cdot \frac{44 \sin^2 \alpha + \sin^2 2\alpha + 32 (\cos 2\alpha - \cos \alpha)}{\sin \alpha (2 + \cos^2 \alpha) - 3\alpha \cos \alpha}.$$

28. Find the centre of gravity of the surface of the paraboloid formed by the revolution about the axis of x of the parabola

$$y^2 = 4mx,$$

a denoting the height of the paraboloid.

$$\bar{x} = \frac{1}{5} \cdot \frac{(3a - 2m)(a+m)^{\frac{3}{2}} + 2m^{\frac{5}{2}}}{(a+m)^{\frac{3}{2}} - m^{\frac{3}{2}}}.$$

29. Find the centre of gravity of the surface generated by the revolution of a semi-cycloid about its axis, the equations of the curve being

$$x = a(1 - \cos \psi), \quad \text{and} \quad y = a(\psi + \sin \psi).$$

$$\bar{x} = \frac{2a}{15} \cdot \frac{15\pi - 8}{3\pi - 4}.$$

30. Find the centre of gravity of the surface generated by the revolution about its axis of one of the loops of the lemniscata

$$r^2 = a^2 \cos 2\theta.$$

$$\bar{x} = \frac{2 + \sqrt{2}}{6} a.$$

31. A cardioid revolves about its axis; find the centre of gravity of the surface generated, the equation of the cardioid being

$$r = a(1 - \cos \theta).$$

$$\bar{x} = \frac{50a}{63}.$$

32. A ring is generated by the revolution of a circle about an axis in its own plane; c being the distance of the centre of the circle from the axis, and a the radius, determine the volume and surface generated.

$$V = 2\pi^2 c a^2, \text{ and } S = 4\pi^2 c a.$$

33. A triangle revolves about an axis in its plane; a_1, a_2, and a_3, denoting the distances of its vertices from the axis, determine the volume generated.

$$V = \frac{2\pi A}{3}(a_1 + a_2 + a_3).$$

34. Find the volume of a frustum of a cone, the radii of the bases being a_1 and a_2, and the height h.

$$V = \frac{\pi h}{3}(a_1^2 + a_1 a_2 + a_2^2).$$

35. Find the volume and surface generated by the revolution of a cycloid about its base.

$$V = 5\pi^2 a^3, \text{ and } S = \frac{64\pi a^2}{3}.$$

XV.

Moments of Inertia.

188. When a body rotates about a fixed axis, the velocity of a particle at a distance r from the axis is

$$r\frac{d\omega}{dt},$$

in which ω is the angle of rotation. The force which acting for a unit of time would produce this motion in a mass m is measured by the momentum

$$mr\frac{d\omega}{dt}.$$

The moment of this force about the axis is therefore

$$mr^2\frac{d\omega}{dt}.$$

The sum of these moments for all the parts of a rigid system is

$$\frac{d\omega}{dt}\Sigma \triangle mr^2,$$

since the angular velocity, $\frac{d\omega}{dt}$, is constant. In the case of a continuous body this expression becomes

$$\frac{d\omega}{dt}\int r^2\,dm,$$

in which dm is the differential of the mass. The factor

$$\int r^2\,dm,$$

which depends upon the shape of the body, is called its *moment of inertia*, and is denoted by *I*.

189. When the body is homogeneous, dm is to be taken equal to the differential of the line, area, or volume, as the case may be. For example, in finding the moment of inertia of a straight line whose length is $2a$, about an axis bisecting it at right angles, we let x denote the distance of any point from the axis; then $dm = dx$, hence we have

$$I = \int_{-a}^{a} x^2\,dx = \frac{2a^3}{3} = \frac{(2a)^3}{12}.$$

Again, in finding the moment of inertia of the semi-circle in figure 25, about the axis of y, let $dm = 2y\,dx$; then, since every point of the generating line is at the distance x from the axis, the moment of inertia is

$$I = 2\int_0^a y x^2\,dx = 2\int_0^a \sqrt{(a^2 - x^2)}\, x^2\,dx.$$

Putting $x = a \sin \theta$, we have

$$I = 2a^4 \int_0^{\frac{1}{2}} \cos^2\theta \sin^2\theta\,d\theta = \frac{\pi a^4}{8}.$$

The Radius of Gyration.

190. If the whole mass of the body were situated at the distance k from the axis, its moment of inertia would be $k^2 m$. Now, if k is so determined that *this moment* shall be equal to the actual moment of inertia of the body, the value of k is *the radius of gyration* of the body with reference to the given axis. Hence

$$k^2 = \frac{\text{Moment of inertia}}{\text{Mass}}.$$

Thus, for the radius of gyration of the line $2a$, whose moment of inertia is found in the preceding article, we have

$$k^2 = \frac{a^2}{3}, \qquad \text{or} \qquad k = \frac{a}{\sqrt{3}};$$

and for the radius of gyration of the semi-circle, whose area is $\frac{1}{2}\pi a^2$,

$$k^2 = \frac{a^2}{4}, \qquad \text{or} \qquad k = \frac{a}{2}.$$

It is evident that this expression is also the radius of gyration of the whole circle about a diameter, for the moment of inertia of the circle is evidently double that of the semi-circle, and its area is also double that of the semi-circle.

191. It is sometimes convenient to use modes of generating the area or volume, other than those involving rectangular coordinates. For example, let it be required to find the radius of gyration of a circle whose radius is a, about an axis passing through its centre and perpendicular to its plane. This circle may be generated by the circumference of a variable circle whose radius is r, while r passes from 0 to a. The differential of the area is then $2\pi r\, dr$, and the moment is

$$I = 2\pi \int_0^a r^3\, dr = \frac{\pi a^4}{2}.$$

Dividing by the area of the circle, we have

$$k^2 = \frac{a^2}{2}.$$

192. Again, to find the radius of gyration of a sphere whose radius is a about a diameter. In order that all points of the elements shall be at the same distance from the axis,

we regard the sphere as generated by the surface of a cylinder whose radius is x, and whose altitude is $2y$. The surface of this cylinder is therefore $4\pi xy$. The differential of the volume is $4\pi xy\,dx$, and the moment of inertia is

$$I = 4\pi \int x^3 y\,dx = 4\pi \int \sqrt{(a^2 - x)}\,x^3\,dx.$$

Putting $x = a \sin \theta$,

$$I = 4\pi a^5 \int_0^{\frac{\pi}{2}} \sin^3 \theta \cos^2 \theta\,d\theta = \frac{8\pi a^5}{15}.$$

Dividing by $\dfrac{4\pi a^3}{3}$, the volume of the sphere, we have

$$k^2 = \frac{2a^2}{5}.$$

Radii of Gyration about Parallel Axes.

193. *The moment of inertia of a body about any axis exceeds its moment of inertia about a parallel axis passing through the centre of gravity, by the product of the mass and the square of the distance between the axes.*

Let h be the distance between the axes. Pass a plane through the element dm perpendicular to the axes, and let r and r_1 be the distances of the element from the axes. Then, r, r_1, and h form a triangle; let θ be the angle at the axis passing through the centre of gravity, then

$$r^2 = r_1^2 + h^2 - 2r_1 h \cos \theta. \quad \ldots \ldots \quad (1)$$

§ XV.] *RADII OF GYRATION ABOUT PARALLEL AXES.* 223

The moment of inertia is therefore

$$\int r^2 dm = \int r_1^2 dm + h^2 m - 2h \int r_1 \cos \theta \, dm. \quad . \quad . \quad (2)$$

Now r_1 and θ are the polar coordinates of dm, in the plane which is passed through the element; hence the last integral in equation (2) is equivalent to

$$- 2h \int x \, dm.$$

But $\int x \, dm$ is the statical moment of the body about the axis passing through the centre of gravity. Now from the definition of the centre of gravity, this moment is zero; hence, equation (2) reduces to

$$\int r^2 dm = \int r_1^2 dm + h^2 m. \quad . \quad . \quad . \quad . \quad (3$$

Introducing the radii of gyration, we have also

$$k^2 = k_1^2 + h^2. \quad . \quad \quad \quad . \quad . \quad (4)$$

194. As an application of this result, we shall now find the moment of inertia of a cone whose height is h, and the radius of whose base is a, about an axis passing through its vertex perpendicular to its geometrical axis. Taking the origin at the vertex of the cone, the axis of x coincident with the geometrical axis, and a circle perpendicular to this axis as the generating element, we have for the area of this element πy^2, and for its radius of gyration about a diameter parallel to the given axis, $\dfrac{y^2}{4}$.

The distance between these axes being x, the proposition proved in the preceding article gives an expression for the radius of gyration of the element about the given axis; viz., $x^2 + \dfrac{y^2}{4}$. Replacing r^2, in the general expression for I (Art. 188), by this expression, and substituting for dm the differential $\pi y^2\, dx$, we have

$$I = \pi \int \left(x^2 + \frac{y^2}{4} \right) y^2\, dx,$$

in which $y = \dfrac{ax}{h}$. Therefore

$$I = \frac{\pi a^2}{h^2} \int_0^h \left(1 + \frac{a^2}{4h^2}\right) x^4\, dx = \frac{\pi a^2 h^3}{5} \left(1 + \frac{a^2}{4h^2}\right),$$

and since

$$V = \frac{\pi a^2 h}{3},$$

$$k^2 = \frac{3}{20}\left(a^2 + 4h^2\right).$$

To find the square of the radius of gyration about a parallel axis through the centre of gravity, we have

$$k_0^2 = \frac{3}{20}\left(a^2 + 4h^2\right) - \left(\frac{3h}{4}\right)^2$$

$$= \frac{3}{80}\left(4a^2 + h^2\right).$$

To find the moment of inertia of a right cone about its geometrical axis we employ the same generating element as before; but in this case the square of the radius of gyration is $\dfrac{y^2}{2}$. Hence

$$I = \frac{\pi}{2} \int y^4\, dx = \frac{\pi a^4}{2h^4} \int_0^h x^4\, dx :$$

therefore
$$I = \frac{\pi a^4 h}{10}, \quad \text{whence} \quad k^2 = \frac{3a^2}{10}.$$

Polar Moments of Inertia.

195. In the case of a plane area, when the axis of rotation passes through the origin, we have

$$r^2 = x^2 + y^2, \qquad \text{hence} \int r^2\,dm = \int (x^2 + y^2)\,dm,$$

therefore
$$I = \int x^2\,dm + \int y^2\,dm;$$

that is, *the sum of the moments of inertia of a plane area about two axes in its own plane at right angles to each other is equal to the moment of inertia about an axis through the origin perpendicular to the plane.* I in the above equation is called *the polar moment of inertia.*

In the case of the circle, since the moment is the same about every diameter, the polar moment is twice the moment about a diameter; that is, denoting the former by I_p and the latter by I_a, we have

$$I_p = 2 I_a = \frac{\pi a^4}{2}.$$

See Art. 191.

Examples XV.

1. Find the radius of gyration of a circular arc ($2s$) about a radius passing through its vertex.

Solution :—

Taking the origin at the centre, and the axis of x bisecting the arc, and denoting by 2α the angle subtended by $2s$, we have

$$mk^2 = \int_{-s}^{s} y^2\, ds = a^3 \int_{-\alpha}^{\alpha} \sin^2 \theta\, d\theta.$$

$$m = 2a\alpha \qquad \therefore \qquad k^2 = \frac{a^2}{2}\left(1 - \frac{\sin 2\alpha}{2\alpha}\right)$$

2. Find the radius of gyration of the same arc about the axis of y, and thence about a perpendicular axis through the centre of the circle. $\qquad k = a.$

3. Find the radius of gyration of the same arc about an axis through its vertex perpendicular to the plane of the circle.
See Ex. XIV., 14, *and denote by c the subtending chord.*

$$k^2 = 2a^2\left(1 - \frac{c}{2s}\right).$$

4. Find the moment of inertia of the chord of a circular arc, in terms of the diameter parallel to it, and its angular distance from this diameter.

See Arts. 189 *and* 193. $\qquad I = \dfrac{d^2}{24}(3\cos\alpha - \cos 3\alpha).$

5. Find the radius of gyration of an ellipse about an axis through its centre perpendicular to its plane.
Find the radius of gyration about the major axis and about the minor axis, and apply Art. 195.

$$k^2 = \tfrac{1}{4}(a^2 + b^2).$$

6. Find the radius of gyration of an isosceles triangle about a perpendicular let fall from its vertex upon the base ($2b$).

$$k^2 = \frac{b^2}{6}.$$

§ XV.] EXAMPLES. 227

7. Find the radius of gyration about the axis of the curve, of the area enclosed by the two loops of the lemniscata

$$r^2 = a^2 \cos 2\theta.$$

$$k^2 = \frac{a^2}{48}(3\pi - 8).$$

8. Find the radius of gyration of a right triangle, whose sides are a and b, about an axis through its centre of gravity perpendicular to its plane

$$k^2 = \frac{a^2 + b^2}{18}.$$

9. Find the radius of gyration of a portion of a parabola bounded by a double ordinate perpendicular to the axis, about a perpendicular to its plane passing through its vertex.

$$k^2 = \tfrac{3}{7} x^2 + \tfrac{1}{5} y^2.$$

10. Find the radius of gyration of a cylinder about a perpendicular that bisects its geometrical axis, $2l$ being the length of the cylinder, and a the radius of its base.

$$k^2 = \frac{a^2}{4} + \frac{l^2}{3}.$$

11. Find the radius of gyration of a concentric spherical shell about a tangent to the external sphere, the radii being a and b.

$$k^2 = \frac{7a^5 - 5a^2b^3 - 2b^5}{5(a^3 - b^3)}.$$

12. Find the radius of gyration of a paraboloid of revolution about its axis, in terms of the radius (b) of the base.

$$k^2 = \frac{b^2}{3}.$$

13. Find the moment of inertia of an ellipsoid about one of its principal axes.

$$I = \frac{4\pi abc}{15}(b^2 + c^2).$$

14. Find the radius of gyration of a symmetrical double convex lens about its axis, a being the radius of the circular intersection of the two surfaces, and b the semi-axis.

$$k^2 = \frac{b^4 + 5a^2b^2 + 10a^4}{10(b^2 + 3a^2)}.$$

15. Find the radius of gyration of the same lens about a diameter to the circle in which the spherical surfaces intersect.

$$k^2 = \frac{10a^4 + 15a^2b^2 + 7b^4}{20(b^2 + 3a^2)}.$$

THE END.

www.ingramcontent.com/pod-product-compliance
Lightning Source LLC
Chambersburg PA
CBHW022140300426
44115CB00006B/280